VOLUME THREE

STRATIGRAPHY & TIMESCALES

Cyclostratigraphy and Astrochronology

VOLUME THREE

STRATIGRAPHY & TIMESCALES

Cyclostratigraphy and
Astrochronology

Edited by

MICHAEL MONTENARI

Department of Earth Sciences and Geography
Keele University
Newcastle
United Kingdom

ACADEMIC PRESS

An imprint of Elsevier

Academic Press is an imprint of Elsevier
125 London Wall, London EC2Y 5AS, United Kingdom
525 B Street, Suite 1650, San Diego, CA 92101, United States
50 Hampshire Street, 5th Floor, Cambridge, MA 02139, United States
The Boulevard, Langford Lane, Kidlington, Oxford OX5 1GB, United Kingdom

First edition 2018

Notices
Knowledge and best practice in this field are constantly changing. As new research and
experience broaden our understanding, changes in research methods, professional practices,
or medical treatment may become necessary.

Practitioners and researchers must always rely on their own experience and knowledge in
evaluating and using any information, methods, compounds, or experiments described
herein. In using such information or methods they should be mindful of their own safety and
the safety of others, including parties for whom they have a professional responsibility.

To the fullest extent of the law, neither the Publisher nor the authors, contributors, or editors,
assume any liability for any injury and/or damage to persons or property as a matter of
products liability, negligence or otherwise,or from any use or operation of any methods,
products, instructions, or ideas contained in the material herein.

ISBN: 978-0-12-815098-6
ISSN: 2468-5178

For information on all Academic Press publications visit our
website at https://www.elsevier.com/books-and-journals

Publisher: Zoe Kruze
Acquisition Editor: Jason Mitchell
Editorial Project Manager: Collett Joanna
Production Project Manager: Vignesh Tamil
Cover Designer: Alan Studholme

Typeset by TNQ Technologies

CONTENTS

7. The Down-dip Preferential Sequence Record of Orbital Cycles in Greenhouse Carbonate Ramps: Examples From the Jurassic of the Iberian Basin (NE Spain)   **285**

Beatriz Bádenas and Marcos Aurell

8. Astronomical Calibration of the Tithonian – Berriasian in the Neuquén Basin, Argentina: A Contribution From the Southern Hemisphere to the Geologic Time Scale   **327**

Diego A. Kietzmann, Maria Paula Iglesia Llanos and Melisa Kohan Martínez

9. Paleocene-Eocene Calcareous Nannofossil Biostratigraphy and Cyclostratigraphy From the Neo-Tethys, Pabdeh Formation of the Zagros Basin (Iran) **357**

Seyed Hamidreza Azami, Erik Wolfgring, Michael Wagreich and Mohamad Hosein Mahmudy Gharaie

CONTRIBUTORS

Moujahed Al-Husseini
Gulf PetroLink-GeoArabia, Manama, Bahrain

Marcos Aurell
Departamento de Ciencias de la Tierra-IUCA, Universidad de Zaragoza, Zaragoza, Spain

Seyed Hamidreza Azami
University of Vienna, Department of Geodynamics and Sedimentology, Vienna, Austria

Beatriz Bádenas
Departamento de Ciencias de la Tierra-IUCA, Universidad de Zaragoza, Zaragoza, Spain

Annette E. Götz
University of Portsmouth, School of Earth and Environmental Sciences, Portsmouth, United Kingdom

Linda A. Hinnov
Department of Atmospheric, Oceanic, and Earth Sciences, George Mason University, Fairfax, VA, United States

Chunju Huang
State Key Laboratory of Biogeology and Environmental Geology, School of Earth Sciences, China University of Geosciences, Wuhan, China

Maria Paula Iglesia Llanos
Universidad de Buenos Aires, Facultad de Ciencias Exactas y Naturales, Departamento de Ciencias Geológicas, Ciudad Universitaria, Pabellón 2, Ciudad Autónoma de Buenos Aires, Argentina; CONICET-Universidad de Buenos Aires, Instituto de Geociencias Básicas, Ambientales y Aplicadas de Buenos Aires (IGeBA), Ciudad Universitaria, Pabellón 2, Ciudad Autónoma de Buenos Aires, Argentina

Diego A. Kietzmann
Universidad de Buenos Aires, Facultad de Ciencias Exactas y Naturales, Departamento de Ciencias Geológicas, Ciudad Universitaria, Pabellón 2, Ciudad Autónoma de Buenos Aires, Argentina; CONICET-Universidad de Buenos Aires, Instituto de Geociencias Básicas, Ambientales y Aplicadas de Buenos Aires (IGeBA), Ciudad Universitaria, Pabellón 2, Ciudad Autónoma de Buenos Aires, Argentina

Mohamad Hosein Mahmudy Gharaie
Ferdowsi University of Mashhad, Department of Geology, Mashhad, Iran

Mathieu Martinez
Univ Rennes, CNRS, Géosciences Rennes, Rennes, France

Melisa Kohan Martínez
Universidad de Buenos Aires, Facultad de Ciencias Exactas y Naturales, Departamento de Ciencias Geológicas, Ciudad Universitaria, Pabellón 2, Ciudad Autónoma de Buenos Aires, Argentina; CONICET-Universidad de Buenos Aires, Instituto de Geociencias Básicas, Ambientales y Aplicadas de Buenos Aires (IGeBA), Ciudad Universitaria, Pabellón 2, Ciudad Autónoma de Buenos Aires, Argentina

André Strasser
Department of Geosciences, University of Fribourg, Fribourg, Switzerland

Ákos Török
Budapest University of Technology and Economics, Department of Engineering Geology and Geotechnics, Budapest, Hungary

Michael Wagreich
University of Vienna, Department of Geodynamics and Sedimentology, Vienna, Austria

Erik Wolfgring
University of Vienna, Department of Geodynamics and Sedimentology, Vienna, Austria

PREFACE

Present Status and Future Directions in Cyclostratigraphy and Astrochronology

Cyclostratigraphy and Astrochronology are compared to other stratigraphic disciplines relatively young. Although "cyclic or semicyclic" sedimentation patterns had been noticed and described before, it was not until the seminal publication by Milutin Milanković that a deeper understanding of certain astronomical parameters and their effects on the Earth started to develop. Nowadays it is well established that the cyclic and semicyclic changes of the Earth's orbit, its relation to other planets of the solar system, and its tilt relative to the sun have a profound impact on the solar insolation and hence the amount of energy the Earth receives from the Sun.

The present book on *Cyclostratigraphy and Astrochronology* aims to provide an update and outlook on both stratigraphic disciplines. In the following the chapters and their abstracts will be briefly introduced. In the first chapter, **Linda A. Hinnov** highlights that "after more than a century of doubt and equivocation, cyclostratigraphy—the stratigraphic record of astronomically forced paleoclimatic change—has come to be widely accepted in the geosciences. Cyclostratigraphy depicts a paleoclimate system that is intimately connected with and pervasively tuned to the variations of co-occurring astronomical and geophysical parameters that have affected the Earth through geologic time. Cyclostratigraphy provides a proprietary record of planetary orbital motions and is the sole repository of evidence for interplanetary orbital resonance and chaotic interaction throughout Solar System history. Cyclostratigraphy has also recorded the evolution of Earth's rotation rate and mass distribution, and by consequence, the orbital dynamics of the Moon. For the most recent 10 million years, cyclostratigraphy has been precisely correlated to quantitative astronomical models and, for hundreds of millions of years prior to that, has been matched to models with a progressively decreasing but still significant measure of confidence. This has led to the rise of astrochronology, which assigns cyclostratigraphy to a specific time scale based on its correlation to astronomical solutions. Modern astronomical solutions are highly consistent, indicating that the theory of Solar System dynamics has been described as completely as possible for 0 to at least 50 million years ago. This indicates that cyclostratigraphy can be correlated to the full Earth orbital eccentricity and/or inclination solution over 0–50 Ma, to contribute a high-precision astrochronology for the geologic time scale. For times prior to 50 Ma, astronomical models deviate substantially from one another as predicted by the Lyapunov time of the Solar System and depict chaotic behavior, for example, between the Earth and Mars. On the other hand, specific planetary orbits, notably those of the Jupiter and Saturn, theoretically remain stable over much longer periods of time; these have led to the recognition of 'metronomes' in Earth's orbital motions, specifically $g_2 - g_5$ and $s_3 - s_6$, which when identified in cyclostratigraphy, can provide practical extensions to the Astronomical Time Scale for hundreds of millions of years into the past." This general chapter is followed by the contribution of **Chunju Huang** who establishes that "the astronomical theory and its application to sediment cyclostratigraphy have been widely studied over the past 30 years. Especially during the past decade, these have been successfully applied for a continuous high-resolution calibration of portions of the geologic time scale. The method of astronomical tuning enhances the traditional geological dating methods, such as paleontology, paleomagnetism, and radioisotope dating. In *The Geologic Time Scale 2012* (GTS2012), most of the Cenozoic Era was directly calibrated to the absolute astronomical time scale (ATS); however, only a few intervals of the Mesozoic–Paleozoic were scaled by cyclostratigraphy in discontinuous floating segments. Since then, other cyclostratigraphy studies have been published using the most stable 405-kyr long eccentricity cycle to calibrate portions of the Mesozoic geological time scale. Other than some gaps within the mid-Triassic and Middle Jurassic, it is possible to compile a nearly complete ATS status for the Mesozoic. This chapter compiles and recalibrates an extensive suite of selected paleoclimate proxies series and applies a tuning to the master astronomical scale based on the stable 405-kyr long eccentricity period. This enables the synthesis of a nearly complete Mesozoic ATS to the base of the Triassic." The chapter by **André Strasser** explores "the sedimentary record of ancient shallow-marine carbonate platforms commonly displays a stacking of different facies, which reflects repetitive changes of depositional environments through time. These changes can be induced by external factors such as cyclical changes in climate and/or sea level, but also by internal factors such as lateral migration of sediment bodies and/or changes in the ecology of the carbonate-producing organisms. If it can be demonstrated that the facies changes formed in tune with the orbital (Milankovitch) cycles of known duration, then a high-resolution time framework can be established. This demonstration is not an easy task because the orbital signal may be too weak to be recorded, or it may be distorted and/or overprinted by local or regional processes. The limitations of the cyclostratigraphical approach are discussed, but a case study from the Oxfordian of the Swiss Jura Mountains also shows its potential. A well-established chrono- and sequence-stratigraphic framework and detailed facies analysis allow identification of elementary, small-scale, and medium-scale depositional sequences that formed in tune with the precession, the short eccentricity, and the long eccentricity cycles, respectively. In the best case, a depositional sequence attributed to the precession cycle with a duration of 20,000 years can be interpreted in terms of sequence stratigraphy. This then allows estimating rates of sea-level change and sedimentation within a relatively narrow time window, thus facilitating comparisons between ancient carbonate platforms and Holocene or recent shallow-marine environments where such rates are well quantified." This is followed by the study of **Mathieu Martinez** in which he provides a detailed account on "Eccentricity cycles often exert strong influence on the sedimentary series, while they only have weak powers in the insolation series. I analyze here three case studies previously published in Early Cretaceous hemipelagic marl-limestone alternations of the Tethyan area to suggest mechanisms of transfer of power from the precession to the eccentricity band. In all cases, the sedimentation rates vary cyclically, following the 405-kyr and the 2.4-myr eccentricity cycles. Maximums of sedimentation rates occur in more clayey intervals, suggesting a strong eccentricity forcing on detrital supply and consequently on sedimentation rates. In all sections, proxies related to purely detrital input show an overwhelming influence of the 405-kyr and the 2.4-myr eccentricity cycle compared to the other Milankovitch cycles. Proxies including the carbonate production show lower amplitudes of the 405-kyr cycle and enhanced 100-kyr eccentricity cycle, which can dominate the power in the Milankovitch band. Memory effects of erosional and pedogenetic processes appear to be powerful mechanisms of transfer of amplitude from the precession to the eccentricity band. Highly evolved pedogenesis under more humid climates has longer memory effects, which favors the 405-kyr and the 2.4-myr cycle in the sedimentary series. Carbonate production eventually changes its phasing with the insolation series, which suppresses the low frequencies and favors the 100-kyr eccentricity cycle.

Clay-dominated marl-limestone alternations are associated to higher sedimentation rates and higher amplitudes of the long eccentricity cycles. Conversely, carbonate-dominated marl-limestone alternations are associated to lower sedimentation rates and higher amplitudes of the 100-kyr eccentricity cycles." In the fifth chapter **Moujahed Al-Husseini** presents a new and novel approach to Arabian Orbital Sequences, highlighting "the orbital scale of glacio-eustasy adopts *stratons* as the fundamental time-rock units of sequence stratigraphy, numbers them as consecutive integers, and approximates their duration as 405 kyr. It predicts 12 stratons form periodic *dozons* (4.86 myr) and 3 dozons an *orbiton* (14.58 myr), and introduces a concise alphanumeric code for naming them. Importantly the scale predicts the ages of major and minor sequence boundaries (SB), lowstands, and flooding intervals (MFI) within these long-period sequences using arithmetical formulas. The scale is based on a simplified model of orbital-forcing of glacioeustasy, which was calibrated with Arabian transgressive-regressive (T-R) sequences. The calibration indicated Orbiton 1 was deposited between SB 1 at 16.166 Ma and SB 0 at 1.586 Ma, and SB 37 predicted at 541.046 Ma (1.586 + 37 × 14.58), correlates to the Precambrian/Cambrian Boundary. In many Phanerozoic intervals dozon- and orbiton-scale SBs correlate closely to age estimates of the maximum values in major global positive $\delta^{13}C$ excursions, and negative $\delta^{13}C$ cycles resemble predicted T-R sequences. In this chapter, examples of Early Paleozoic, Early Triassic, Late Cretaceous, and Miocene Arabian sequences are compared to global $\delta^{13}C$ patterns and both data sets are interpreted as glacioeustatic proxies and dated with the orbital scale." This is followed by the contribution of **Annette E. Götz and Ákos Török** who analyzed in detail the "Anisian Muschelkalk ramp cycles are well documented from the northwestern Tethys shelf and its northern Peri-Tethys Basin by detailed studies in Germany, the Netherlands, Poland, and Hungary. High-resolution correlation based on cyclostratigraphy has been attempted in recent years but is still hampered by precise age control. The recent approach of Tethys–Peri-Tethys correlation of the eccentricity-cycle-scaled Anisian carbonate series integrating global climate signatures improves the so far established small-scale cycle correlation using high-frequency eustatic signatures in the Milankovitch frequency band. Cyclostratigraphic calibration of Lower Muschelkalk ramp deposits of different palaeogeographic settings by astronomical cycle-tuning of gamma-ray, isotopic and magnetostratigraphic data, and integration of global climate signatures is the next step toward refined correlation." In the seventh chapter **Beatriz Bádenas and Marcos Aurell** establish that "high-frequency sequences of different scales were recorded in the distinct carbonate ramps developed in the Iberian Basin (Spain) in the Jurassic times. This work reviews and compares the sedimentary record of these sequences in four Iberian carbonate ramps: (1) inner to proximal outer ramp areas of a nonskeletal-dominated homoclinal ramp (upper Sinemurian-lowermost Pliensbachian); (2) middle to proximal outer ramp areas of a skeletal-dominated homoclinal ramp (upper Pliensbachian); (3) shallow to relatively deep domain of a microbial/siliceous sponge-dominated distally steepened ramp (Bajocian); and (4) inner to open ramp domains of a homoclinal ramp with coral-microbial reefs (upper Kimmeridgian). The comparative review reveals that the sequences were recorded differently from shallow to deep ramp areas. In shallow ramp areas there is a preferential record of eccentricity-related sequences, whereas high-frequency (precession) sequences are ubiquitous in relatively deep ramp settings. This preferential preservation was likely controlled by the interplay between accommodation changes and internal processes controlling accumulation above and below fair-weather and storm wave base levels, rather than by climate (warm *vs.* cold greenhouse climate modes) and type of carbonate production. Comparison with cyclostratigraphic analysis performed in similar settings indicates that this variable down-dip sequence record seems to be a common feature in greenhouse carbonate ramps: eccentricity-related, meter-thick facies sequences are dominant in shallow ramp areas, whereas the deep outer ramp-basin is characterized by preservation of precession or sub-Milankovitch cycles. It is emphasized that the middle to outer ramp sedimentary domain (located around the storm wave base) has the greatest potential for cyclostratigraphic analysis because this domain is more likely to record both eccentricity- and/or precession-related sequences." This is followed by the contribution of **Diego A. Kietzmann** et al., who demonstrate that "detailed cyclostratigraphical analyses have been made from five Tithonian–Berriasian sections of the Vaca Muerta Formation, exposed in the Neuquén Basin, Argentina. The Vaca Muerta Formation is characterized by decimeter-scale rhythmic alternations of marlstones and limestones, showing a well-ordered hierarchy of cycles, where elementary cycles, bundles of cycles, and superbundles have been recognized. According to biostratigraphic data, elementary cycles have a periodicity of $\sim$21 ky, which correlates with the precession cycle of Earth's axis. Spectral analysis based on time series of elementary cycle thickness allows us to identify frequencies of

~400 ky and ~90–120 ky, which we interpret as the modulation of the precessional cycle by the Earth's orbital eccentricity. Correlation between studied sections allowed us to estimate a minimum duration for each Andean ammonite zone. Moreover, cyclostratigraphic data allowed us to build the first continuous floating astronomical time scale for the Tithonian–Berriasian, which is anchored to the geological time scale through magnetostratigraphy. We estimated a minimum duration of 5.67 myr for the Tithonian and 5.27 myr for the Berriasian. The resulted durations of some polarity chrons are also different with respect to the GTS2016; however, such differences could be due to condensation or discontinuities not detected in the studied sections." The ninth chapter by *Seyed Hamidreza Azami* **et al.** highlights that "the Pabdeh Formation in the Zagros Basin, Iran, records cyclic pelagic sedimentation from the middle Paleocene to the middle Eocene in a Tethyan setting. The cyclic successions consist of deeper-water pelagic to hemipelagic shale, marl(stone) and limestone, with a predominantly shaley lower part, and a marl-limestone upper part. Nannofossil biostratigraphy indicates standard zones CNP8 - NP6 to CNE8 - NP14. The Paleocene-Eocene Thermal Maximum (PETM) interval is indicated by a distinct nannofossil assemblages and a significant negative carbon isotope excursion. Sediment accumulation rates are in general around 6–68 mm/ka. Cyclic signals investigated include fluctuations in carbonate content. Power spectra using LOWSPEC (Robust Locally-Weighted Regression Spectral Background Estimation) and EHA (Evolutive Harmonic Analysis) indicates the presence of twenty-one 405 ka cycles from the base of the PETM up to the top of the studied section. Orbital tuning to the established Laskar target curve is in accordance with the general Paleocene-Eocene cyclostratigraphy and yields insights into Paleogene chronostratigraphy."

Michael Montenari
Keele University
United Kingdom

CHAPTER ONE

Cyclostratigraphy and Astrochronology in 2018

Linda A. Hinnov

Department of Atmospheric, Oceanic, and Earth Sciences, George Mason University, Fairfax, VA, United States
E-mail: lhinnov@gmu.edu

Contents

Stratigraphy & Timescales, Volume 3
ISSN 2468-5178
https://doi.org/10.1016/bs.sats.2018.08.004

Abstract

After more than a century of doubt and equivocation, cyclostratigraphy — the stratigraphic record of astronomically forced paleoclimatic change — has come to be widely accepted in the geosciences. Cyclostratigraphy depicts a paleoclimate system that is intimately connected with and pervasively tuned to the variations of co-occurring astronomical and geophysical parameters that have affected the Earth through geologic time. Cyclostratigraphy provides a proprietary record of planetary orbital motions, and is the sole repository of evidence for interplanetary orbital resonance and chaotic interaction throughout Solar System history. Cyclostratigraphy has also recorded the evolution of Earth's rotation rate and mass distribution, and by consequence, the orbital dynamics of the Moon. For the most recent ten million years, cyclostratigraphy has been precisely correlated to quantitative astronomical models, and for hundreds of millions of years prior to that, has been matched to models with a progressively decreasing but still significant measure of confidence. This has led to the rise of astrochronology, which assigns cyclostratigraphy to a specific time scale based on its correlation to astronomical solutions. Modern astronomical solutions are highly consistent, indicating that the theory of Solar System dynamics has been described as completely as possible for 0 to at least 50 million years ago (Ma). This indicates that cyclostratigraphy can be correlated to the full Earth orbital eccentricity and/or inclination solution over 0–50 Ma, to contribute a high-precision astrochronology for the geologic time scale. For times prior to 50 Ma, astronomical models deviate substantially from one another as predicted by the Lyapunov time of the Solar System, and depict chaotic behavior, for example, between Earth and Mars. On the other hand, specific planetary orbits, notably those of Jupiter and Saturn, theoretically remain stable over much longer periods of time; these have led to the recognition of "metronomes" in Earth's orbital motions, specifically, $g_2 - g_5$ and $s_3 - s_6$, which when identified in cyclostratigraphy can provide practical extensions to the Astronomical Time Scale for hundreds of millions of years into the past.

1. INTRODUCTION

The Earth's stratigraphic record of astronomically forced climate change originates from variations in the Earth's orientation with respect to the Sun imposed by gravitational attraction from the Moon, Sun and other planets. The motions of the attracting bodies are described by astrodynamical theory, and their interactions with the Earth invoke geodynamical theory. The coupling of these variations with the incoming solar radiation, known as insolation, is embodied in the Milankovitch theory of climate.

Planetary orbital motions are modeled as variations in orbital eccentricity and inclination (Section 1.1). A prediction of chaotic interactions between Earth and Mars during Cretaceous time has recently been validated by cyclostratigraphy. The Earth also experiences precession and obliquity variations due to its axial tilt and gravitational attraction from the Moon and Sun, and to a minor extent the other planets (Section 1.2). Tidal dissipation in the Earth-Moon system results in lunar recession, deceleration in Earth's rotation, and decreasing frequencies in the Earth's precession and obliquity. These effects are now being quantified by powerful new analyses of the cyclostratigraphic record. Earth's precession and obliquity variations control the amount of solar energy incident on the Earth, and influence paleoclimate change (Section 1.3).

Recovery of the cyclostratigraphic record of Earth's astronomical parameters has grown rapidly over the past several decades, first focusing on the Cenozoic Era, but advancing progressively into the Mesozoic Era, and most recently, the Paleozoic Era and the Precambrian (Section 2). These advances are attended by new problems, including a severe paucity of geochronologic constraints for cyclostratigraphic data series, and lack of knowledge about the deep-time evolution of the planetary orbits or Earth-Moon system evolution. The time calibration of cyclostratigraphy, or astrochronology, thus involves numerous strategies, most recently with the advent of extremely fast computers, brute-force optimization and Bayesian techniques (Section 3). The goal of astrochronology is to construct an "astronomical time scale" with a resolving power that exceeds and/or maintains the accuracy and precision of geologic time provided by radioisotope geochronology (Section 4).

1.1 Astrodynamics — Orbital Eccentricity and Inclination

The evolution of planetary orbits has been an active line of inquiry in astronomy for centuries, with Laplace and Lagrange describing in the late 18th century the equations of motion that are used for solutions to this day (historical perspective in Laskar, 2013). The original goal for accurate reconstruction of planetary motion has broadened to consider the more general problem of Solar System stability, which, due to the long timescales involved, requires observational data from the cyclostratigraphic record (Laskar et al., 2011a, 2011b). The basic solution variables are described (1.1.1), followed by a summary of legacy and current astronomical solutions and their limitations (Section 1.1.2), and ending with examples of cyclostratigraphic observations that have been used to validate one astronomical solution in particular (Section 1.1.3).

1.1.1 Orbital Eccentricity and Inclination Variables

Classical secular perturbation theory of the Solar System for $n = 1, 2, 3, \ldots,$ $N = 9$ planets (including Pluto) in order of distance from the Sun involves the solution of orbital eccentricity and inclination variables (e.g., Eqs. (2) and (5) in Bretagnon, 1974):

$$z_n(t) = e_n(t)\exp(i\Pi_n t) = h_n(t) + ik_n(t) \approx \sum_{j=1}^{N} A_{n,j} \exp\left(ig_j t\right)$$

$$\zeta_n(t) = \sin\left(\frac{I_n(t)}{2}\right)\exp(i\Omega_n t) = p_n(t) + iq_n(t) \approx \sum_{j=1}^{N} B_{n,j} \exp\left(is_j t\right)$$

(1)

The orbital elements e, Π, I and Ω are eccentricity, longitude of perihelion, inclination and longitude of ascending node, respectively, and variables $h_n(t)$, $k_n(t)$, $p_n(t)$, and $q_n(t)$; $A_{n,j}$ and $B_{n,j}$ are amplitudes, and g_j and s_j are the fundamental secular frequencies of the planets (e.g., Table 3 in Laskar et al., 2004).

Earth's orbital eccentricity is (Fig. 2A and B; Appendix A):

$$e_3(t) = \sqrt{h_3^2(t) + k_3^2(t)}$$

(2)

for which the g_j are the contributing frequencies of the planetary orbital perihelia. Values for $h_3(t)$ and $k_3(t)$ for time t in 1 kyear (kyr) steps

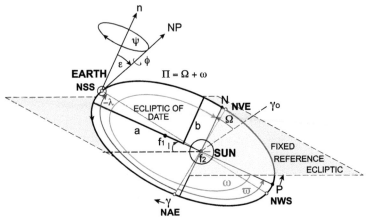

Figure 1 Earth's astronomical parameters viewed from above the Earth's Northern Hemisphere. Earth is positioned near modern-day northern Summer Solstice. The Earth's orbit is elliptical with major axis a and minor axis b, and two foci (f_1, f_2). The Sun's position is at f_2. The orbital elements are eccentricity $e = \sqrt{(a^2 + b^2)}/a$, longitude of perihelion (Π), inclination (I) and longitude of the ascending node (Ω). The plane of the Earth's orbit is the "ecliptic of date" and is inclined at an angle I relative to a "fixed reference ecliptic." The Earth's orbit intersects the reference ecliptic at a longitude Ω at point N, the ascending node, relative to a fixed vernal point γ_0. The orbital perihelion point P, measured relative to γ_0 as longitude of perihelion $\Pi = \Omega + \omega$, moves slowly anticlockwise. The argument of the perihelion ω measures the angular distance between P and N. The angle $\varpi = \Pi + \psi$ tracks the moving vernal point γ relative to P, which moves clockwise along the orbit due to the luni-solar precession ψ. The true longitude of the Earth λ varies from 0 to 360° throughout the year with respect to γ, in the clockwise direction, i.e., in the direction of the Earth's path; here for clarity it is drawn in the opposite direction. The Earth has an anticlockwise rotation rate ϕ and an obliquity angle ε that precesses clockwise at a rate ψ. NP is the Earth's rotation axis ("North Pole"), and n is the normal to the ecliptic of date. Note: today, perihelion P is $\omega = 103°$ from γ, i.e., it occurs on calendar day January 3.

(negative t for time before present) are provided for the 2010 Laskar solutions. Earth's orbital eccentricity is otherwise customarily reported as a frequency decomposition, e.g., Table 6 of Laskar et al. (2004) lists the 20 leading trigonometric terms.

Earth's orbital inclination is (Fig. 2C and D; Appendix A):

$$I_3(t) = 2 \sin^{-1}\left(\sqrt{p_3^2(t) + q_3^2(t)} \right) \tag{3}$$

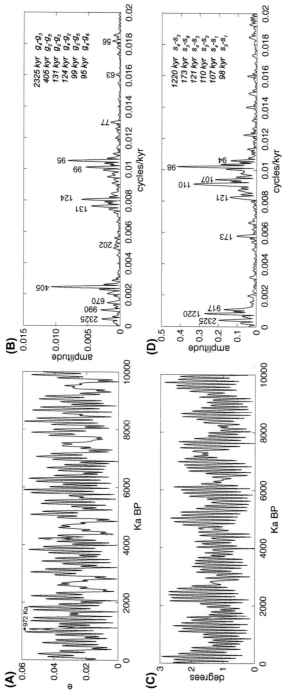

Figure 2 Earth's orbital parameters from the La2010d astronomical solution (Laskar et al., 2011a). (A) The orbital eccentricity variation, 0–10 Ma. (B) Periodogram of the orbital eccentricity variation shown in (A). (C) The orbital inclination variation relative to the invariable plane, 0–10 Ma (Note: Fig. 6 in Laskar et al. (2004) shows La2004 orbital inclination relative to the *ecliptic* plane). (D) Periodogram of the orbital inclination variation shown in (C). Details for how to calculate A and C are provided in Appendix A.

for which the s_j are the contributing frequencies of the planetary orbital nodes. Values for $p_3(t)$ and $q_3(t)$ are explicitly provided for the La2010 solution (see Appendix A). For the La2004 solution, the modeled (h,k,p,q) values must be extracted using *insola.f* (subroutine Telor and its dependencies) (see URL in Appendix A). Orbital eccentricity has long been a major focus in cyclostratigraphy (e.g., Section 1.4), orbital inclination far less so until recently (e.g., Section 3.2.2).

1.1.2 Astronomical Solutions
Originally, astronomical solutions were carried out analytically with trigonometric series approximations (e.g., Le Verrier, 1855; Stockwell, 1873; Brouwer and van Woerkom, 1950; Sharaf and Budnikova, 1967; Bretagnon, 1974). The advent of computers enabled solutions to be carried out by numerical integration, first for the outer planets (Jupiter to Pluto), which can be integrated over large steps, e.g., 40 days for Jupiter. Later the addition of the inner planets (Mercury to Mars) required integration over much smaller steps, e.g., 0.5 days (Laskar, 1993), and for a much larger number of combinations (Ito and Tanikawa, 2007). The first full Solar System solution (8 planets) involving numerical integration used averaged orbits and an integration step-size of 500 years (La1990 of Laskar, 1990). The La2004 solution was a "direct numerical integration of the gravitational equations" of the entire Solar System (8 planets plus Pluto) using a step-size of 0.005 years (=1.82625 days) (Laskar et al., 2004). It is valid (provides a unique solution) for 0 to 41 Ma. The La2010 solution (Laskar et al., 2011a) improved on La2004 by using a new, high-precision planetary ephemeris INPOP (Intégration Numérique Planétaire de l'Observatoire de Paris). It is valid for 0 to 50 Ma. Incorporating the asteroids Ceres, Pallas, Vesta, Iris and Bamberga (Laskar et al., 2011b) has led to the La2011 solution, also valid from 0 to 50 Ma (Laskar et al., 2012). New modeling by Zeebe (2017) provides unique solutions for 0 to 54 Ma. The time limitation of ~ 54 Ma is governed by the Lyapunov time of the Solar System, i.e., the timescale at which uncertainty in the planetary orbits increases by a factor of e, estimated to be approximately 5 Myr (Laskar, 1989).

1.1.3 Paleo-Astrodynamics
The question of whether the motions of the planets have been stable through time has led to the discovery of chaotic zones (Laskar, 1990), to which the orbits of the inner planets are particularly susceptible (Laskar, 2015). One set of motions of quantifiable interest has been the secular

resonance between Earth and Mars, between their rotating orbital perihelia ($g_4 - g_3$) and inclinations ($s_4 - s_3$). Laskar's numerical models predict that for 0 to at least 50 Ma, the solutions for these two motions are in 2:1 resonance, and are manifested in the Earth's orbital eccentricity and obliquity variations as long-period modulations of 2.4 Myr and 1.2 Myr, respectively (Fig. 3A). This means that the Earth–Mars resonance state should be observable in the cyclostratigraphic record. The 1.2 Myr obliquity

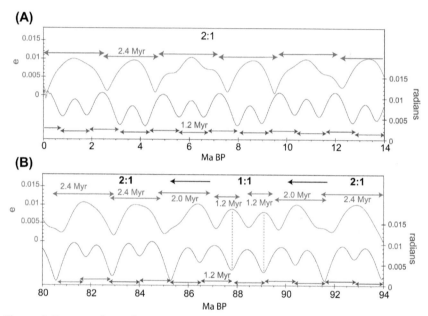

Figure 3 Excerpts from the 0–250 Ma record of $g_4 - g_3$ and $s_4 - s_3$ modulations from the La2004 solution highlighting a chaotic interval between 80 Ma to 94 Ma. The $g_4 - g_3$ modulations were extracted from the La2004 orbital eccentricity solution (Laskar et al., 2004) by Taner bandpass filtering (Kodama and Hinnov, 2015) with a lower cutoff frequency of 0.095 cycles/kyr, center frequency of 0.01025 cycles/kyr and upper cutoff frequency 0.0110 cycles/kyr; the filtered series was Hilbert-transformed to obtain the amplitude envelope, i.e., the modulations of interest. The $s_4 - s_3$ modulations were extracted from the La2004 obliquity solution by Taner bandpass filtering with a lower cutoff frequency of 0.023 cycles/kyr, a center frequency of 0.0275 cycles/kyr and upper cutoff frequency of 0.032 cycles/kyr; the filtered series was Hilbert transformed to obtain the modulations. (A) The modulations for 0 to 14 Ma show a stable 2:1 secular resonance between the orbital motions of Earth and Mars, i.e., 2.4 Myr:1.2 Myr. (B) The modulations for 80 to 94 Ma are chaotic: $g_4 - g_3$ modulations transition from a 2.4 Myr periodicity to a 1.2 Myr periodicity for two cycles, so that Earth-Mars orbital resonance to 1:1, then subsequently transition back to 2.4 Myr periodicity and 2:1 resonance.

modulation was discovered in Cenozoic cyclostratigraphy from ODP Leg 154 for a 10-Myr long interval spanning the Oligocene-Miocene boundary (Shackleton et al., 1999); later, the 2.4 Myr orbital eccentricity modulation was detected in the same data confirming that the Earth and Mars orbits were in the stable 2:1 resonance pattern as predicted by both La1993 and La2004 (Pälike et al., 2004).

The La2004 solution computed back to 250 Ma predicts that the most recent major chaotic transition took place from 83 Ma to 94 Ma. The modulation pattern for this interval (Fig. 3B) indicates that the $g_4 - g_3$ modulation shortens starting at 92 Ma, first to a single 2 Myr cycle, then to 1.2 Myr cycling for two repetitions before transitioning back first to a single 2.0 Myr cycle and then to 2.4 Myr at 85 Ma. Importantly the $s_4 - s_3$ modulation maintains a 1.2 Myr periodicity through the entire episode. This chaotic event has recently been detected in cyclostratigraphy of the Niobrara Formation (Western Interior Seaway) (Ma et al., 2017), and in the continental Songliao Basin (Wu et al., 2018a, 2018b), with a near-perfect fit to the characteristics of La2004 solution shown in Fig. 3B (The La2004 solution indicates another similar event between 78 Ma and 80 Ma, but involving only one 1.2 Myr cycle in the $g_4 - g_3$ modulation.) It should be emphasized that astronomers are unconvinced that any of the astronomical solutions, including La2004, could be accurate much further back in time from 55 Ma, and so the Cretaceous empirical evidence may not constitute a validation of the La2004 model specifically. Nonetheless, the discovery of this ancient chaotic interval in cyclostratigraphy is a significant step forward for paleo-astrodynamics.

One of the lessons of the Cretaceous example is that tracking $g_4 - g_3$ modulations only in cyclostratigraphy may be sufficient to identify Earth-Mars chaotic episodes. The stratigraphic amplification of the orbital eccentricity frequencies in the Triassic Newark Series and Inuyama Sequence (Section 2.2.3) brings to light very long-period 1.6 Myr to 1.8 Myr cycles ascribed as a direct contribution of $g_4 - g_3$ to the orbital eccentricity variation (not as modulator of the short orbital eccentricity) (Olsen and Kent, 1999; Ikeda and Tada, 2013). A 1.75 Myr cycle persists throughout the Newark Series, which suggests that the Earth-Mars orbits were in 1:1 resonance for an extended time of at least 22 Myr during the Late Triassic Period (see Fig. 16 in Olsen, 2010).

Other recent attempts to isolate possible $g_4 - g_3$ and $s_4 - s_3$ modulations have been made for Paleozoic cyclostratigraphy. The 13 Myr-long Late Permian section from Meishan and Shangsi (South China) do not reveal clearly recognizable modulations; slowly shifting depositional environments may have interfered with the rock magnetic proxies used in the study (Wu et al., 2013). Early-Middle Permian cyclostratigraphy at Shangsi (China) shows signs of 2:1 resonance, but both interpreted $g_4 - g_3$ and $s_4 - s_3$ periodicities are slightly shorter than 2.4 Myr and 1.2 Myr (Fang et al., 2017). In the Chinese Carboniferous, a 2:1 to 1:1 resonance transition from the Serpukovian to Moscovian was detected in joint $g_4 - g_3$ and $s_4 - s_3$ modulations from magnetic susceptibility measurements of the Luokun section (South China; Fang et al., 2018). These reports are being made whenever very long cyclostratigraphic sections are at hand, and so are opportunistic in nature. That said, the most critical geologic periods to investigate for Solar System chaos are those closest to the terminus of the astronomical model accuracy, i.e., back in time from 55 Ma, and in light of the Cretaceous evidence above, from 55 Ma to 80 Ma. Finally, the establishment of firm times of resonance transitions from cyclostratigraphy will be the key to identifying the best astronomical solution (Zeebe, 2017).

1.2 Geodynamics —Precession and Obliquity

1.2.1 Luni-Solar Precession

The Earth's luni-solar precession arises from the angle ε of the Earth's rotational axis with respect to the ecliptic plane, its equatorial bulge due to rotational effects on the non-rigid Earth, and by gravitational attraction from the Moon and Sun and the other planets on the equatorial bulge (Fig. 4). Astronomers refer to the Earth's obliquity as "inclination of the ecliptic to the equator", or "obliquity of the ecliptic", and the Earth's luni-solar precession as the "luni-solar precession of longitude" or (the major component of) "precession of the equator" (Dehant and Capitaine, 1997). The annually averaged luni-solar precession rate is (see Appendix C for variable definitions):

$$p_{\odot+\mathbb{C}} = \frac{3}{2}\left(\frac{G}{\phi}\right)H\left[\frac{m_\odot}{a^3}\left(1-e^2\right)^{-3/2}\right.$$
$$\left. + \frac{m_\mathbb{C}}{a_\mathbb{C}^3}\left(1-e_\mathbb{C}^2\right)^{-3/2}\left(1-\left(\frac{3}{2}\right)sin^2 I_\mathbb{C}\right)\right]cos\varepsilon \tag{4}$$

The variable $p_{\odot+\mathbb{C}}/cos\varepsilon$ is designated as the "precession constant" α in Eq. (8) of Laskar et al. (2004). Other smaller contributions to the total "precession of the equator" are summarized in Fig. 1 of Dehant and Capitaine (1997) and Table 3 of Williams (1994). For the International

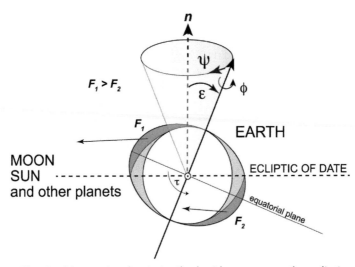

Figure 4 The Earth's rotational axis is tilted with respect to the ecliptic of date normal n at an obliquity angle ε; Earth's rotation Ω is anti-clockwise. The rotation raises a permanent equatorial bulge at right angles to the rotation axis (shown in green and pink). The equatorial bulge experiences gravitational forces F_1 and F_2 from the attracting bodies of the Moon, Sun and (to a much lesser extent) other planets (Note: Moon's orbit is presently inclined by $I_C = 5.15°$ relative to the ecliptic.). The pink areas line up in the direction of gravitational forcing, experience equal but opposite torques, and cancel out. The green areas, however, are offset from the direction of forcing; the bulge closest to the attracting bodies experiences greater forcing so that $F_1 > F_2$. This inequality leads to a net torque τ along a line in the equatorial plane that is normal to both the forcing direction and Earth's rotation axis. This torque generates an additional component of angular momentum on the Earth that is parallel to τ, changing the direction of the rotation axis. This leads to clockwise precession ψ of the rotation axis, which varies with the changing positions of the attracting bodies with respect to the equatorial bulge (e.g., at Spring Equinox $\tau = 0$), with an annually averaged value of approximately 50 arcseconds/ year (see Appendix C). *From Dehant and Mathews (2015) and Lowrie (2007).*

Astronomical Union 1976 System of Astronomical Constants (IAU76), $\varepsilon = 84381.448$ *arcseconds* $= 23.43929°$, and $p_{\odot+C} = 50.40736050$ arcseconds/year, or 1 cycle/(25,711 years). Astronomers continue to refine the precession model with improvements to the Solar System ephemerides and geophysical observations (Liu and Capitaine, 2017). The La2004 solution gives 50.475838 arcseconds/year, denoted as "p", e.g., Eq. (40) in Laskar et al. (2004), corresponding to 1 cycle/(25,676 years). Owing to the continuously revolving positions of the Moon and Earth with respect to the Sun, current instantaneous $p_{\odot+C}$ ranges from 0 (at the equinoxes) to as high as 115 arcseconds/year, varying with fortnightly and semi-annual tidal periodicities (Fig. 4 in Edvardsson et al., 2002).

1.2.2 Precession and Obliquity Equations

Planetary perturbations to the ecliptic plane, thus to the Earth's precession ψ and obliquity ε, are accounted for by the coupled precession and obliquity equations (De Surgy and Laskar, 1997, Eq. (6) in Laskar et al., 2004):

$$\frac{dX}{dt} = L\left(1 - \frac{X^2}{L^2}\right)^{\frac{1}{2}} \left(\mathcal{B}(t)sin\psi - \mathcal{A}(t)cos\psi\right)$$

$$\frac{d\psi}{dt} = \frac{\alpha X}{L} - \frac{X}{L\left(1 - \frac{X^2}{L^2}\right)^{\frac{1}{2}}}\left(\mathcal{A}(t)sin\psi + \mathcal{B}(t)cos\psi\right) - 2\mathcal{C}(t) \tag{5}$$

where $X = cos\varepsilon$ and $L = C\phi$, $\alpha = p_{\odot+\complement}/cos\varepsilon$, and $\mathcal{A}(t)$, $\mathcal{B}(t)$ and $\mathcal{C}(t)$ give time-dependent variations to the ecliptic plane caused by planetary perturbations, modeled by the astronomical solution (Appendix C).

The "solution of order zero" for the Earth's precession due to the Moon and Sun with no planetary disturbances, hence a constant obliquity ε, is (Eq. 27 in Laskar et al., 2004): $\psi = \psi_0 + pt$, where $p = p_{\odot+\complement}$. Higher order solutions to account for the planetary contributions take the form of (Eq. 34 in Laskar et al., 2004):

$$\psi = \psi_0 + pt + p_1 t^2 + other\ terms \tag{6}$$

where $\psi_0 = 49086\ arcsec$ and $p_1 = -13.526564\ arcsec/year^2$.

1.2.3 Tidal Dissipation and Dynamical Ellipticity

The Earth's rotation rate depends on tidal dissipation and mass redistribution (Lambeck, 1980). Tidal dissipation arises from anelastic response of the Earth to the tidal forces generated by the Sun and Moon (p. 290, Lambeck, 1980). The fluid (ocean) and anelastic (solid Earth) tidal responses result in a time delay in the net position of the tidal bulge relative to the Earth-Moon axis, the bulge leads the axis by a small angle δ (e.g., Fig. 4 in Wahr, 1988). A torque in the opposite direction of the Earth's rotation is raised between the Moon and the leading bulge, causing continuous rotational deceleration. Evidence from paleontology and tidalites confirms that the Earth's rotation rate has decreased over the past 2 Ga, and also that the deceleration was variable, apparently even reversing sign over at least one interval (Fig. 5).

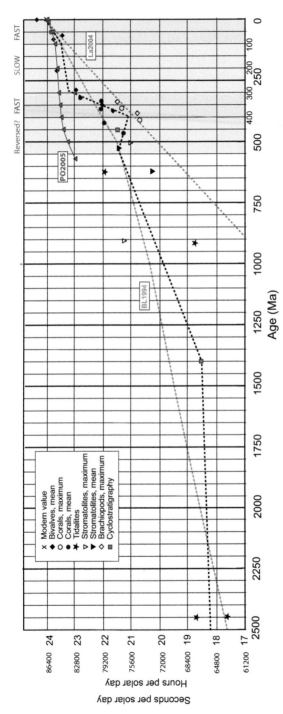

Figure 5 Evolution of Earth's rotation. Data are redrawn from Williams (2000) who compiled data from Scrutton (1978), Berry and Barker (1968), Wells (1970), Mazzullo (1971), Pannella (1972), and Vanyo and Awramik (1985). Data from cyclostratigraphy are from Meyers and Malinverno (2018). Results from the paleo-tidal model of Poliakow (2005) are also shown (purple triangles and line). The red dashed line indicates the Lunar Laser Ranging based empirical model assumed by Laskar et al. (2004) for the La2004 solution, and is extrapolated back in time until the model indicates 17 h LOD; the green dashed line indicates the paleontological model assumed by Berger and Loutre (1994), which averages through the paleontological and tidalite data.

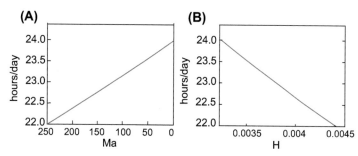

Figure 6 Dynamical ellipticity and rotation rate. (A) Lunar Laser Ranging measurements of 3.82 cm/year lunar recession (Dickey et al., 1994), biased slightly to 3.89 cm/year (Laskar et al., 2004), corresponds to a decrease in length-of-day of 2.68 milliseconds/century extrapolated back in time to 250 Ma. (B) Dynamical ellipticity H obtained by Eq. (6).

The Earth's shape is ellipsoidal and dynamic, commonly referred to as "dynamical ellipticity" (or "dynamical oblateness"), and depends on the Earth's rotation rate, internal density and the consequent gravitational and centrifugal force balances. It is characterized largely by rotational flattening of the poles, and has been approximated as (Eq. 9 in Berger and Loutre, 1994; related discussion on hydrostatic flattening on p. 28 in Lambeck, 1980) (Fig. 6):

$$H = (C - A)/\, C = 6.094 \; x \; 10^5 \; x \; \phi^2 \qquad (7)$$

where A and C are Earth's equatorial and polar moments of inertia, and ϕ is Earth's rotation rate (see Appendix C for values). H can also be affected by mass redistribution, e.g., ice sheets that can result in "climate friction", which contributes to changes in rotation rate and could potentially even lead to resonance with the orbital parameter related to $s_6 - g_6 + g_5$ (Laskar et al., 1993). H is fast-varying, contributing time-variable changes to p on the order of 1% per 100,000 years (Thomson, 1990). ϕ is slow-varying, contributing deep-time, secular changes in p on the order of 1% per 50 million years (Denis et al., 2002).

1.2.4 Paleo-Geodynamics

The importance of Earth's rotational dynamics in Milankovitch cycles rose to prominence during early efforts to estimate length-of-day (LOD) and Earth–Moon distance from ancient corals, mollusks and stromatolites, and sedimentary tidalites (Fig. 5). In a groundbreaking paper, Walker and Zahnle (1986) invoked an astronomical origin, namely the lunar nodal cycle, to explain the origin of 23:1 iron-silica lamination bundling patterns in the 2.5 Ga Weeli Wolli Formation of Australia (Trendall, 1973), for which each lamination pair was assumed to represent a varve. At 2.5 Ga, the lunar nodal cycle would have been significantly longer than its present-day 18.6-year period due to a shorter 0.86 of present-day Earth-Moon distance. They further proposed that an independent determination of their estimate could be made from coeval Milankovitch cycles, which they predicted would have had periodicities of 17 kyr for obliquity and 13 kyr for precession. 2.5 Ga Milankovitch cycles have not yet been firmly established (although see Section 2.4.3), but a recent study of Proterozoic Milankovitch cycles estimates 0.887 of present-day Earth-Moon distance at 1.4 Ga (Section 2.4.2; Meyers and Malinverno, 2018; see also Fig. 5). The Weeli Wolli Formation was later reassessed by Williams (1989) to have 28−30 laminae per bundle, and indicative of tidalites, but the suggestion by Walker and Zahnle (1986) that Milankovitch cycles could serve as a validation tool has remained intact.

The dependence of the precession and obliquity equations (Eq. 5) on $p_{\odot+c}$ (Eq. 4), which is defined by Earth's rotation rate and shape, was merged with the paleontological and tidalite evidence for shorter LOD in the geologic past, an Earth flattening model (Eq. 9), and lunar recession to estimate Milankovitch cycle periodicities from 0−500 Ma (Berger et al., 1989; Loutre et al., 1990; Berger et al., 1992) and later to 2.5 Ga (Berger and Loutre, 1994). This was followed by inclusion of adjustable tidal dissipation (TD) and dynamical ellipticity (E_D) variables in the La1993 and La2004 astronomical solutions (Laskar et al., 2004), while the La2004 solution adopted values based on present-day Lunar Laser Ranging observations of lunar recession. These variables were exploited in testing the power of Milankovitch cycles to solve these two unknowns for the Cenozoic (Pälike and Shackleton, 2000; Lourens et al., 2001; Zeeden et al., 2014). The results suggested that the La2004 model conforms well with Cenozoic cyclostratigraphy, suggesting that over the past 25 Myr, paleo-geodynamics was operating with present-day magnitudes (La2004 red dashed line in Fig. 5).

More recently an effort was made to estimate uncertainties imposed by paleo-geodynamics on Milankovitch cycles back to 700 Ma (Waltham, 2015); this work was motivated by a much lower tidal dissipation energy than present-day required by conditions surrounding the time of origin of the Moon. This requirement is supported by models of the early Eocene Ocean indicating much lower tidal dissipation compared with present-day

(1.44 TW for early Eocene vs. 2.78 TW for present day) (Green and Huber, 2013), with additional modeling for the Cretaceous (2.1 TW) and Permo-Triassic transition (0.9 TW) (Green et al., 2017), and by Poliakow (2005). Together with the observational evidence (black dashed line in Fig. 6) the modeling indicates a more slowly decreasing LOD from 250–50 Ma.

The Milankovitch record over this time, by contrast, appears to indicate obliquity cycle periodicities, in particular, that conform to the faster La2004 TD-E_D model, indicating 35 kyr obliquity at 245 Ma (Anisian; Li et al., 2018a), 32.7–32.9 kyr obliquity at 250–251 Ma (Griesbachian; Li et al., 2016); and a slight lengthening to 34-kyr at 252–260 Ma (Wuchiapingian-Changhsingian; Wu et al., 2013). In a new extraordinary 34 Myr-long section from the Chinese Carboniferous (Shangruya, Huashiban, Dala and Maping formations), there is evidence for an obliquity cycle evolving from a 30.5–30.7 kyr period at the base of the sequence, in the Visean-Bashkirian (333–315 Ma), to a 31.4 kyr period at the top of the sequence, in the Gzhelian (300 Ma) (Wu et al., 2018a, 2018b). A short 2 Myr long section from the late Bashkirian Copacabana Formation (Bolivia) analyzed with ASM (Section 3.5.2) supplemented with an optimization on k, indicates k = 58.7 arcsec/kyr (Ma et al., 2018). Assuming present-day $s_4 = -17.755$ arcsec/year and $s_3 = -18.850$ arcsec/year yields main obliquity periodicities of 31652 years and 32522 years, and a 22 h LOD. These periodicities are slightly longer than those estimated from the Chinese data; application of the same optimization procedure developed by Ma et al. might reconcile the differences. Finally, a more detailed optimization on k and g_i ("timeOptMCMC", Section 3.5.4) for the 1.4 Ga Xiamaling Formation (China) yielded k(=p) = 85.790450 arcsec/year, corresponding to an 18.68-h LOD (Meyers and Malinverno, 2018).

1.3 Milankovitch Cycles and Paleoclimate Responses

The promise of cyclostratigraphy to yield information about Solar System dynamics has in recent years, in the rush simply to identify astronomical signals, deemphasized research into the paleoclimatic response to astronomical forcing. Understanding the paleoclimatic response begins with the insolation equation: in particular, the insolation has very weak direct contributions from the Earth's obliquity and orbital eccentricity variations. There is no precession variation inherent to global insolation; nonlinear and threshold responses of the climate are required to generate precession index signals in insolation-forced climate change. Additional filtering takes place when climatic change is registered in the stratigraphic record (Sections 1.3.2 and 1.4).

1.3.1 The Insolation Equation

The solar irradiance has a major daily cycle due to the Earth's rotation, and a strong annual cycle due to the Earth's revolution around the Sun. The solar irradiance reaching any geographic location on Earth (see Fig. 6 in Hinnov, 2013) is given by the "insolation equation":

$$W = \left(S_0/r^2 \right) \cos z = \left(S_0/r^2 \right) \left(\sin\theta \sin\delta + \cos\theta \cos\delta \cos\hbar \right) \qquad (8)$$

Earth-Sun distance r depends on the orbital eccentricity and time of year, and solar declination angle δ depends on the obliquity; S_0 is the solar constant (in W/m^2), z is angle of the Sun in the local sky relative to the local zenith point, θ is the geographic latitude of the location, and $\hbar$ is the hour angle of the Sun in the local sky, where $\hbar = 0$ is at the meridian (local midday).

Eq. (8) was introduced by Meech (1856), who implemented elliptic integrals to accurately position the Earth on its orbit. Meech also considered obliquity and orbital eccentricity effects (based on Le Verrier) on insolation for specific values, concluding that both were insufficient to drive significant climate change. Pilgrim (1904) provided a 1-myr long calculation of the combined orbital eccentricity, obliquity and precession on insolation, based on the Stockwell (1873) astronomical solution (Appendix B). Finally, a comprehensive theory of astronomically forced insolation was completed, first using Pilgrim's values (Milankovitch, 1920), and later comparing this with the (Le Verrier-based) astronomical solution of Miskovitch (1931) in (Milankovitch, 1941) (see reproduction in Fig. 2 of Loutre, 2003).

FORTRAN codes for computing insolation were introduced by Berger (1978) and Laskar et al. (1993). A myriad of important insolation calculations was discussed in Berger et al. (1993), including instantaneous insolation (irradiance, in W/m^2), integrated insolation (irradiation, in J/m^2), and zenith-class distance. The freeware *Analyseries* (Paillard et al., 1996) calculates insolation with options for three astronomical solutions. Huybers (2006) adapted the Berger (1978) FORTRAN code into the MATLAB script *daily_insolation.m*. The MATLAB GUI *Earth Orbit v2.1* enables detailed visualization of the astronomical parameters and insolation (Kostadinov and Gilb, 2014). Berger et al. (2010) compared elliptical integral versus summation methods for accurate integrating insolation over time intervals along the Earth's orbit, with code recently made available in *palinsol* for R (Crucifix, 2016).

Calculating Eq. (8) with hourly dt obviously resolves the diurnal cycle; hourly dt would be needed when evaluating zenith-class insolation (Berger et al., 1993) when a climate response is expected for a limited range of solar angles in the sky, e.g., a west-facing mountain glacier. Monthly dt, resolving the seasonal cycle of insolation, reveals some surprises (Fig. 7). The seasonal (annual) cycle undergoes small interannual modulations forced by different astronomical parameter contributions throughout the year. At 65° North, winter insolation is modulated by the obliquity, spring (or fall) insolation is modulated by the precession index only, and summer insolation by a combination of obliquity and precession index variations. Moreover, from month to month, the phase of the precession index shifts by 30° increments, the consequence of each month occurring at a different point along the Earth's orbit. Other significant implications of this annually resolved insolation calculation are discussed at length in Section 1.3.2 below.

Inter-annually sampled insolation (for dt ≫ 1 year) resolves the astronomical parameters only, notably the latitudinal dependence of obliquity forcing as shown for June 21 at different Northern latitudes (Fig. 8). Obliquity forcing increases poleward, jumping up dramatically from 60° N to 80° N (i.e., across the Arctic Circle at 66° N), although never exceeding the amplitudes of the precession index. Also, the summer insolation increases from an average of 400 W/m^2 at the Equator to 525 W/m^2 at 80° N, due to more daylight hours with increasing latitude (up to 24 h north of the Arctic Circle). Finally, the Earth's orbital eccentricity plays a miniscule role in the direct forcing of insolation (see Section 1.4 for discussion on the dominance of orbital eccentricity signals in cyclostratigraphy).

1.3.2 Paleoclimate Responses

The annually resolved insolation calls attention to a long-standing dilemma in Milankovitch theory: the seasonal cycle of insolation (at 65° N) has an amplitude of more than ±200 W/m^2 (Fig. 7C), whereas the Milankovitch cycles have amplitudes of up to ±15 W/m^2 (for the obliquity) and up to 20 W/m^2 for the precession index (Fig. 8B), and occasionally at most, at 80° N, ±75 W/m^2 (peak-to-peak) amplitudes (Fig. 8A). By this measure, the amplitude of the astronomically forced insolation variations are only about a 10th of the amplitude of the seasonal insolation. Thus, the

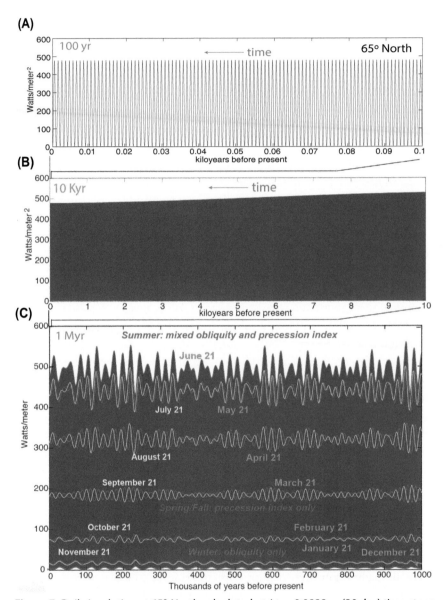

Figure 7 Daily insolation at 65° North calculated at $\Delta t = 0.0822$ yr (30 day) time steps using the MATLAB script *daily_insolation.m* (Huybers, 2006), with modifications (The mean value of k is maintained throughout the year in this calculation, and not changed as required by the lunar orbit, as mentioned in Section 1.2.1). (A) The most recent 100 years, showing a strong seasonal cycle in insolation. (B) The most recent 10 kyr, revealing a long-term change in the insolation cycle maxima. (C) The most recent 1 Myr, which shows astronomically forced maxima and minima (June 21 and December 21, respectively) in the Northern Hemisphere insolation cycles.

characterization of insolation at the interannual scale only (as in Fig. 8) does not acknowledge that the majority of insolation power is in the seasonal cycle.

Therefore, the modulations of the seasonal cycle of insolation must be the source of astronomically forced climate change. A simple model of how astronomically forced climate variations could come to dominate the paleoclimate record is shown in Figs. 9 and 10:

(1) The seasonal insolation, taken as the driver of the climate system and shown for the past 5 years (Fig. 9Aa), has an annual-band spectrum (Fig. 9B) with a major peak at 1 cycle/year, and small side-band peaks at $1 \pm 1/(54 \text{ kyr})$, $1 \pm 1/(41 \text{ kyr})$ and $1 \pm 1/(29 \text{ kyr})$ from the modulation by the obliquity variation, and a very small side-band peaks at $1-1(23\text{kyr})$ and $1-1/(9\text{kyr})$ from modulations by the precession index, which are one-sided due to the uni-directional counter-clockwise motion of the precession along the annual orbit (Fig. 1). The power spectrum across all frequencies (Fig. 10A) confirms that the annual spectral peak contains the majority of the power; there is a well-defined, but very low-power peak in the obliquity band, approximately 4 orders of magnitude lower than the annual power.

(2) A "threshold climate" response to the seasonal insolation would involve a 'threshold' insolation below which the climate system does not respond (is "insensitive"); above this threshold, the climate responds linearly to the insolation (Fig. 9C). For example, there may be a minimum insolation required to trigger marine productivity or warm the ocean surface layer. This produces a paleoclimate proxy that is a "half-rectified" version of the insolation. The annual-band spectrum shows a slight strengthening of the obliquity- and precession-induced side-bands with respect to the annual peak (Fig. 9D). The power spectrum across all frequencies (Fig. 10B) shows an annual peak that is slightly diminished in power (reflecting the near-halving of the annual cycle amplitude by the threshold), but now in addition to obliquity power there is precession index power at a comparable level,

---

The magenta and light orange curves indicate daily insolation for specific calendar days in the year, revealing the familiar astronomical variations, with a mix of precession index and obliquity variations affecting insolation maxima (June 21), precession index only at spring (or fall) equinox (March 21 or September 21), and obliquity affecting insolation minima (December 21), and a shifting phase of the precession index variations (see text).

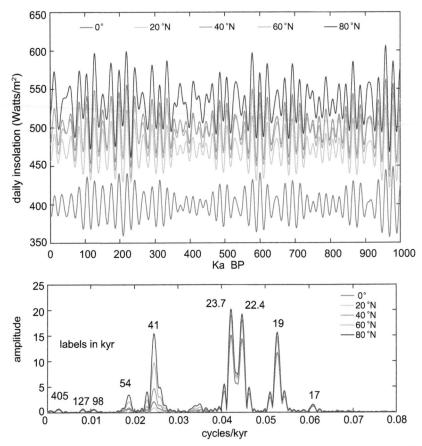

Figure 8 Mean daily insolation at June 21 (northern Summer Solstice) for latitudes 0°, 20° N, 40° N, 60° N and 80° N. (A) Insolation sampled at Δt = 1 kyr for 0 to 1 Ma using the La2004 astronomical solution. (B) 2π multitaper amplitude spectra of the insolation time series shown in (A) All calculations were carried out in *Analyseries 2.0.8* (Paillard et al., 1996).

albeit both have significantly less power than the annual frequency, which is still nearly 4 orders of magnitude higher in power.

(3) A "mixed climate" response might additionally respond non-linearly to the insolation above the threshold level envisioned above (Fig. 9E). For example, there could be disruptions to ocean surface layer processes from ocean–atmospheric dynamics causing delays, or stochastic weather responding rapidly to rising insolation. The annual-band spectrum (Fig. 9F) now shows obliquity and precession index side-bands on both sides of the annual frequency, which maintains dominant power. However, in the power spectrum across all frequencies (Fig. 10C)

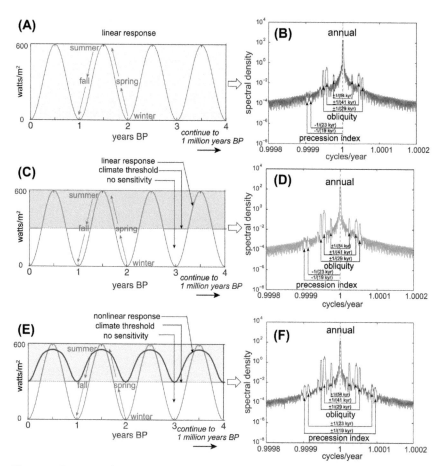

Figure 9 Conceptual climate responses to insolation forcing at 65° North, past 4 years, Δt = 30 days (=0.0822 years). (A) *Linear climate*: Solar irradiance at the top of the atmosphere is the driver of a linearly responding climate system; (B) Power spectrum of a calculated for 0–1 Ma. (C) *Threshold climate*: half-wave rectification (green curve) truncating the seasonal insolation cycles (blue curve) below a threshold (dashed green line) to simulate a threshold insolation required, for example, to trigger marine productivity, or to warm the ocean surface layer. (D) Power spectrum of c calculated for 0–1 Ma. (E) *Mixed climate*: Half-wave rectified seasonal insolation cycles (green curve from B) with amplifications, delays and other nonlinearities (purple curve), for example, to simulate disruptions to ocean surface layer processes from ocean-atmosphere dynamics, weather, etc. (F) Power spectrum of E calculated for 0–1 Ma.

obliquity and precession index power is now only approximately 1 order of magnitude lower than the annual power; also, obliquity power exceeds precession index power, which is normally accomplished with time-integrated insolation (e.g., Huybers, 2006; Berger et al., 2010).

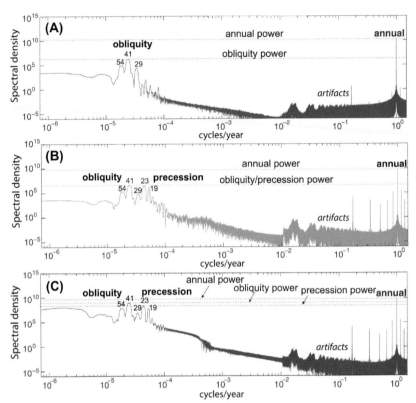

Figure 10 Power spectra of climate responses to 65° N insolation for 0 to 1 Ma presented in Fig. 8. (A) Linear climate response; (B) threshold climate response; (C) mixed climate response. The horizontal dashed lines indicate power level of the annual, obliquity and precession index frequencies; "artifacts" are likely due to time sampling errors (there are no signal components in the spectrum between the annual and astronomical frequency bands).

1.4 Stratigraphic Record of Milankovitch Cycles

While climate is fast-acting and seasonal, its sedimentary record typically involves "filtering" through a much slower system of responses. Some depositional environments have accumulation rates that preserve seasonality in varved sequences, with accumulation rates on the order of a 1–10 mm/year. However, these environments tend not to operate long enough to preserve astronomical timescales, i.e., they fill up quickly. However, there are a few exceptions, e.g., the sequence of 200,000 evaporite varves of the Castile Formation (Ochoan, late Permian) (Anderson, 2011). The vast majority of depositional systems have much lower accumulation rates, on the order of 1 to 10 cm/kyr, which biases the sedimentary record to preservation of the Milankovitch cycles. Below additional transformations of Milankovitch cycles are described, to explain how orbital eccentricity (405 kyr and ~100 kyr) cycles come to dominate cyclostratigraphy.

1.4.1 Rectification Effects

Consider a threshold climate (see Section 1.3.2) that registers only the mean of June and July insolation in the seasonal cycle every year; this would result in an interannual climate forcing signal (Fig. 11A, with a power spectrum showing major power at precession index frequencies, and moderate-low power at the obliquity frequencies (Fig. 11B). Suppose that there is a sedimentation response only when insolation is greater than the mean value (Fig. 11C); this "clipping" imposes a half-rectification on the climate signal that will be recorded in the sediment. Its power spectrum shows the appearance of difference tones in the orbital eccentricity frequencies and addition tones at high frequencies (Fig. 11D). Sedimentation of the climate proxy (which assumes the insolation values) now takes place, with a 1 cm/kyr sedimentation rate whenever insolation is above the mean value, and 0 cm/kyr otherwise, squeezing the signal as it takes on its stratigraphic form (Fig. 11E), which amplifies the difference tones (the orbital eccentricity frequencies) and suppresses the 1/(19 kyr) precession index frequency and the addition tones (Fig. 11F). Also of some surprise is amplification of the 1/(54 kyr) obliquity frequency. Finally, sediment mixing (e.g., bioturbation) is simulated for a 10-cm depth (Fig. 11G), which further amplifies the orbital eccentricity frequencies (Fig. 11H), and for a 20-cm depth (Fig. 11I), which in this case almost completely eradicates the precession index (Fig. 11J). This nonlinear sedimentation and mixing model is similar to the one proposed by Ripepe and Fischer (1991) to explain the strong orbital eccentricity frequencies and near absent precession index frequencies in the Lower Cretaceous pelagic Scisti a Fucoidi (Section 2.2.1).

1.4.2 Variable Accumulation Rates

Stratigraphic time scales are notoriously difficult to determine owing to variable sedimentation rates through time. These rates can have random fluctuations, secular trends, and even cyclic variations, and can defocus and obscure what would otherwise be detected as a clear astronomical signal. The synthetic example shown in Fig. 12 illustrates the strong distorting

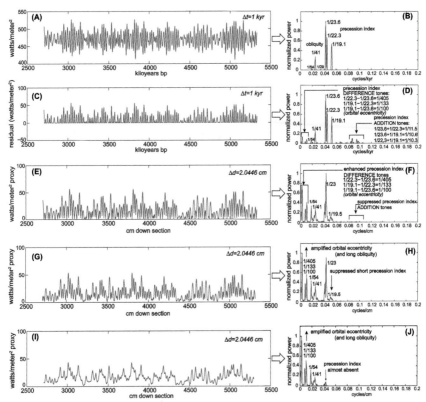

Figure 11 Stratigraphic enhancement of orbital eccentricity in cyclostratigraphy. (A) Mean June–July insolation at $65°$ North with La2004 solution from 2.706 Ma to 5.329 Ma, calculated at dt = 1 kyr using Analyseries (Paillard et al., 1996), (B) 2π multitaper power spectrum of A. (C) Insolation from A, half-rectified to simulate sedimentation response for insolation greater than the mean value of 470 W/m^2, reset to zero in the plot. (D) 2π multitaper power spectrum of C showing redistribution of power into the difference tones and addition tones of the precession index frequencies. (E) Conversion of C into stratigraphy assuming a 1 cm/kyr sedimentation rate and assuming insolation values as the paleoclimatic proxy. To maintain the actual start and stop times of the stratigraphic series (2.706 Ma to 5.329 Ma) the stratigraphic sample rate was set at 2.0446 cm/kyr. (F) 2π multitaper power spectrum of E showing amplification of difference tones in the orbital eccentricity band and 1/(54 kyr) obliquity frequency, and suppression of addition tones and the short 1/(19 kyr) precession index. (G) Simulation of bioturbation with a simple 10-m smoothing window applied to the stratigraphic series in E. (H) 2π multitaper power spectrum of (G) showing further amplification of orbital eccentricity frequencies and 1/(54 kyr) obliquity frequency, and further reduction of addition tones. (I) Simulation of bioturbation with a simple 20-m smoothing window applied to the stratigraphic series in E. (J) 2π multitaper power spectrum of I showing further amplification of orbital eccentricity frequencies and near full suppression of the precession index.

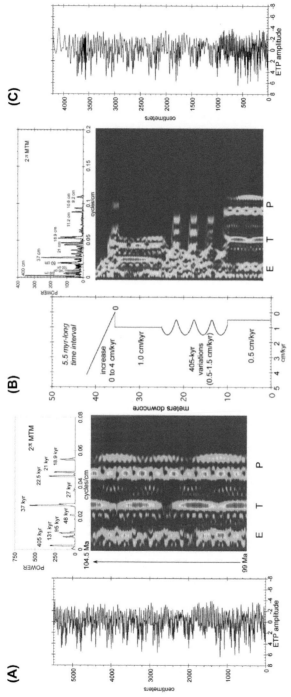

Figure 12 A demonstration of variable accumulation rates and their effects on an astronomical forcing signal, represented as the synthetic combination ETP (defined in Section 3.1.1). (A) Left: ETP signal from 99 Ma to 104.5 Ma from the La2004 solution at http://vo.imcce.fr/insola/earth/online/earth/earth.html, converted to "stratigraphy" assuming a 1 cm/kyr sedimentation rate. Top: 2π power spectrum; bottom: FFT spectrogram computed with a 250-cm window. (B) Variable sedimentation rates to be applied to the same ETP signal that is recorded in A. (C) Right: ETP signal converted to stratigraphy according to the sedimentation rates in B. Top: 2π power spectrum; bottom: FFT spectrogram computed with a 250-cm window.

effects of a series of sedimentation rate variations on an astronomical signal. The example (Fig. 12B) considers a series of sedimentation rates starting with a constant rate of 0.5 cm/kyr, changing to a cyclic variation with a 405-kyr periodicity, followed by a constant 1 cm/kyr rate, a brief (one sample long) hiatus, and ending with an interval with a linear increase in rates from 0 to 4 cm/kyr. While the two stratigraphic intervals characterized by constant rates have an easily identifiable ETP signal, it is far more difficult to detect the ETP signal in the intervals with variable rates (Fig. 12C). This problem is ubiquitous in cyclostratigraphy — especially in the very long, multi-million year-long sequences that are being collected today — and is the main motivation for the practice of "tuning" discussed at length below (Section 3.3).

2. CYCLOSTRATIGRAPHY

Today sedimentologists and stratigraphers are sensitive to an unprecedented degree that astronomical forcing could be recorded in their sections. Researchers with a singular focus on cyclostratigraphy and reconstructing timescales (Section 4) are now collecting multi-million year-long cyclstratigraphic series that are correlated to all available biostratigraphy, magnetostratigraphy, chemostratigraphy and radioisotopic geochronology. Cyclostratigraphy from 0 to 55 Ma can theoretically be matched directly to the astronomical solution (Section 1.1.2), although modeling and testing will continue to be required to determine $p_{\odot+c}$ for all geologic times (Section 1.1.4). Visual identification (Section 3.3.1) of the 405-kyr metronome (Section 3.2.1) has been especially provident for effective alignment of astronomical signals in Mesozoic and Paleozoic cyclostratigraphy (Section 4). Representative examples of well-known cyclostratigraphic sequences are highlighted below. Also notable is that the best cyclostratigraphic evidence almost exclusively comes from drill core data, which provides the required level of high-precision control on stratigraphic position.

2.1 Cenozoic

The cyclostratigraphy that has contributed the most to the Neogene time scale is from the Mediterranean sapropelic sequence, which has been analyzed for biostratigraphy, magnetostratigraphy and oxygen isotopes, and correlated with deep-sea drill core counterparts from the Atlantic and

Pacific oceans (Lourens et al., 2004; Hilgen et al., 2012). The astronomically forced benthic marine $\delta^{18}O$ record for 0—5 Ma has also been correlated and "stacked" across 57 sections from the Atlantic, Pacific, Indian and Southern oceans (Lisiecki and Raymo, 2005), testimony to the strong global nature of the record.

Following on the Cenozoic $\delta^{18}O$ curve compiled by Zachos et al. (2001, 2008), a benthic marine $\delta^{18}O$ "Megasplice" over 0 to 35 Ma was recently stitched together from single ODP core records in series for study of the time evolution of phasing of recorded astronomical parameters (De Vleeschouwer et al., 2017a). This "Megasplice" (Fig. 13A) shows the characteristic Oi1 step at the Eocene-Oligocene boundary, interpreted as the glaciation of Antarctica, followed by strong astronomically forced cycling throughout the Oligocene, ending with a pronounced Mi1 spike at the Oligocene-Miocene boundary (Paleogene-Neogene boundary), a reversal during the mid-Miocene, followed by renewed cooling at Mi3, and in this record, stasis from 13 to 3 Ma, when major glaciations came to dominate the Northern Hemisphere. The astronomical cycles are very pronounced, and reasonably well-aligned by the current tuning, except at precession timescales (detail and power spectrum in Fig. 13A).

Gamma ray (GR) logs from boreholes in the continental Bohai Bay Basin (North China) have been assembled into a complete Paleogene sequence of syn-rift and post-rift sedimentation from 66 Ma to 23 Ma, with a continuous 43-Myr long record of astronomical forcing (Liu et al., 2017a). 405-kyr cycles were identified and used to reconstruct the sedimentation history of the basin, showing a dramatic drop (by a factor of 6) in sedimentation rate at 35.99 Ma just prior to Eocene-Oligocene boundary (and glaciation of Antarctica). Rifting events in the basin, indicated by increases in sedimentation rate, are newly calibrated by the 405-kyr-based timescale, and compared with the subduction of the neighboring Pacific Plate to the east, collision of the Indian and Eurasian plates to the south, and drying episodes in the Late Eocene of eastern Asia.

Elsewhere, Eocene cyclostratigraphy has been correlated to the astronomical solution with core-scanning XRF and carbon isotope data indicating 405-kyr cycles, with researchers chipping away at a longstanding mid-late Eocene cyclostratographic gap (Westerhold et al., 2012, 2014, 2015; Boulila et al., 2018). The timing of the Paleocene/Eocene boundary has been astronomically estimated, using interpreted precessional cycles, as 170 kyr (Röhl et al., 2007). In the Paleocene, cyclostratigraphy from ODP cores and outcrops (notably Zumaia, Spain) has also been used to

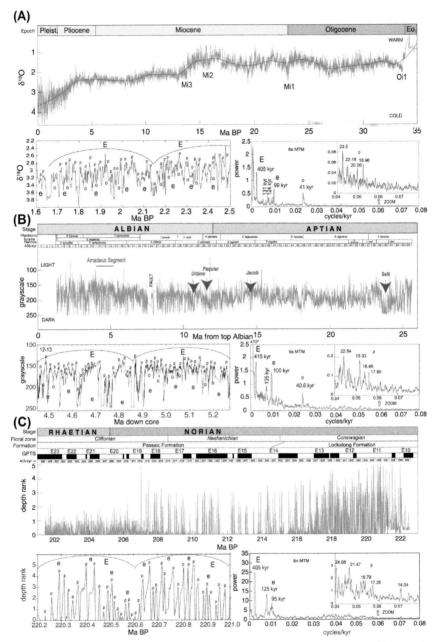

Figure 13 Exceptional cyclostratigraphic sequences from the Cenozoic and Mesozoic eras. For each case, three panels are displayed: top: the tuned cyclostratigraphic proxy; bottom left: a detail; bottom right: 6π multitaper power spectrum, with a vertical zoom of the precession index band. Precession index power is low in part

calibrate the age of the K/Pg boundary, which occurred during an orbital eccentricity minimum (Herbert, 1999), with research progressively adjusting the number of observed 405-kyr cycles through the epoch from 24 to 25 together with new radioisotope geochronology (65.28 Ma or 65.68 Ma, Westerhold et al., 2007, 2008; 65.95 Ma, Kuiper et al., 2008; 66 Ma, Hilgen et al., 2010; 66.043 ± 0.043 Ma, Renne et al., 2013; 66.022 ± 0.040 Ma, Dinarès-Thurell et al., 2014).

2.2 Mesozoic

Over the past decade there has been rapid progress in studies of Mesozoic cyclostratigraphy and astrochronology. A complete account of the significant developments is provided in Huang (this volume), with highlights summarized below.

2.2.1 Cretaceous

The K/Pb boundary calibrated from cyclostratigraphy from below (i.e., from the Maastrichtian) is consistent with the Paleocene calibration from Zumaia and neighboring outcrops (65.56 Ma or 65.97 Ma; Batenburg et al., 2012, 2014; 66 Ma, Dinarès-Thurell et al., 2013) and from ODP Site 762 (66 ± 0.07 Ma, Husson et al., 2011). The estimate of approximately 300 kyr from the K/Pb boundary to the base of C29r (e.g., Table 1 in

◄ _____

due to uncorrected high frequency sedimentation rate variations. Labels: E = 405 kyr long orbital eccentricity, e = ~100 kyr short orbital eccentricity, o = obliquity, and p = precession index cycles. (A) 35 Myr long composite benthic marine "δ^{18}O Megasplice" from late Eocene-Pleistocene ODP drill core records consisting of 12171 measurements with an average $\Delta t = 3$ kyr (De Vleeschouwer et al., 2017a). Epochs are shown at the top; Oi1, Mi1, Mi2 and Mi3 indicate glaciation events. Note: the δ^{18}O scale (y-axis) points down. This "Megasplice" is publicly available at https://doi.pangaea.de/10.1594/PANGAEA.869815. (B) 23.936 Myr long grayscale scan with a 0.8 mm resolution of the 77-m long Aptian-Albian abyssal Scisti a Fucoidi in the Piobbico Core, central Italy, tuned to the indicated 405-kyr cycles, resolving an average $\Delta t = 0.24$ kyr (Grippo et al., 2004; Huang et al., 2010a, 2010b, 2010c). The "Amadeus Segment" is the 8-m segment analyzed by Herbert et al. (1986) and Park and Herbert (1987). The numbers in the rectangles represent 405-kyr cycle number from top of Cenomanian. *Data are courtesy of A. Grippo.* (C) 21.12 Myr long depth rank series of a 3447-m long composite section of drill cores from the Norian-Rhaetian Newark Basin, eastern USA with 4040 points, tuned to the indicated 405-kyr "McLaughlin" cycles, for an average $\Delta t = 5$ kyr (Kent et al., 2017, 2018). The numbers in the rectangles represent 405-kyr cycle number from the present. This dataset is publicly available at https://www.ldeo.columbia.edu/~polsen/nbcp/data.html.

Table 1 Planetary Secular Frequencies from Table 3 in Laskar et al. (2004). The Subscripts 1–6 Represent Each Planet (1 = Mercury; 2 = Venus; ..., 6 = Saturn)

Secular frequencies	Arcseconds/yr	yr
g_2	7.452	173913
g_3	17.368	74620
g_4	17.916	72338
g_5	4.257452	304307
s_2	−7.050	−183830
s_3	−18.850	−68753
s_4	−17.755	−72911
s_6	−26.347855	−49188

Hennebert, 2014) inspired an upward cyclostratigraphic extrapolation from the base of C29r to identify the K-Pb boundary in the continental Songliao Basin of China (Wu et al., 2014), which exhibits cyclostratigraphy through much of its depositional history through the entire Cretaceous Period (Wang et al., 2013). A well-known Tethyan cyclostratigraphic section from Bottacione Gorge, Italy was evaluated, yielding a Santonian duration of 2.94 Myr, and a Coniacian duration of 3.46 Myr (Sprovieri et al., 2013). In the Western Interior Seaway, USA cyclostratigraphy integrated into the regional chronostratigraphy to provide an astronomical timescale for the entire Niobrara Formation with a Santonian duration of 2.3 ± 0.82 Myr (shorter than the Tethyan scale) and a Coniacian duration of 3.26 ± 0.82 Myr (Sageman et al., 2014). This was followed by a transformational analysis of the long-term modulations of the eccentricity ($g_4 - g_3$) and obliquity ($s_4 - s_3$) in the Niobrara Formation, which detected evidence for an orbital resonance transition between Earth and Mars that had been predicted by the La2004 solution from 83 Ma to 90 Ma (Ma et al., 2017) (see Section 1.1.3).

In the Tethyan abyssal realm, in central Italy (not far from Bottacione) a core was drilled through the entire Aptian-Albian section, the "Piobbico Core" (Fischer et al., 1991). Among the many innovative studies that have been carried out on this core, a high-resolution grayscale scan of digitized core photographs was analyzed (Fig. 13B; Grippo et al., 2004; Huang et al., 2010a). The recognition of 405-kyr cycles through the entire section provides a 24 Myr long astrochronology, continuous sampling of the major lower Cretaceous black shales (e.g., the Selli event) and recovery of an explicit record of all three astronomical parameters (detail and power spectrum in Fig. 13B). The grayscale scan (lightness L*) of a drill core through the Volgian Stage (the lowermost Cretaceous to upper Jurassic) in the Norwegian Sea records 13.3 Myr of astronomical cycling, with the potential to solve the long-standing problem of correlating the Volgian to the international chronostratigraphic units (Tithonian and Berriasian stages; Huang et al., 2010b).

2.2.2 Jurassic

The upper Jurassic source rock, the Kimmeridge Clay (Dorset, England) was drilled in 1996 and 1997, providing excellent drill cores for study of its organic shales and cyclostratigraphy (Morgans-Bell et al., 2001). An initial cyclostratigraphic study (Weedon et al., 2004) was revised into a 405-kyr cycle calibration using total organic carbon (TOC) and Formation Micro Scanner (FMS) data (Huang et al., 2010c), indicating a 6.74 Myr long duration for the formation. Studies of shorter cyclostratigraphic intervals have identified 405-kyr cycles in a number of European Aalenian-Oxfordian sections (see Appendix G). The Toarcian cyclostratigraphic sequence of the Sancerre Core (France) has been tuned to 405-kyr cycles for a total duration of 8.3 Myr (Boulila et al., 2014), including the renowned Toarcian Oceanic Anoxic Event, for which the astrochronological interpretation remains controversial and unsettled (reviewed in Boulila et al., 2017). Hand-held XRF measurements of the Pliensbachian Mochras Core (Wales) have revealed pronounced astronomical cycles, with strong 405-kyr cycles indicating a duration of 8.7 Myr (Ruhl et al., 2016). Currently in the lower Jurassic only the Sinemurian is missing complete cyclostratigraphic coverage; the Hettangian-Lower Sinemurian outcrops from St. Audrie's Bay/Quantoxhead (Somerset, UK) provide a 2.7 Myr long astronomical record at the start of the Jurassic Period (Hüsing et al., 2014).

2.2.3 Triassic

The Newark Series is a thick continental sedimentary sequence from a tropical rift system in eastern North America (Olsen, 1997, 2010). Within the Newark Basin a nearly 7 km-thick cyclic lacustrine sequence from 233 Ma to 199 Ma archives the early evolution of the dinosaurs, the rifting of Pangea, and an astronomically forced paleoclimate system (Olsen and Kent, 1996). Depositional cycles are comprised of organic-rich, laminated shales (deep lake) yielding to subaerial-exposed, desiccated mudstones (lake margin) in repeating sequences, documented by facies analysis of drill cores, and compiled as a "depth rank" series (Fig. 13C). This depth rank series shows the hallmarks of precession-eccentricity cycling, with 20-kyr "Van Houten cycles" ('p') bundled into ~100-kyr "short modulating cycles" ('e') and 405-kyr "McLaughlin cycles" ('E') (Kent et al., 1995; Olsen and Kent, 1996; Kent et al., 2017). The latter recently were shown to be precisely phased with the 405-kyr term of the La2004 solution (Table 6 in Laskar et al., 2004; see Appendix F) (Kent et al., 2018; Hinnov, 2018). Coeval to the Newark Series — but a world away — is a remarkable marine bedded chert sequence (Inuyama, Japan), which spans 70 Myr from Lower Triassic to Lower Jurassic (Ikeda et al., 2010; Ikeda and Tada, 2013, 2014; Ikeda et al., 2017) (see Fig. 5 in Huang, 2018). This sequence, with its extremely slow accumulation rate, has a greatly magnified the recorded orbital eccentricity, and modulations with a 1.6 to 1.8 Myr periodicity that may be related to a similar, 1.75 Myr periodicity detected in the Newark depth rank series (Ikeda and Tada, 2013; Olsen and Kent, 1999). Both records differ from the low-frequency orbital eccentricity solutions computed for this time period (i.e., La2004, La2010a—d). The extraordinarily long span of the Inuyama sequence also reveals even longer term, ~8 Myr cycling, which has yet to be assigned an attribution. Finally, Lower Triassic cyclostratigraphy in South China (Li et al., 2016), has recently been correlated with the German Basin (Li et al., 2018a), as well as to the Japanese Inuyama sequence (Huang, 2018).

2.3 Paleozoic

As would be expected, cyclostratigraphy of the Paleozoic Era is relatively unexplored. Since there are no astronomical solutions to provide an accurate time framework, Paleozoic cyclostratigraphy is limited to empirical evidence. An active research initiative is underway to improve knowledge of this evidence (International Geological Correlation Programme 652, http://www.geolsed.ulg.ac.be/IGCP_652/). The recent confirmation that the strong 405-kyr orbital eccentricity cycle in the La2004 solution is precisely phased with the radioisotopically calibrated Newark Series at 215 Ma (Section 2.2.3) raises expectation that as with Cenozoic and Mesozoic, 405-kyr cycles also dominated Paleozoic cyclostratigraphy, and this is affirmed below.

2.3.1 Permian

Intense interest in the end-Permian biotic extinction and its causes has generated high-precision geochronological data (Burgess et al., 2014; Burgess and Bowring, 2015; Burgess et al., 2017; Ramezani and Bowring, 2018) and analysis of late Permian Chinese cyclostratigraphy exhibiting strong 405-kyr cycles (Wu et al., 2013). In the lower North American Ochoan (Late Permian), the extraordinary evaporitic Castile Formation, USA with more than 200,000 carbonate-anhydrite varves provides a rare and fascinating observational window into paleoclimatic variations from annual to orbital eccentricity scale (Anderson, 1982; Anderson, 2011). The sequence of basinal chert-mudstone alternations of the Guadalupian (Middle Permian) Gufeng Formation (Lower Yangze, South China) was tested with a three-step (obliquity/long-precession/obliquity) tuning process (similar to the procedure of Imbrie et al., 1984) sharpen power in the obliquity and precession index bands (Yao et al., 2015). To the west, the coeval, highly cyclic limestone-rich Maokou Formation (Upper Yangze, South China) was measured for magnetic susceptibility and anhysteretic remanent magnetism, revealing astronomical forcing. Tuning to strong 405-kyr cycles aligns a strong 32 kyr cycle interpreted as obliquity (shortened from present-day, see Section 1.2.4), and evidence for orbital resonance transitions (Section 1.1.3) (Fang et al., 2015, 2017).

2.3.2 Carboniferous

The Carboniferous Earth was an icehouse with large and dynamic continental ice sheets enveloping southern Gondwana, and large sea level oscillations repeatedly flooding the continental margins. One of the great legacies of the sea level transgression-regressions are the well-known Carboniferous cyclothems, recognized worldwide, e.g., the Midcontinent Sea (Algeo and Heckel, 2008) and Russian Platform of Euramerica (Eros et al., 2012), eastern Panthalassic Gondwana margin (Grader et al., 2008), South China Block (Feng et al., 1998), and Australia (Fielding et al., 2008). Despite the high repute of the cyclothems, until recently, their origins have not been ascribed to Milankovitch forcing. High-precision radioisotope geochronology was the first line of evidence indicating that cyclothems occurred at 405-kyr intervals (Davydov et al., 2010; Schmitz and Davydov, 2012). The cyclic periplatform slope deposits of the Dian-Qian-Gui-Xiang Platform in South China were recently discovered to contain a 14.6 Myr long record (36 405-kyr cycles) of astronomical forcing (Fang et al., 2018). Another ongoing investigation at Naqing (South China) has documented a 34 Myr-long record of astronomical cycles that is so long that a shift toward lower-period obliquity cycles can be detected, from 30.5 kyr to 31.5 kyr (Wu et al., 2018a, 2018b). Finally, new analysis of Late Bashkirian cyclostratigraphy in the Copacabana Formation of Bolivia (Carvajal et al., 2018) has enabled an estimate of $k(= p_{\odot+c})$ and Carboniferous LOD (Section 1.2.4) (Ma et al., 2018).

2.3.3 Devonian
The Devonian Period was predominantly a greenhouse, except for the latest Devonian (Famennian), during which climatic instabilities swinging between intervals of glaciation and oceanic anoxia occurred repeatedly in advance of the Carboniferous icehouse (Isaacson et al., 2008; McGhee, 2013). The astronomical forcing of the Famennian was investigated comprehensively by Pas et al. (2018), who analyzed drill core from the deep shelf deposits of the Illinois Basin, USA, finding a 13.5 Myr-long record of strong 405-kyr cycling, accompanied by ∼100 kyr and 34 kyr cycles, ascribed to forcing from the orbital eccentricity and obliquity, with stratigraphic effects presumably suppressing what had to have included strong precession index forcing (see Section 1.4.1). Following on modeling by Le Hir et al. (2011) demonstrating that combined continental drifting and land plant expansion caused the significant CO_2 drawdown during the Devonian, De Vleeschouwer et al. (2014) used the Hadley Centre GCM to simulate Devonian climate conditions for different combinations of the astronomical parameters. The results indicate that for the Frasnian greenhouse, confluent obliquity and orbital eccentricity maxima produce the highest mean annual temperatures (27 °C) and confluent minima produce the lowest (19.5 °C) temperatures (This is consistent with Mesozoic and Cenozoic responses to these parameter configurations; e.g., Hinnov, 2018.). This pattern was explored further to explain the timing of the Late Devonian extinctions during the Frasnian-Famennian transition (De Vleeschouwer et al., 2017b). Cyclostratigraphy of outer ramp carbonate facies of the Belgian Givetian was analyzed by De Vleeschouwer et al. (2015) using a magnetic susceptibility proxy, documenting 405-kyr and ∼100 kyr cycles (orbital eccentricity), 33 kyr cycles (obliquity) and 18 kyr cycles (precession index), and an estimated duration (based on the 405-kyr cycles) of 4.35 ± 0.45 Myr for the Givetian Stage. The 33 kyr periodicity assigned to the

obliquity suggests a slower LOD than for the Carboniferous, for which a substantially faster 30.5 kyr to 31.5 kyr obliquity periodicity has been estimated (Section 2.3.2). Moroccan cyclostratigraphy of the Eifelian Stage based on low-resolution magnetic susceptibility measurements suggests a duration of 6.28 Myr for the stage (Ellwood et al., 2015). Astronomical forcing of Early Devonian cyclostratigraphy of the Czech Republic was identified with ASM modeling (Section 3.5.2) assuming the modeled Devonian (400 Ma) astronomical parameter periodicities of Berger et al. (1992) (orbital eccentricity: 400 kyr, 131 kyr, 123 kyr, 100 kyr and 95 kyr; obliquity: 38.7 kyr and 31.6 kyr; precession index: 19.7 kyr and 16.7 kyr) (Da Silva et al., 2016; power spectra showing obliquity and precession index bands appear in their Appendix.)

2.3.4 Silurian

While there is a long-standing interest in Silurian cyclostratigraphy, practically no studies have yet attempted to identify whether astronomical forcing played a role. In their study of the Lochovian GSSP, Crick et al. (2001) measured the magnetic susceptibility across the Silurian-Devonian boundary in the Czech Republic and Morocco, and considered the possibility that the sedimentary cyclicity common to both localities had been forced by obliquity or orbital eccentricity. The evidence for global sea level cycles throughout the Silurian at the Myr-scale (Calner, 2008) should be sampled in greater stratigraphic detail — ideally by coring (e.g., by the Swedish Deep Drilling Program; Lorenz, 2010) — in order to resolve astronomical forcing timescales. Short shallow marine sections of the Estonian Wenlockian have been described as having shallowing-upward cycles (Nestor et al., 2001); recently, examination of Upper Homerian (Sheinwoodian) cyclostratigraphy of Lithuania suggests astronomical forcing, with nested "fourth" and "fifth" order cycles, plus an unnamed higher order of cycling (Radzevičius et al., 2017).

2.3.5 Ordovician

A growing number of Milankovitch-forced successions have been reported from the Late Ordovician, presumed to involve a strong glacioeustatic component during the Katian. These include the Juniata Formation in the eastern Appalachians indicating an obliquity periodicity of 30 kyr (Hinnov and Diecchio, 2015), glacial deposits in northern Africa (Loi et al., 2010), the Mallowa Salt in northwestern Australia with an estimated obliquity periodicity of 31.3 ± 3 kyr (Williams, 1991), and the Kope Formation of eastern-central North America (Ellwood et al., 2013). The Vauréal, Ellis Bay and Becscie formations of Anticosti Island, eastern Canada (Desrochers et al., 2010; Elrick et al., 2013) have recently been sampled at high resolution, which has the potential to reveal astronomical variations (Sinnesael et al., 2017a, 2017b). In the Sandbian, the lower slope deposits of the Pingliang Formation (Guanzhuang, China; Fang et al., 2015) and the Arnestad Formation of southern Norway (Svensen et al., 2015) both show strong organized sedimentary cyclicity indicative of orbital eccentricity, obliquity, and precession index forcing, the obliquity with shorter periodicities (Pingliang: 30.6 kyr; Arnestad: 30.3 kyr). The Llanvirn (Middle Ordovician) Krivaya Luka section of Siberia has strong sedimentary cycles with magnetic susceptibility variations suggestive of astronomical forcing (Rodionov et al., 2003). The Darwillian (Middle Ordovician) of France has revealed promising patterns indicative of astronomical forcing, but lacking in sufficient age constraints (Dabard et al., 2015; Sinnesael et al., 2017a, 2017b).

In the Lower Ordovician, periodic sedimentation has been found in the Ibexian-Whiterockian marginal marine shelf deposits of the Wah and Juab formations, Utah, with hierarchical thickness bundling meter-scale cycles suggestive of astronomical forcing (Gong and Droser, 2001). The Tremadocian-Arenigian thick Dumugol Formation (Korea) is comprised of a succession of meter-scale basinal, ramp and subtidal cycles, each thought to represent a 96 kyr periodicity (Kim and Lee, 1998). The 115 shallow-marine carbonate cycles of the (mostly Arenigian) El Paso Group in the Franklin Mountains, Texas have cycle thicknesses bundling patterns with unfamiliar ratios (14:1, 9:1, 6.25:1 and 4.17:1) and supposed not due to astronomical forcing (Goldhammer et al., 1993). This conclusion was based on subjective facies analysis at the outcrop; for objective measurements of magnetic susceptibility or other geophysical/geochemical proxy, there might be a different result (Kodama and Hinnov, 2015).

2.3.6 Cambrian

The Cambrian is well known for its worldwide thick sequences of cyclic shallow marine carbonate platforms (and their continuation into the early Ordovician, see Section 2.3.5) (Derby et al., 2012). Most of the analysis of these cyclic peritidal sequences measured cycle thicknesses as a basis for assessing whether astronomically forced sea level oscillations could explain their stacking patterns, e.g., Olseger and Read (1991). Similarly, Bazykin and Hinnov (2002) analyzed thickness bundling frequencies in the Cambro-Ordovician of Kazahkstan and compared them with those of astronomical models. An interesting development was the "gamma method" (Section 3.5.5) to estimate sedimentation rates of cyclic sub-facies of middle Cambrian carbonate platform cycles (e.g., subtidal, intertidal and supratidal) by Bond et al. (1991). The gamma age model of the middle Cambrian Trippe Limestone sequence that they studied realigned power into the frequencies that would be expected for astronomical forcing. The gamma method was later abandoned due to its predisposition to assign more time to thicker sub-facies, when for shallow marine carbonates, sub-tidal units tend to accumulate faster than intertidal and supratidal units (due to accommodation space).

2.4 Precambrian

While representing the lion's share of Earth history, the Precambrian system remains the least explored of geology, especially in terms of cyclostratigraphy. This state of affairs is rapidly changing, however, with the advent of high-precision geochronology and chemostratigraphy providing the chronostratigraphic framework in the absence of biostratigraphy. Now, it is evident that cyclostratigraphy can also play a role. However, this depends on whether the Solar System operated similarly, e.g., is the Earth's orbital eccentricity still recognizable? How different was the Earth's rotation? There is pre-existing data and modeling to help anticipate the answers, which are that the orbital eccentricity variations were similar, with 405-kyr, ~ 100 kyr cycles, and that the obliquity and precession index cycles were shorter, as indicated by evidence for shorter LOD (Fig. 5). Below is a selection of notable studies.

2.4.1 Neoproterozoic

At the start of the Ediacaran Period when the Earth exited its "snowball" phase, in succession it experienced a series of three very pronounced negative marine carbon isotope excursions (CIEs), the emergence of the Ediacaran biota, traces and burrows, and at the end of the period, the occurrence of shelly faunas. The Shuram Excursion (SE), the last and largest of the CIEs immediately preceded the explosion of life (Grotzinger et al., 2011). The timescale of the SE is unknown, with an onset date sometime between 560 Ma to 580 Ma. In some localities, it occurs within a thick sequence of cyclic deposits, hypothesized as astronomically forced. In the Johnnie Formation (Death Valley, USA) the SE is estimated to have lasted 8.2 Myr, with an 818 kyr long deep phase near the start of the event (Minguez et al., 2015); in the Doushantuo Formation (Donghahe, South China) cyclostratigraphy indicates that the SE lasted an estimated 9.1 Myr (Gong et al., 2017).

The Cryogenian Period, with its repeated snowball Earth glaciations, is characterized by extended interglacial deposits separating the glacial deposits associated with the snowball events. Between the Sturtian and Marinoan glaciations, cyclostratigraphy of a 292 m-long ZK1909 drill core of the Datangpo Formation (South China) reveals a remarkable series of astronomical cycles lasting 9.8 Myr from 660 Ma to 650.2 Ma (Bao et al., 2018), with 405-kyr tuned obliquity and precession index periodicities consistent with a model based on $k = p_{\odot+C} = 64.6$ arcseconds/year (650 Ma; Williams, 2000).

2.4.2 Mesoproterozoic

The cyclostratigraphy of the 1.4 Ga Xiamaling Formation (Xiahuayuan, North China) consists of alternating basinal iron oxide-rich and iron oxide-poor mudstones and black shales with cm to m scale bedding, interpreted as astronomically forced cycles with obliquity periodicities of 21 kyr, 27 kyr and 30 kyr, and precession index periodicities of 12 kyr, 14 kyr and 16 kyr (Zhang et al., 2015). A short segment of Cu/Al data, a productivity/redox proxy, representing approximately 600 kyr, was further analyzed to optimize estimates of $k(= p_{\odot+C})$, g_2, g_3, g_4 and g_5, (Table 2 in Meyers and Malinverno, 2018), which supports the Zhang et al. interpretation, with an estimated $k = 85.79$ $arcsec/year$ and precession index periodicities of 12.5 kyr to 14.4 kyr. Further down in the Chinese Mesoproterozoic section is the 1.6 Ga Wumishan Formation (Tianjin, North China), a 3-km thick carbonate platform with a stack of 626 m-scale peritidal cycles, each consisting of lagoonal to supratidal flat subfacies (Mei et al., 2001). Observations of widespread 4:1 bundling of cycle thicknesses, and local 1:5 to 1:8 bundling, led to speculations of Milankovitch forcing (Mei and Tucker, 2013). Studies are ongoing with magnetic susceptibility measurements to completely document the astronomically forced cyclicity of this extraordinary sequence, which is so long that it spans an aragonite/calcite sea transition with a change from cement-dominated to micrite-dominated stromatolites, and records processes at the Wilson cycle scale (Hinnov et al., 2015).

2.4.3 Paleoproterozoic

The Lower Proterozoic (1.89 Ga) Rocknest Formation (Northwest Territories, Canada) consists of a stack of up to 160 m-scale peritidal carbonate platform cycles suggestive of, but not yet assessed for astronomically forced sea levels (Grotzinger, 1986). The documentation of these cycles at the sub-facies level should make it possible to reconstruct a "depth-rank" facies (analogous to that of the Newark Series) for spectral analysis, and ultimately, perhaps, a new field campaign to reinvestigate the formation with objective technologies (e.g., magnetic susceptibility).

Most intriguing of all are the banded iron formations (BIFs), which count among Earth's oldest sedimentary deposits, and coming into prominence at the start of the Proterozoic (e.g., Fig. 2 in Klein, 2005). The most intensely studied of BIFs in terms of cyclic deposition is the 2.5 Ga Dales Gorge Member of the Brockman Iron Formation (Hamersley Basin, Australia), the first from Trendall (1965, 1966). The origins of the pervasive and multiscale cycling of iron oxide and silica remain enigmatic, stymied by lack of adequate geochronology, and by the lack of proxies that can be sampled adequately through sections to resolve the multiple scales. Grayscale scanning presently suffices as a crude proxy for characterizing iron-rich vs. silica-rich lithologies, and in the upper part of the Dales Gorge Member reveals affinities in the cycling patterns that would be expected for astronomical forcing at that time. Namely, precession index cycles would be expected to have a 11.3 kyr to 12.7 kyr periodicity (Berger and Loutre, 1994), and assuming that orbital eccentricity cycles maintained a ~ 100 kyr periodicity at the time, a ~ 10:1 bundling of cycles would be expected, as observed in the 16-cm thick "mesobands" in the upper part of the Dales Gorge (Fig. 2 in Rodrigues et al., 2018).

2.4.4 Archean

The oldest cyclostratigraphic record described thus far is a sequence of shallowing-upward carbonate cycles from the Archean (2.65 Ga) Cheshire Formation (Zimbabwe) (Hofmann et al., 2004). A 10:1 bundling of the cycle thicknesses is consistent with precession index forcing with an 11.6 kyr mean periodicity, bundled into ~ 100 kyr orbital eccentricity cycles.

3. ASTROCHRONOLOGY

In the 19th century, without the knowledge of radioisotopes, geologists turned to cyclic sedimentary deposits for ways to determine geologic time. This included the renowned Charles Lyell who, influenced by the astronomical theory of ice ages by Adhémar and Croll, calibrated glacial deposits form the Last Ice Age (now known to be 20 Ka) to the most recent maximum orbital eccentricity at 850 Ka, for which the modern solution indicates 972 Ma (Fig. 2A) (Hilgen, 2010). Geologists continued to flirt with the use of an astronomical-forced paleoclimate record as a means for timescale calibration, until radioactivity was discovered in 1895, leading to the radioisotope geochronology revolution. Nonetheless, the later characterization of astronomical forcing as "the insolation canon" by Milankovitch (1941) and the idea that planetary orbital influences on Earth's surface processes could be highly deterministic was never fully struck from geological doctrine. The seminal report on Pleistocene evidence for astronomical forcing by Hays et al. (1976) was the shot in the arm needed to motivate geologists to initiate a wide-ranging search for astronomical signals in stratigraphy (Fischer, 1986; Fischer et al., 1988), leading to the founding of "cyclostratigraphy" and declarations such as: "Milankovitch cycles are like varves and are proper time units, at least within the accuracy of geological measurements." (p. 197, Schwarzacher, 1993). The following sections review the profusion of methods that have been devised to link cyclostratigraphic proxy series to the astronomical parameters in the attempt to define timescales.

3.1 Astronomical Targets

3.1.1 Time Domain

For explaining the great Pleistocene ice ages, summer insolation at 65° North was marked as the best indicator for ice sheet expansion when insolation was too low to allow ice to melt during the summers (Milankovitch, 1941, and other earlier works). Thus, deepest insolation minima from 0—600 Ka (Fig. 14A) were matched to the Würm I, II, and III, Riss I and II, Mindel I and II and Günz I and II European ice ages (review in Loutre, 2003). This time domain approach has persisted ever since in countless research publications. Among these the Late Neogene astronomical time scale was constructed using the same 65° N summer insolation, matching its maxima to the sedimentary record of Mediterranean sapropels (+3 kyr shift, required by radiocarbon-dated Holocene Sapropel S1 for 0 to 2.7 Ma) (Hilgen et al., 1993; Lourens et al., 2004). Astronomically forced glacial models (Fig. 14B) attempting to replicate the strong sawtooth variations and time delays in Pleistocene paleoclimate records (e.g., Imbrie and Imbrie, 1980) have also been used as tuning targets, e.g., the marine oxygen isotope stack of Lisiecki and Raymo (2005). As more data have been collected, it has become evident that simple insolation models do not fit data very well, especially the amplitudes of the variations (notably the

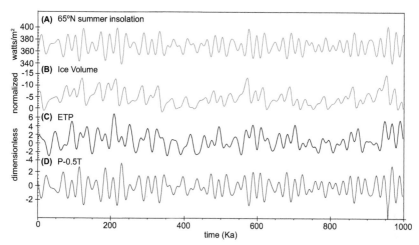

Figure 14 Time-domain astronomical targets, all sampled with Δt = 1 kyr from 0 to 1 Ma, all using the La2004 solution in *Analyseries 2.0.8* (Paillard et al., 1996). (A) Classic summer half-year insolation at 65° North. (B) Glaciation model according to Imbrie and Imbrie (1980), calculated in *Analyseries 2.0.8* (Basic Series menu) with the default settings. (C) Sum of standardized orbital eccentricity, obliquity and precession index, or ETP. (D) Standardized precession index (P) minus 0.5 of the standardized obliquity (T) (similar to mid-latitude summer insolation).

high amplitude orbital eccentricity signals that are not predicted by insolation). This has led to simplified astronomical models, such as the sum of standardized orbital eccentricity, obliquity, and precession index, known as ETP (or EOP) (Fig. 14C), devised by Imbrie et al. (1984). Another useful target is P-0.5T (Fig. 14D), developed for ease in testing different TD and E_D models (e.g., Lourens et al., 1996; Lourens et al., 2001, Section 1.2.4).

3.1.2 Frequency Domain

Astronomical targets have increasingly moved from the time domain, where incremental manipulation of cyclostratigraphic data alters both frequencies and phasing of the cycling, and heightens the circularity problem in the practice of tuning, to the frequency domain. Since cycle phasing has important information, e.g., the phasing of the precession index can indicate season of forcing (Fig. 8). Therefore, one goal of tuning should be to preserve phasing within the data as much as possible. Minimal tuning (e.g., tuning to 405 kyr cycles only) is one way to achieve this (see Section 3.3.2 below). Another way is by searching for a sedimentation rate that maximizes power in the astronomical frequency bands. This is the approach taken by the objective techniques offered by the average spectral misfit (ASM) and correlation coefficient (COCO) methods (see Section 3.3.5 below). Finally, evolutionary spectrograms reveal shifting and drifting of frequencies, as shown in the synthetic example of Fig. 12, and can be used to correct sedimentation rates by tracking a single spectral component that changes frequency through the spectrogram (e.g., Preto et al., 2001; Yao et al., 2015).

3.2 Astronomical Metronomes

3.2.1 Venus and Jupiter Longitude of Perihelia and Earth's Orbital Eccentricity g_2-g_5

The most prominent of the Earth's orbital eccentricity terms is $g_2 - g_5 = 3.199279''/year$ which completes a cycle in 405,091 years, determined by a recurrent alignment of the perihelia of Venus and Jupiter (Fig. 2A and B; Tables 1, 2). The cycle is very stable due to the large mass of Jupiter; over a 250 Myr long solution, g_2 indicates a variability of 13.4%, but g_5 has a vanishingly small variability of only 0.0007% (Table 3 and Fig. 9 of Laskar et al., 2004). Thus, it is an excellent candidate for use as a high-precision "metronome". It is readily observed in a majority of cyclostratigraphy, which is marked by enhanced orbital eccentricity cycles (Section 1.4.1; Fig. 11). The phase of $g_2 - g_5$ of the La2004 solution was recently verified by high-precision radioisotopic dating of the 200 Ma to 215 Ma continental Newark Series (Fig. 13G–I; Kent et al., 2018). The definition of the La2004 $g_2 - g_5$ metronome is given in Fig. 15, for which the first maximum in past time occurs at 192.125 Ka. It should be noted that the recent work on the Newark Series by Kent et al. (2017, 2018) assumed the less accurate maximum of 216 Ka based on the full orbital

Table 2 Main Terms Used to Determine Earth's Orbital Eccentricity, Obliquity and Climatic Precession (Precession Index) Periodicities According to Laskar et al. (2004). The g_i and s_i are the Planetary Secular Frequencies; Subscripts 1–6 Represent Each Planet (1 = Mercury; 2 = Venus; ..., 6 = Saturn) and k is Earth's Precession Constant. For Present-day Earth, $k = 50.475838$ Arcseconds/year (Laskar et al., 2004)

Orbital eccentricity		Obliquity		Climatic precession	
E	$g_2 - g_5$	o_1	$k + s_6$	p_1	$k + g_5$
e_1	$g_3 - g_2$	o_2	$k + s_3$	p_2	$k + g_2$
e_2	$g_4 - g_2$	o_3	$k + s_4$	p_3	$k + g_3$
e_3	$g_3 - g_5$	o_4	$k + s_2$	p_4	$k + g_4$
e_4	$g_4 - g_5$				

(A)

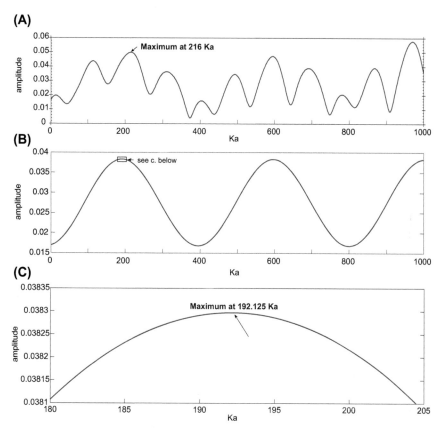

Figure 15 The $g_2 - g_5$ 405-kyr metronome. (A) La2004 orbital eccentricity variation (Laskar et al., 2004) with apparent maximum at 216 Ka. (B) $s_3 - s_6$ over the same interval based on Table 6 in Laskar et al. (2004). (C) The first $s_3 - s_6$ 405 kyr cycle indicates a maximum at 192.15 Ka (see Appendix F for MATLAB script).

eccentricity solution (Fig. 15A) for their metronome formula: the age of kth 405-kyr cycle maximum = $0.216 + (k-1) \times 0.405$ Ma (Eq. 1 in Kent et al., 2017), or a difference of 23875 years. For the Arabian Orbital Time Scale, the age of the Mth 405 kyr Straton base = $0.371 + (M-1) \times 0.405$ Ma (Al Husseini, 2015), which also differs slightly from the La2004 definition of 192.125 Ka + 405/2 kyr = 394.625 Ka, or a difference of 23625 years.

3.2.2 Earth and Saturn Precession of Nodes and Earth's Orbital Inclination s_3-s_6

The Earth's orbital inclination plays a major role in the modulations of the obliquity variation. One of these modulations involves Saturn's precession of

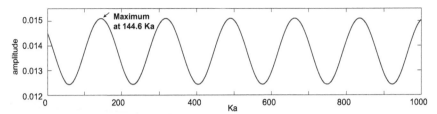

Figure 16 The $s_3 - s_6$ 173-kyr metronome, computed using values from Table 5 and Eq. (26) in Laskar et al. (2004) (see Appendix F for MATLAB script).

nodes and its interaction with those of the Earth, $s_3 - s_6$; it has a periodicity of 173 kyr (Fig. 2D). As with $g_2 - g_5$, the $s_3 - s_6$ is unusually stable because of the large mass of Saturn. Thus, it can be used as a metronome in sequences characterized by dominant obliquity forcing and low precession index contributions (i.e., low or no $g_2 - g_5$). The definition of the La2004 $s_3 - s_6$ metronome is given in Fig. 16, for which the first maximum in past time occurs at 144.6 Ka. This metronome has only very recently been described and used to tune a 10-Myr long Eocene deep-sea sequences characterized by dominant obliquity forcing, from 38 Ma to 48 Ma (Boulila et al., 2018). The joint recovery of $g_2 - g_5$ and $s_3 - s_6$ metronomes from two frequency bands in cyclostratigraphy promises to enhance our confidence in astronomically derived timescales.

3.3 Classical Tuning

3.3.1 Visual Tuning

Simply looking at a plot of a cyclostratigraphic proxy series, with assistance from a "proviso" time scale when possible, is a crucial first step to establish its potential for providing an astronomical forcing record. In some (albeit mostly rare) cases, the affinity of data with an astronomical signal (i.e., target) is unambiguous, as in Fig. 4 of Hilgen et al. (2014), where it is irresistible to correlate each of the peaks between the data and astronomical target. This is the essence of visual tuning, and this has been in practice from the very beginning, to develop depth-to-time transformations for cyclostratigraphy. Visual tuning is the basis for the "Linage" and "Splinage" functions in Analyseries (Paillard et al., 1996) which plots the data and target side by side and invites the user to use a mouse to select tie points (An equivalent procedure is now also available in the "linage" function of *astrochron*; Meyers, 2014.) However, as discussed in Section 3.3.5, circular reasoning is an ever-present risk in tuning procedures that "force" data-target matches based on visual patterns, which may vary according to one's "personal equation." Thus, additional measures should be taken to control this problem; some remedies are discussed below.

3.3.2 Minimal Tuning

One way to reduce circular reasoning is to tune data to a single astronomical frequency and assess the success or failure of the tuning in aligning other astronomical terms in a procedure known as "minimal tuning" (Muller and MacDonald, 2000). 405-kyr cycle tuning is the most common implementation of minimal tuning, and is extremely successful. For example, the Cretaceous and Triassic cyclostratigraphic data depicted in Fig. 13B and C were visually tuned to 405-kyr cycles (see numbered rectangles). Amazingly, even individual spectral peaks in the precession band are as well defined (although at different frequencies) as those in the Cenozoic spectrum of Fig. 13A, which involved another minimal tuning approach, namely limiting the tuning tie points of all of the contributing records to ~100 kyr intervals (Zeeden et al., 2015). This strategy for the Cenozoic data led to the discovery of a change in precession phasing, interpreted as evidence for astronomical forcing of the global deep ocean temperatures as predominantly in the Northern Hemisphere during the late Oligocene, then in the Southern Hemisphere during the Middle Miocene Climate Optimum, and switching back to the Northern Hemisphere during the Quaternary Period (De Vleeschouwer et al., 2017a). In another example, Kodama et al. (2010) tuned a 4 Myr long ARM proxy of a late Eocene flysch deposit to the La2004 orbital eccentricity solution (see their Fig. 9) which preserved an October-November precession band phasing (see their Fig. 11), interpreted as evidence for a Mediterranean climate responding to astronomical forcing.

3.3.3 Tune and Release

Another tuning strategy involves tuning to one astronomical frequency, then "releasing" it and tuning to another astronomical frequency. The approach was pioneered by Imbrie et al. (1984), who iteratively tuned bandpass filtered versions of the SPECMAP stack in the precession and obliquity bands to the BER1978 solution for a total of 120 iterations. The procedure, however, never became popular and has not been used since then. Recently, Yao et al. (2015) tested a similar approach, bandpass filtering their Late Permian chert-mudstone rank series to isolate an obliquity signal, tuning it to 32 kyr (predicted for the Late Permian), then bandpass filtering the results for the long precession — which is actually a multiple frequency component — and tuning that to 20.2 kyr, and finally bandpass filtering those results for the obliquity band and retuning to the 32 kyr obliquity to restore any multiple components in the long precession (procedure demonstrated in Fig. 3 of Yao et al., 2015).

3.4 Sedimentological Tuning

While the classical tuning techniques discussed above manipulate data in a variety of ways to "fit" the astronomical target, there are other non-manipulative procedures based completely on sedimentological criteria —and currently severely underutilized, or not utilized at all — that have been shown to focus astronomical frequencies in cyclostratigraphy. These are as follows.

3.4.1 Aluminum Tuning

In marine sediment (weight percent) aluminum concentrations are typically extremely low, and their fluctuations along a section are mostly the result of variable accumulation of the remaining sediment. Thus, correcting aluminum concentrations to a constant value by adjusting the time scale can be used to estimate sedimentation rates along a section. Application of this approach to the Pleistocene SPECMAP V28–238 $\delta^{18}O$ record significantly sharpened the power spectrum at the obliquity and precession index frequencies (Kominz et al., 1979). However, the procedure has not been used since then. "Aluminum tuning" (or a modernized or equivalent version of it) could be reinstated with potential great success given the rise of XRF core scanning (Croudace and Rothwell, 2015).

3.4.2 Carbonate Tuning

Herterich and Sarnthein (1984) assumed a constant sedimentation rate for the noncarbonate fraction and a variable sedimentation rate for the carbonate fraction in their carbonate-rich Brunhes section to model an incremental timescale along the section. Application of the model increased obliquity and precession spectral power, and resolved 23 kyr and 19 kyr spectral peaks in the precession band. A similar procedure is described in Herbert et al. (1986) for estimating carbon fluxes according to the Albian Piobbico core.

3.4.3 Depth-Derived Time Scales

Depth-derived time scale modeling considers cyclostratigraphy from multiple sections for a given time interval, and the availability of independent time control, e.g., radioisotopic data (Huybers and Wunsch, 2004; Huybers, 2007). Common stratigraphic features, e.g., magnetic reversals, are taken as time-correlative control points among the sections. Independent time control is used to develop a preliminary sedimentation rate model for each section, and to assign ages to the control points. The mean age of each control point across all sections and its uncertainty is then estimated. The goal is to retain variability in the time-calibrated signal that might be due to nonlinear climate change. The variant of the method was applied to a basin-wide set of sections from the Eocene Green River Formation, USA (Aswasereelert et al., 2013), leading to new insights: alluvial sediments alternating with carbonate-rich lacustrine sediments recorded ~100 kyr cycles, while shorter cycles were recorded preferentially by lacustrine sediments.

3.4.4 Turbidite, Conglomerate Removal
Cyclostratigraphy depends on the assumption that sedimentation rates are stable (constant), or slowly varying, such as a pelagic rain in a marine basin. A turbidite represents an instantaneous depositional event, in which a thick (cm to m) and sometimes relatively coarse sedimentary layer is deposited in a matter of minutes to hours. Thus, while the normal sedimentation mode at a depositional site might involve very slow sedimentation rates, e.g., 1 cm/kyr, a 1 cm turbidite represents less than a day. Turbidites therefore "contaminate" a basinal section, and if included can severely distort timescale estimates. The solution is simply to remove the turbidite(s) prior to analyzing a cyclostratigraphic section for astronomical forcing. This was done in a study of the Triassic basinal Buchenstein beds, which dramatically reduced the number of spectral peaks in the turbidite-free version of the cyclostratigraphic proxy series (Maurer et al., 2004). In a related procedure, thick conglomerate intervals were removed from the gamma ray log of an extended, cyclic lacustrine-fluvial-alluvial sequence from the Lower Cretaceous of Songliao Basin (China), which clarified a strong orbital eccentricity signal in the log (Liu et al., 2017b).

3.5 Statistical Tuning
3.5.1 Average Spectral Misfit (ASM)
The average spectral misfit (ASM) method was developed for cyclostratigraphy with uncertain timescales (Meyers and Sageman, 2007). The method relies on preliminary spectral analysis of cyclostratigraphic data to identify statistically significant "line" frequencies with multitaper F-testing (Thomson, 1982, 2009). Detected data lines exceeding a pre-determined

significance level (e.g., 95%) are input to ASM, which computes the difference between the set of data line frequencies converted from stratigraphic (e.g., cycles/cm) to time domain (e.g., cycles/kyr) with a test sedimentation rate, and the set of target line frequencies of an astronomical model. The procedure is repeated over a range of test sedimentation rates to find the minimum ASM.

Additionally, for each sedimentation rate, Monte Carlo simulation is performed by randomizing the data line frequencies (in cycles/kyr) for large number of trials (e.g., 10,000) and computing the ASM value for each trial. The number of trials with ASM values indicating a difference of zero between the sets of data and target lines is used as an estimate of the significance level for rejection of the null hypothesis (H0: no astronomical signal) for the original ASM value. H0 below a critical significance level is taken as evidence for frequencies in the data that are consistent with those of the astronomical model. This has been suggested to be the inverse of the total number of investigated test sedimentation rates (e.g., 0.02 for a run with 500 sedimentation rates). The optimal sedimentation rate is the one with the lowest ASM value and H0 with the highest significance level and lowest P value. This method is provided in the "asm" function of the *astrochron* package in R (Meyers, 2014). For cases where significant variations in sedimentation rate are suspected along a cyclostratigraphic section, an "evolutionary" version of ASM is available, that calculates the ASM procedure along a running window through the section, provided as "eAsm" in *astrochron*.

3.5.2 Bayesian Tuning

Malinverno et al. (2010) developed a "Bayesian Monte Carlo" approach to search for the sedimentation rate in a cyclostratigraphic proxy series that maximizes spectral power at the astronomical frequencies. The Monte Carlo method is used to develop uncertainties for the estimated sedimentation rate. This can be applied with a running window to estimate sedimentation rates through sections. However, since its introduction and application to the Aptian Selli Event interval, it has not been used.

Lin et al. (2014) developed a procedure for the alignment of stratigraphic records that applies Bayesian inference and hidden Markov modeling to develop the probability distribution for the alignment, which is then used for estimating its uncertainty. Their set of MATLAB scripts is available at: http://ccmbweb.ccv.brown.edu/cgi-bin/download_HMM_Match.pl.

3.5.3 TimeOpt and TimeOptMCMC

The time optimization (timeOpt) method was developed with the same goals as ASM. However, in this case cyclostratigraphy is analyzed for the sedimentation rate that optimizes orbital eccentricity frequency characteristics of the precession index amplitude envelope, and power at the precession index and orbital eccentricity frequencies (Meyers, 2015). For each test sedimentation rate, the data are fitted to model precession index and orbital eccentricity time series in two steps: (1) the data are band-pass filtered and Hilbert transformed to isolate the precession index amplitude envelope. This envelope is then linearly regressed on a model time series with orbital eccentricity frequencies, and the result is reported together with the squared model-data Pearson correlation coefficient $r_{envelope}^2$; and (2) the data are linearly regressed on a model time series with orbital eccentricity and precession index frequencies, and the result is reported together with the squared model-data Pearson correlation coefficient r_{power}^2.

The product $r_{opt}^2 = r_{envelope}^2 \cdot r_{power}^2$ indicates the fraction of power shared by the precession index envelope and the orbital eccentricity plus precession index time series. For a perfect data-model fit, and perfect synchronization, as would be the case between the precession index, its amplitude envelope and the orbital eccentricity, $r_{opt}^2 = 1$. However, r_{opt}^2 is typically much lower than 1 due to data filtering and noise; Monte Carlo simulation is required to evaluate the statistical significance of the computed r_{opt}^2 maximum (Meyers, 2015). This method is available in the "timeOpt" and "timeOptSim" functions in the *astrochron* package in R (Meyers, 2014).

An important extension - time optimization with Markov Chain Monte Carlo simulation - was recently introduced in order to search additionally for optimal k ($= p_{\odot + c}$) (Eq. 4) and g_1 to g_5 for the model. This was applied to two cyclostratigraphic sections, one from the 1.4 Ga Xiamaling Formation (China) and the other from the 55 Ma Walvis Ridge ODP Core 1262, for joint optimization of sedimentation rate, k, and g_1 to g_5 (Meyers and Malinverno, 2018). The procedure is available in the function "timeOptMCMC" in *astrochron 0.8*.

3.5.4 Correlation Coefficient (COCO)

As with ASM (Section 3.5.2), the correlation coefficient (COCO) method of Li et al. (2018b) evaluates two problems simultaneously: (1) a potential astronomical signal in cyclostratigraphy, and (2) stratigraphic distortion due to variable sedimentation rates. COCO estimates the correlation coefficient between the power spectra of an astronomical target signal and paleoclimate proxy series across a range of test sedimentation rates. As with ASM, a null hypothesis of no astronomical forcing is evaluated using Monte Carlo simulation. When applied using a sliding stratigraphic window, "eCOCO" estimates sedimentation rate variations along cyclostratigraphic sections, and other insights into astronomical forcing.

3.5.5 Other Assorted Techniques

Other techniques that have been developed for establishing statistically based fits between cyclostratigraphy and astronomical targets include: linear inversion (Martinson et al., 1982), varimax norm demodulation (Schiffelbein and Dorman, 1986), a minimal cost function method (Bruggemann, 1992), dynamic optimization (Yu and Ding, 1998), dynamic programming (Lisiecki and Lisiecki, 2002), CORRELATOR (Olea, 1994, 2004), precession-eccentricity optimization (Zeeden et al., 2015), dynamic time warping (Kotov and Pälike, 2017), and automatic correlation optimization (Zeeden et al., 2018).

Finally, for stratigraphic cycles defined by repeating sub-facies, "gamma analysis" was designed to estimate the sedimentation rates of individual subfacies contributing to a cycle (Kominz and Bond, 1990; Kominz et al., 1991). In each cycle, the thickness of the ith subfacies c_i has an unknown inverse sedimentation rate γ_i, and so the periodicity of the cycle with N subfacies is: $T_{cy} = \sum_{i=1}^{N} \gamma_i c_i$. If the cycle period is unknown, $T_{cy} = 1$ is assumed. There is an equation for each cycle, so for a total of N cycles each occurring with a period T_{cy}, the set of N equations is inverted to solve for γ_i.

3.6 Astronomical Time Scale

The Astronomical Time Scale (ATS) is based on astronomical solutions matched with cyclostratigraphy that is correlated with global chronostratigraphy. There are "anchored" and "floating" versions of the ATS, the former anchored to the present day, and the latter to high-precision geochronologic dates in the GTS. For 0 to 5 Ma, insolation models are typically used for matching; earlier than 5 Ma, the effects of shorter LOD on precession need to be included, astronomical targets such as "P-0.5T La2004$_{1,1}$" (e.g., Zeeden et al., 2014) may become standard targets for early Neogene

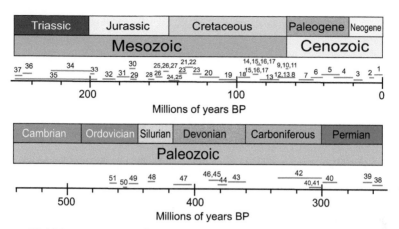

Figure 17 Major sequences of $g_2 - g_5$ cycles in the cyclostratigraphic record. A detailed list of the numbered sources (1–50) appears in Appendix G. Additional information for the Mesozoic is provided in Huang (2018).

and Paleogene cyclostratigraphy. At all times, identification of the 405-kyr metronome is a priority. Fig. 17 shows the current 405-kyr metronome coverage in the cyclostratigraphic record (detailed list in Appendix G). The formal presentation of the ATS should have a format such as presented in the figures of Huang (2018).

4. CONCLUSIONS

In 2018, cyclostratigraphy can be said to have entered a "golden age", with multi-million year-long sequences now routine in data collecting, many of these exceeding 10 Myr in duration. These new datasets represent major contributions to the geologic time scale with dominantly deterministic astronomical metronomes readily recognized in the cyclic successions. The principal metronome originates from the interactions of the orbital perihelia of Venus and Jupiter, $g_2 - g_5$ with a periodicity of 405-kyr and is commonly found in cyclostratigraphy dominated by precession and orbital eccentricity forcing (see Huang, this volume). The La2004 405-kyr metronome defined by the +5 Myr to −15 Myr integration (Table 6 in Laskar et al., 2004) is valid as far back is 215 Ma (Kent et al., 2018). A second metronome originating from the orbital inclinations of Earth and Saturn, $s_3 - s_6$ with a periodicity of 173-kyr occurs in cyclostratigraphy dominated by obliquity forcing. In the future, the assessment of jointly occurring $g_2 - g_5$ and $s_3 - s_6$ metronomes will provide greater confidence in astrochronology than either metronome on its own.

The accuracy of astrochronology is sufficient for high quality intercalibration with radioisotope geochronology. The integration of both information sources is expected to lead to improved time scales that are continuous and precise at an unprecedented level (achieving 0.05% age uncertainty).

The application of optimization methods to solve for Earth's precession, and planetary secular frequencies for times prior to 50 Ma represents a major advance in cyclostratigraphy, and will provide detailed constraints for Solar System modeling in the very near future.

The next line of inquiry in cyclostratigraphy will involve investigation of Solar System resonance state and chaotic events through geologic time. This goes hand in hand with the long sequences now being assembled, in light of the million-year period modulations that need to be recovered for this purpose.

Finally, in the high stakes race to discover astronomical forcing signals in cyclostratigraphy, understanding the pathway from insolation to stratigraphy has been severely shortchanged. Thus, the other future major line of inquiry will involve modeling insolation forcing of the paleoclimate system, and the responses of the depositional systems that have inherited the astronomical signal.

APPENDIX A — CALCULATION OF EARTH'S ORBITAL ECCENTRICITY AND INCLINATION

A.1 La2010 Solution Files

At: http://vo.imcce.fr/insola/earth/online/earth/earth.html, e.g., in the "Data files here" link under "Solutions La2010 for Earth orbital elements from −250 Myr to the present," the file "La2010d_alkhqp3L.dat" provides values for La2010d:

t: the time from J2000 (in kyr) (negative back in time)

a: semi-major axis

l: mean longitude (expressed in radians)

k: $e \cos (\Pi)$

h: $e \sin (\Pi)$

q: $\sin(I/2) \cos (\Omega)$

p: $\sin(I/2) \sin (\Omega)$

(Note: In the main text, the subscript "3" is used to designate the Earth's orbital elements.)

A.2 Orbital Eccentricity

For time t, given: $k = e\cos\Pi$ *and* $h = e\cos\Pi$:

$$k^2 + h^2 = e^2(\cos^2\Pi + \sin^2\Pi) = e^2$$

$$e = \sqrt{(k^2 + h^2)}$$

A.3 Orbital Inclination

For time t, given: $q = \sin(I/2)\cos\Omega$ and $p = \sin(I/2)\sin\Omega$:

$$q^2 + p^2 = \sin^2(I/2)(\cos^2\Omega + \sin^2\Omega) = \sin^2(I/2)$$

$$\sin(I/2) = \sqrt{q^2 + p^2}$$

$$I = 2\sin^{-1}\left(\sqrt{q^2 + p^2}\right)$$

APPENDIX B — ASTRONOMICAL SOLUTIONS TRANSFORMATIVE TO CYCLOSTRATIGRAPHY

B.1 Analytical Solutions

LEVER1855 (Le Verrier, 1855) orbital eccentricity and obliquity solution.
STOCK1873 (Stockwell, 1873) originally completed in 1870.
LEVMIS1931 (Miskovitch, 1931) solution based on LEVER1855.
BROVW1960 (Brouwer and van Woerkom, 1950) includes terms up to the sixth degree in the eccentricities and inclinations of Jupiter and Saturn, and terms related to the "great inequality", i.e. the near equality to the 5:2 mean motion resonance between Jupiter and Saturn up to the second order in planetary masses.
SHABU1967 (Sharaf and Budnikova, 1967, 1969) (corrections to BROVW1960)
VERNE1972 (Vernekar, 1972) (BROVW1960 corrected by SHABU1967)
BRETA1974 (Bretagnon, 1974) includes terms up to the fourth degree in planetary eccentricities and inclinations, and up to second order in planetary masses.

BER78 (Berger, 1978) based on BRETA1974, with FORTRAN code to compute insolation.

B.2 Numerical Integrations

La1990 (Laskar, 1990) includes terms up to the sixth degree in planetary eccentricities and inclinations, and up to second order in planetary masses.
La1993 (Laskar et al., 1993) includes adjustable parameters for Earth rotation and dynamical ellipticity with the La1990 solution, valid from 0 to 35 Ma.
La2004 (Laskar et al., 2004) is a numerical solution for full Solar System with an Earth precession model based on LLR observations, orbital solution valid from 0 to 40 Ma.
La2010 (Laskar et al., 2011a) incorporates new INPOP planetary ephemeris, no Earth precession model, 4 solutions, valid from 0 to 50 Ma.
La2011 (Laskar et al., 2011b; Laskar et al., 2012) inclusion of asteroid bodies, no Earth precession model, valid from 0 to 50 Ma.
ZB2017 (Zeebe, 2017) new integrator algorithms, separate Moon, asteroids, no Earth precession model, 17 solutions, valid from 0 to 54 Ma.

La1990−2010a-d models are at: http://vo.imcce.fr/insola/earth/online/earth/earth.html.
La2011 model is provided in the 'getLaskar' function of the astrochron for R package (Meyers, 2014).
Zeebe models are at: https://www.soest.hawaii.edu/oceanography/faculty/zeebe_files/Astro.html.

APPENDIX C − VARIABLES OF EARTH'S PRECESSION
C.1 − Luni-Solar Precession

The annually averaged net torque due to gravitational forcing from the Sun is (p. 712, Williams, 1994):

$$\tau_\odot = \frac{3 G m_\odot (C - A) sin\varepsilon\ cos\varepsilon}{2a^3 \left(1 - e^2\right)^{3/2}}$$

This results in retrograde precession (see Fig. 3) of the Earth's rotation axis:

$$\frac{d\psi_\odot}{dt} = \frac{\tau_\odot}{C\phi\ sin\varepsilon} = \frac{3 G m_\odot (C - A) cos\varepsilon}{2C\phi\ a^3 \left(1 - e^2\right)^{\frac{3}{2}}} = 2.450183 \times \frac{10^{-12}}{s}$$

$$= 15.948788\ \text{arcseconds/year}$$

Similarly, gravitational forcing from the Moon, with an added factor to account for the lunar orbital inclination I_C with respect to the Earth, results in:

$$\frac{d\psi_C}{dt} = \frac{\tau_C}{C\phi \, sin\varepsilon} = \frac{3Gm_C(C-A)cos\varepsilon}{2C\phi \, a_C^3(1-e_C^2)^{\frac{3}{2}}} \left(1 - \frac{3}{2}sin^2 I_C\right)$$

$$= 5.334529 \times \frac{10^{-12}}{s} = 34.723638 \text{ arcseconds/year}$$

The sum gives the luni-solar total of 50.672426 "/year and in general:

$$\frac{d\psi_\odot}{dt} + \frac{d\psi_C}{dt} = \frac{3Gm_\odot(C-A)cos\varepsilon}{2C\phi \, a^3(1-e^2)^{3/2}}$$

$$+ \frac{3Gm_C(C-A)cos\varepsilon}{2C\phi a_C^3(1-e_C^2)^{3/2}} \left(1 - \frac{3}{2}sin^2 I_C\right)$$

$$= \frac{3Gm_\odot(C-A)cos\varepsilon}{2C\phi \, a^3(1-e^2)^{3/2}} + \frac{3Gm_C(C-A)cos\varepsilon}{2C\phi a_C^3(1-e_C^2)^{3/2}} \left(1 - \frac{3}{2}sin^2 I_C\right)$$

$$\frac{d}{dt}(\psi_\odot + \psi_C) = \frac{3GH}{2\phi} \left[\frac{m_\odot}{a^3(1-e^2)^{\frac{3}{2}}} + \frac{m_C\left(1 - \frac{3}{2}sin^2 I_C\right)}{a_C^3(1-e_C^2)^{\frac{3}{2}}}\right] cos\varepsilon = p_{\odot+C}$$

This appears as Eq. (4) in the main text (This derivation benefitted from online notes by Mclemore and Koonce, 2016).

C.2 — Orbital Contributions to Precession and Obliquity

Given p and q for the Earth (Appendix A above), the following expressions provide the planetary perturbation contributions to the precession and obliquity equations (Eq. 5 in the main text):

$$\mathcal{A}(t) = \frac{2}{\sqrt{1-p^2-q^2}} \, [\dot{q} + p(q\dot{p} - p\dot{q})]$$

$$\mathcal{B}(t) = \frac{2}{\sqrt{1-p^2-q^2}} \, [\dot{p} - q(q\dot{p} - p\dot{q})]$$

$$\mathcal{C}(t) = q\dot{p} - p\dot{q}$$

with

$$\dot{p} = \frac{dp}{dt} \text{ and } \dot{q} = \frac{dq}{dt}$$

C.3 — Variable Definitions

ϕ — Earth rotation rate (7.2921150×10^{-5} rad/s at J2000, IERS)
G = gravitational constant (6.67408×10^{-11} m^3/(kg s^2))
H — Earth dynamical ellipticity = $(C-A)/C$ (present-day: 0.003273787)
A — Earth's equatorial moment of inertia (8.008×10^{37} kg/m^2)
C — Earth's polar moment of inertia (8.034×10^{37} kg/m^2)
a — semi-major axis of Earth's orbit (1.4959802×10^{11} m at J2000, IERS)
e — Earth orbital eccentricity (0.016708634 at J2000, IERS)
$a_\mathbb{C}$ — semi-major axis of Moon's orbit (3.833978×10^{8} m at J2000, IERS)
$e_\mathbb{C}$ — eccentricity of the Moon's orbit (0.05554553 at J2000, IERS)
$I_\mathbb{C}$ — inclination of Moon orbit on the ecliptic (5.156690 at J2000, IERS)
$m_\mathbb{C}$ — mass of the Moon ($7.34767309 \times 10^{22}$ kg)
$m_\odot$ — mass of the Sun (1.98855×10^{30} kg) ($=27068510\ m_\mathbb{C}$)
$Gm_\odot = 1.3271244 \times 10^{20}$ m^3/s^2 (at J2000, IERS)
$Gm_\mathbb{C} = 4.902799 \times 10^{12}$ m^3/s^2 (at J2000, IERS)
ε — obliquity angle of Earth (23.43928 at J2000, IERS)
$\odot$ — symbol for the Sun
$\mathbb{C}$ — symbol for the Moon
$\oplus$ — symbol for the Earth

APPENDIX D — LUNAR RECESSION AND EARTH DECELERATION

In the Earth-Moon system, tidal dissipation results in a transfer of angular momentum from the Earth to Moon. The Earth's rotational angular momentum is $L_E = C\phi$, and the Moon's orbital angular momentum is $L_M = \kappa L_E$, where present-day $\kappa = 4.93$. Assuming that tidal friction caused by Moon and Sun is independent and proportional to the square of the ratio of their tidal amplitudes (0.46) (Eq. 2 in Deubner, 1990):

$$dL_E = -dL_M \left[1 + (0.46)^2 \left(\frac{a_0}{a}\right)^6 \right]$$

Integrating and rearranging Eq. (3) in Deubner (1990) gives:

$$\frac{\phi}{\phi_0} = 6.00967 - 4.93\left(\frac{a}{a_0}\right)^{\frac{1}{2}} - 0.080245231\left(\frac{a}{a_0}\right)^{\frac{13}{2}}$$

where ϕ_0 is present-day ϕ (=7.2921150 × 10−5 rad/s) and a_0 is present-day lunar semi-major axis (=3.833978 × 10^8 m). A detailed treatment of the problem is provided in Chapter 10 of Lambeck (1980).

APPENDIX E — EARTH-MOON ROCHE LIMIT

The Roche Limit is the distance between two attracting bodies at which the smaller of the two bodies experiences disruptive forces that are sufficient to break it part. The Roche Limit (in km) for the Earth-Moon system, assuming rigid bodies is (p. 53, Lowrie, 2007):

$$d_R = R_\oplus \left[2\left(\frac{\rho_\oplus}{\rho_C}\right)\right]^{\frac{1}{3}} = 1.4887 \; x \; 6371 \; km = 9484 \; km$$

where $R_\oplus$ = Earth radius = 6371 km $\rho_\oplus$ = Earth mean density = 5.51 g/cm^3 ρ_C = Moon mean density = 3.34 g/cm^3

APPENDIX F — CALCULATING THE g_2-g_5 AND s_3-s_6 METRONOMES WITH LA2004

The following two MATLAB scripts calculate the $g_2 - g_5$ and $s_3 - s_6$ metronomes for 0−1 Ma:

```
% This calculates g2-g5
% using the formula in Table 6 of Laskar et al., 2004;
% negative time t in years BP (into the past)
dt − -1.; % 1 year sample rate (change to −1000 for kyr)
t = 0:dt:−1000000; % 0 to 1 million years BP (negative time)
% e0, mu, b and faz from Laskar et al., 2004, Table 6
e0 = 0.0275579;
mu = 3.199279*pi/(3600*180); % convert to radians
b = 0.010739;
faz = 170.739*pi/180.; % convert to radians
```

```
ecc405 = e0+b*cos(mu*t + faz);
figure; plot(t,ecc405);


% This calculates s3—s6
% using s3 and s6 from Table 5 and Eq. 26 in Laskar et al., 2004;
% negative t in years BP (into the past)
dt = −1.; % 1 year sample rate (change to −1000 for kyr)
t = 0:dt:−1000000; % 0 to 1 million years BP (negative time)
t = t';
s3a = 0.01377449*cos((−18.845116*pi/(3600*180))*t−pi*111.31/180);
s3b = 0.01377449*sin((−18.845116*pi/(3600*180))*t−pi*111.31/180);
s6a = 0.00133215*cos((−26.34788*pi/(3600*180))*t + pi*127.306/180);
s6b = 0.00133215*sin((−26.34788*pi/(3600*180))*t + pi*127.306/180);
s36a = s3a−s6a;
s36b = s3b−s6b;
s36 = sqrt(s36a.^2 + s36b.^2);
figure; plot(t,s36);
```

APPENDIX G — MAJOR SEQUENCES OF g_2–g_5 CYCLES IN THE GEOLOGIC RECORD

The following is a list of publications that describe $g_2 - g_5$ cycles in cyclostratigraphic sections. Coverage in geologic time is summarized in Fig. 17; bold numbers refer to the labels in the figure; approximate time intervals (and durations) are also given. For the Mesozoic, additional detailed information is provided in Huang (2018).

Cenozoic

Neogene
Pliocene-Pleistocene: (1) Nie, 2017: 0—5.5 Ma (5.5 Myr)
Miocene: (2) Hilgen et al., 1999: 6.6—9.5 Ma (3.1 Myr); (3) Miller et al. (2017): 14—20 Ma (6 Myr)
Oligocene: (4) Pälike et al., 2006: 23—36 Ma (13 Myr)

Paleogene
Eocene: (5) Westerhold et al., 2014: 31—43 Ma (12 Myr): (6) Westerhold et al., 2015: 41—49 Ma (8 Myr)

Paleocene: (7) Westerhold et al., 2012: 48—58 Ma (10 Myr); (8) Hilgen et al., 2015: 61—66 Ma (5 Myr)

K/Pg boundary: (9) Kuiper et al., 2008; (10) Renne et al., 2013; (11) Hennebert, 2014.

Mesozoic

Cretaceous

Maastrichtian: (12) Batenburg et al., 2014: 66—72 Ma (6 Myr); (13) Wu et al., 2014: 66—72.1 Ma (6.1 Myr)

Campanian: (13) Wu et al., 2014: 70.5—83.4 Ma (12.9 Myr)

Santonian: (14) Thibault et al., 2016: 84—87 Ma (3 Myr); (15) Wu et al., 2013: 83.6—86.3 Ma (3.3 Myr); (16) Sageman et al., 2014: 84.19—86.49 Ma (2.3 Myr); (17) Locklair and Sageman, 2008:

Coniacian: (15) Wu et al., 2013: 86.3—89.6 Ma (3.6 Myr); (16) Sageman et al., 2014: 86.49—89.75 Ma (3.26 Myr)

Turonian: (18) Eldrett et al., 2015: 89.75—94.1 Ma (4.35 Myr)

Cenomanian: (18) Eldrett et al., 2015: 94.1—98 Ma (4.1 Myr; *upper part*)

Albian: (19) Grippo et al., 2004: 101—113 Ma (12 Myr)

Aptian: (20) Huang et al., 2010a: 113—126 Ma (13 Myr)

Barremian: (21) Hinnov et al., 2008: 126—131.67 Ma (5.67 Myr) (based on (22) Fiet and Gorin, 2000)

Hauterivian: (23) Martinez et al., 2015: 129.5 to 135.25 Ma or 126.1 to 132 Ma (5.265 Myr)

Valanginian: (23) Martinez et al., 2015: 135.25—140.25 Ma or 132 to 137 Ma (5.08 Myr)

Berriasian: (24) Huang et al., 2010c: 137.8 to 144.7 Ma (6.9 Myr; Upper Volgian); (25) Kietzmann et al., 2015: 139.9—145 Ma (5.1 Myr)

Jurassic

Tithonian: (25) Kietzmann et al., 2015: 145—149 Ma (4 Myr; *upper part*) (26) Huang et al., 2010b: 147.2—150.52 Ma (2.5 Myr; *lower part*) (27) Rameil, 2005: 6.8 Myr.

Kimmeridgian: (26) Huang et al., 2010b: 147.2—153.93 Ma (6.73 Myr)

Oxfordian: (28) Boulila et al., 2010: 157.2—161.2 Ma (4 Myr; *lower-middle part*)

Callovian: Incomplete data

Bathonian: Incomplete data
Bajocian: (29) Sucheras-Marx et al., 2013: ~168–172 Ma (4.1 Myr; *lower part*)
Aalenian: (30) Huret et al., 2008a, 2008b: 170.85–174.7 Ma (3.85 Myr); (29) Sucheras-Marx et al., 2013: (0.94 Myr; *upper part*)
Toarcian: (31) Boulila et al., 2014: 174.7–183 Ma (8.3 Myr)
Pliensbachian: (32) Ruhl et al., 2016: 183.8–192.5 Ma (8.7 Myr); Ikeda and Tada, 2014: xxx-yyy (zzz)
Sinemurian-Hettangian: (33) Hüsing et al., 2014: 197.3–200 Ma (2.7 Ma; *lower Sinemurian*)

Triassic
Rhaetian: (34) Kent et al., 2017: 205.5–201.4 Ma (4.1 Myr); (35) Ikeda and Tada, 2014: 208.5–201.4 Ma (7.1 Myr)
Norian: (34) Kent et al., 2017: 227–205.5 Ma (21.5 Myr); Ikeda and Tada, 2014: 225–208.5 Ma (16.5 Myr)
Carnian: (35) Ikeda and Tada, 2014: 235.2–225 Ma (10.2 Myr)
Ladinian: Ikeda and Tada, 2014: xxx–235.2 Ma (xxx Myr)
Anisian: (36) Li et al., 2018a: 241–247 Ma (6 Myr)
Induan-Olenekian: (37) Li et al., 2016: 252–247 Ma (5 Myr)

Paleozoic
Permian
Wuchiapingian-Changhsingian: (38) Wu et al., 2013: 260–251 Ma (9 Myr)
Wordian-Capitanian: (39) Fang et al., 2017: 262–269 Ma (7 Myr)
Asselian-Artinskian: (40) Schmitz and Davydov, 2012: 288–299 Ma (11 Myr; *lower Artinskian*)

Carboniferous
Moscovian-Gzhelian: (41) Davydov et al., 2010; (40) Schmitz and Davydov, 2012: 299–309 Ma (10 Myr; *upper Moscovian*)
Visean-Gzhelian: (42) Wu et al., 2018a, 2018b: 299–333 Ma (34 Myr; *upper Visean*)
Tournaisian: Incomplete data

Devonian
Famennian: (43) Pas et al., 2018: 358.9–372.2 Ma (13.5 Myr)
Frasnian: (44) De Vleeschouwer et al., 2012: 376.7–383.6 Ma (6.9 Myr)

Givetian: (45) De Vleeschouwer et al., 2015: 383.6—387.95 Ma (4.35 Myr)
Eifelian: (46) Ellwood et al., 2015: 387.95—394.16 Ma (6.21 Myr)
Emsian: No data
Lockovian-Pragian: (47) Da Silva et al., 2016: 404—418 Ma (14 Myr)

Silurian
Pridoli: No data
Ludfordian: No data
Gorstian: No data
Homerian: Incomplete data
Sheinwoodlian: No data
Telychian: (48) Gambacorta et al., 2018: 433.4—438.86 Ma (5.46 Myr)
Aeronian: No data
Rhuddanian: Incomplete data

Ordovician
Katian-Hirnantian: (49) Hinnov and Diecchio, 2015: 443—450 Ma (7 Myr)
Sandbian: (50) Fang et al. (2016): 453—456 Ma (3 Myr; *upper part*)
Darwillian-Dapingian: (51) Rasmussen et al., 2018: 463—468.5 Ma (5.5 Myr; *upper Dapingian, lower Darwillian*)
Floian: No data
Tremadocian: No data.
Cambrian: No data.

ACKNOWLEDGMENTS

My deep gratitude goes to Alfred G. Fischer (*December 12, 1920 - †July 2, 2017), who supported my effort to understand cyclostratigraphy for more than 30 years. Al's scientific leadership and meticulous documentation of the cyclic sedimentary record, and his ability to synthesize complex concepts throughout his life inspired countless geologists in their careers. Al was responsible for the term "cyclostratigraphy" and he quickly adapted new ideas into his research, for example, the recognition of 405-kyr cycles and their use as a metronome. He was responsible for the 405-kyr cycle definitions in the Aptian-Albian Piobbico core record highlighted in this paper. I thank Jim Ogg, who has tirelessly supported and promoted cyclostratigraphic research for decades, carefully integrating cyclostratigraphy into the International Geologic Time Scale 2004, 2008, 2012, 2016, and now the upcoming GTS 2020. I am grateful to my long-time colleague David Florkowski, who through many years has never failed to alert me about new developments in signal processing, Solar System modeling, geodynamics, and much, much more. Finally, I thank all of my long-suffering, wonderful colleagues and co-authors for their graciousness in always sharing their data and including me in their research, and reviewer Christian Zeeden, who provided incisive and valuable comments that greatly improved this presentation.

REFERENCES

Al Husseini, M.I., 2015. Arabian orbital stratigraphy revisited — AROS 2015. GeoArabia 20 (4), 183—216.

Algeo, T.J., Heckel, P.H., 2008. The late Pennsylvanian Midcontinent Sea of North America: a review. Palaeogeogr. Palaeoclimatol. Palaeoecol. 268, 205—221. https://doi.org/ 10.1016/j.palaeo.2008.03.049.

Anderson, R.Y., 2011. Enhanced climate variability in the tropics: a 200 000 yr annual record of monsoon variability from Pangea's equator. Clim. Past 7, 757—770. https://doi.org/ 10.5194/cp-7-757-2011.

Anderson, R.Y., 1982. A long geoclimatic record from the Permian. J. Geophys. Res. 87 (C9), 7285—7294. https://doi.org/10.1029/JC087iC09p07285.

Aswasereelert, W., Meyers, S.R., Carroll, A.R., Peters, S.E., Smith, M.E., Feigl, K.L., 2013. Basin cyclostratigraphy of the Green River formation. Geol. Soc. Am. Bull. 125, 216—228. https://doi.org/10.1130/B30541.1.

Bao, X., Zhang, S., Jiang, G., Wu, H., Li, H., Wang, X., An, Z., Yang, T., 2018. Cyclostratigraphic constraints on the duration of the Datangpo Formation and the onset age of the Nantuo (marinoan) glaciation in South China. Earth Planet. Sci. Lett. 483, 52—63. https://doi.org/10.1016/j.epsl.2017.12.001.

Batenburg, S.J., Gale, A.S., Sprovieri, M., Hilgen, F., Thibault, N., Boussaha, M., Oruc-Etxebarria, X., 2014. An astronomical time scale for the Maastrichtian based on the Zumaia and Sopelana sections (Basque country, northern Spain). J. Geol. Soc. 171, 165—180. https://doi.org/10.1144/jgs2013-015.

Batenburg, S.J., Sprovieri, M., Gale, A.S., Hilgen, F.J., Hüsing, S., Laskar, J., Liebrand, D., Lirer, F., Orue-Etxebarria, X., Pelosi, N., Smit, J., 2012. Cyclostratigraphy and astronomical tuning of the late Maastrichtian at Zumaia (Basque country, Spain). Earth Planet. Sci. Lett. 359—360, 264—278. https://doi.org/10.1016/j.epsl.2012.09.054.

Bazykin, D.A., Hinnov, L.A., 2002. Orbitally-driven depositional cyclicity of the Lower Paleozoic Aisha-Bibi seamount (Malyi Karatau, Kazakstan): integrated sedimentological and time series study. In: Zempolich, W.G., Cook, H.E. (Eds.), Paleozoic Carbonates of the Commonwealth of Independent States (CIS): Subsurface Reservoirs and Outcrop Analogs, pp. 19—41. SEPM Special Publication No. 75.

Berger, A., 1978. A Simple Algorithm to Compute Long Term Variations of Daily or Monthly Insolation. Contribution No. 18. Institut d'Astronomie et de Geophysique Georges Lemaitre, Universite Catholique de Louvain, Louvain-la-Neuve, Belgique, 50 p.

Berger, A., Loutre, M.-F., 1994. Astronomical forcing through geologic time. In: DeBoer, P.L., Smith, D.G. (Eds.), Orbital Forcing and Cyclic Sequences: International Association of Sedimentologists Special Publication, vol. 19, pp. 15—24.

Berger, A., Loutre, M.-F., Tricot, C., 1993. Insolation and Earth's orbital periods. J. Geophys. Res. 98 (D6), 10341—10362. https://doi.org/10.1029/93JD00222.

Berger, A., Loutre, M.-F., Yin, Q.Z., 2010. Total irradiation during any time interval of the year using elliptic integrals. Quat. Sci. Rev. 29, 1968—1982. https://doi.org/10.1016/ j.quascirev.2010.05.007.

Berger, A., Loutre, M.-F., Laskar, J., 1992. Stability of the astronomical frequencies over the Earth's history for paleoclimatic studies. Science 255, 560—566. https://doi.org/ 10.1126/science.255.5044.560.

Berger, A., Loutre, M.-F., Dehant, V., 1989. Influence of the changing lunar orbit on the astronomical frequencies of pre-Quaternary insolation patterns. Paleoceanography 4, 555—564. https://doi.org/10.1029/PA004i005p00555.

Berry, W.B., Barker, R.M., 1968. Fossil bivalve shells indicate longer month and year in Cretaceous than present. Nature 217, 938–939.

Bond, G.C., Kominz, M.A., Beavan, J., 1991. Evidence for orbital forcing of Middle Cambrian peritidal cycles: Wah Wah range, south-central Utah. Kansas Geol. Survey Bull. 233, 294–317.

Boulila, S., Vahlenkamp, M., De Vleeschouwer, D., Laskar, J., Yamamoto, Y., Pälike, H., Kirtland Turner, S., Sexton, P.E., Westerhold, T., Röhl, U., 2018. Towards a robust and consistent middle Eocene astronomical timescale. Earth Planet. Sci. Lett. 486, 94–107. https://doi.org/10.1016/j.epsl.2018.01.003.

Boulila, S., Hinnov, L.A., Galbrun, B., 2017. A review of tempo and scale of the early Jurassic Toarcian OAE: implications for carbon cycle and sea level variations. Newslett. Stratigr. 50, 363–389. https://doi.org/10.1127/nos/2017/0374.

Boulila, S., Galbrun, B., Huret, E., Hinnov, L.A., Rouget, I., Gardin, S., Huang, C., Bartolini, A., 2014. Astronomical calibration of the Toarcian Stage: implications for sequence stratigraphy and duration of the early Toarcian OAE. Earth Planet. Sci. Lett. 386, 98–111.

Boulila, S., Galbrun, B., Hinnov, L.A., Collin, P.-Y., Ogg, J.G., Fortwengler, D., Marchand, D., 2010. Milankovitch and sub-milankovitch forcing of the oxfordian (late jurassic) terres Noires formation (SE France) and global implications. Basin Res. 22, 712–732. https://doi.org/10.1111/j.1365-2117.2009.00429.x.

Bretagnon, P., 1974. Termes à longues périodes dans le système solaire. Astronomy Astrophys. 30, 141–154.

Brouwer, D., van Woerkom, A.J.J., 1950. The secular variations of the orbital elements of the principal planets. Astronomical Pap. Prep. Use Am. Ephemer. Naut. Alm. 13 (2), 81–107.

Brüggemann, W., 1992. A minimal cost function method for optimizing the age-depth relation of deep-sea sediment cores. Paleoceanography 7, 467–487. https://doi.org/10.1029/92PA01235.

Burgess, S.D., Muirhead, J.D., Bowring, S.A., 2017. Initial pulse of Siberian Traps sills as the trigger of the end-Permian mass extinction. Nat. Commun. 8, 164. https://doi.org/10.1038/s41467-017-00083-9.

Burgess, S.D., Bowring, S.A., 2015. High-precision geochronology confirms voluminous magmatism before, during, and after Earth's most severe extinction. Sci. Adv. 1 (7) https://doi.org/10.1126/sciadv.1500470 e1500470 1–14.

Burgess, S.D., Bowring, S.A., Shen, S.Z., 2014. A new high-precision timeline for Earth's most severe extinction. Proc. Natl. Acad. Sci. 111, 3316–3321. https://doi.org/10.1073/pnas.1317692111.

Calner, M., 2008. Silurian global events—at the tipping point of climate change. In: Elewa, A.M.T. (Ed.), Mass Extinctions. Springer-Verlag, Berlin, pp. 21–58.

Carvajal, C.P., Soreghan, G.S., Isaacson, P.E., Ma, C., Hamilton, M.A., Hinnov, L.A., Dulin, S.A., 2018. Atmospheric dust from the Pennsylvanian Copacabana Formation (Bolivia): a high-resolution record of paleoclimate and volcanism from northwestern Gondwana. Gondwana Res. 58, 105–121. https://doi.org/10.1016/j.gr.2018.02.007.

Crick, R.E., Ellwood, B.B., Hladil, J., El Hassani, A., Hrouda, F., Chlubac, I., 2001. Magnetostratigraphy susceptibility of the pridolian-lochkovian (Silurian-Devonian) GSSP (Klonk, Czech Republic) and a coeval sequence in anti-atlas Morocco. Palaeogeogr. Palaeoclimatol. Palaeoecol. 167, 73–100. https://doi.org/10.1016/S0031-0182(00)00233-9.

Crucifix, M., 2016. Palinsol: Insolation for Palaeoclimate Studies, R Package. https://CRAN.R-project.org/package=palinsol.

Croudace, I.W., Rothwell, R.G., 2015. Micro-XRF Studies of Sediment Cores, Developments in Paleoenvironmental Research, vol. 17. Springer, Dordrecht. https://doi.org/10.1007/978-94-017-9849-5, 656 p.

Da Silva, A.C., Hladil, J., Chadimova, L., Slavik, L., Hilgen, F.J., Babek, O., Dekkers, M.J., 2016. Refining the early devonian time scale using Milankovitch cyclicity in lochkovian—pragian sediments (prague Synform, Czech Republic). Earth Planet. Sci. Lett. 455, 125—139. https://doi.org/10.1016/j.epsl.2016.09.009.

Dabard, M.P., Loi, A., Paris, F., Ghienne, J.F., Pistis, M., Vidal, M., 2015. Sea-level curve for the middle to early late ordovician in the armorican massif (western France): icehouse third-order glacio-eustatic cycles. Palaeogeogr. Palaeoclimatol. Palaeocol. 436, 96—111. https://doi.org/10.106/j.palaeo.2015.06.038.

Davydov, V.I., Crowley, J.L., Schmitz, M.D., Poletaev, V.I., 2010. High-precision U-Pb zircon age calibration of the global Carboniferous time scale and Milankovitch band cyclicity in the Donets Basin, eastern Ukraine. G-cubed 11, Q0AA04. https://doi.org/10.1029/2009GC002736.

Dehant, V., Mathews, P.M., 2015. Precession, Nutation and Wobble of the Earth. Cambridge University Press, Cambridge, 536 p.

Dehant, V., Capitaine, N., 1997. On the precession constant: values and constraints on the dynamical ellipticity; link with Oppolzer terms and tilt-over-mode. Celest. Mech. Dyn. Astronomy 65, 439—458. https://doi.org/10.1007/BF00049506.

Desrochers, A., Farley, C., Achab, A., Asselin, E., Riva, J.F., 2010. A far-field record of the end ordovician glaciation: the Ellis Bay formation, Anticosti Island, eastern Canada. Palaeogeogr. Palaeoclimatol. Palaeoecol. 296, 248—263. https://doi.org/10.1016/j.palaeo.2010.02.017.

Deubner, F.-L., 1990. Discussion on Late Precambrian tidal rhythmites in South Australia and the history of the Earth's rotation. J. Geol. Soc. London 147, 1083—1084.

De Vleeschouwer, D., Da Silva, A.C., Sinnesael, M., Chen, D., Day, J.E., Whalen, M.T., Guo, Z., Claeys, P., 2017b. Timing and pacing of the Late Devonian mass extinction event regulated by eccentricity and obliquity. Nat. Commun. 8, 2268. https://doi.org/10.1038/s41467-017-02407-1.

De Vleeschouwer, D., Vahlencamp, M., Crucifix, M., Paelike, H., 2017a. Alternating Southern and Northern Hemisphere climate response to astronomical forcing during the past 35 m.y. Geology 45 (4), 375—378. https://doi.org/10.1130/G38663.1.

De Vleeschouwer, D., Boulvain, F., Da Silva, A.-C., Pas, D., Labaye, C., Claeys, P., 2015. The astronomical calibration of the givetian (middle devonian) timescale (dinant Synclinorium, Belgium). In: Da Silva, A.C., Whalen, M.T., Hladil, J., Chadimova, L., Chen, D., Spassov, S., Boulvain, F., Devleeschouwer, X. (Eds.), Magnetic Susceptibility Application: A Window onto Ancient Environments and Climatic Variations, vol. 414. Geological Society, London, pp. 245—256. https://doi.org/10.1144/SP414.3. Special Publications.

De Vleeschouwer, D., Crucifix, M., Bounceur, N., Claeys, P., 2014. The impact of astronomical forcing on the Late Devonian greenhouse climate. Glob. Planet. Change 120, 65—80. https://doi.org/10.1016/j.gloplacha.2014.06.002.

De Vleeschouwer, D., Whalen, M.T., Day, J.E.J., Claeys, P., 2012. Cyclostratigraphic calibration of the Frasnian (Late Devonian) time scale (western Alberta, Canada). Geological Society of America Bulletin 124 (5/6), 928—942. https://doi.org/10.1130/B30547.1.

Denis, C., Schreider, A.A., Varga, P., Zavoti, J., 2002. Despinning of the earth rotation in the geological past and geomagnetic paleointensities. J. Geodyn. 34, 667–685. https://doi.org/10.1016/S0264-3707(02)00049-2.

De Surgy, N.O., Laskar, J., 1997. On the long term evolution of the spin of the Earth. Astronomy Astrophys. 318, 975–989.

Derby, J., Fritz, R., Longacre, S., Morgan, W., Sternbach, W., 2012. The great american carbonate bank: the geology and economic Resources of the cambrian-ordovician Sauk megasequence of laurentia. AAPG Memo. 98, 528 p.

Dickey, J.O., Bender, P.L., Faller, J.E., Newhall, X.X., Ricklefs, R.L., Ries, J.G., Shelus, P.J., Veillet, C., Whipple, A.L., Wiant, J.R., Williams, J.G., Yoder, C.F., 1994. Lunar laser ranging: a continuing legacy of the Apollo Program. Science 265, 482–490. https://doi.org/10.1126/science.265.5171.482.

Dinarès-Thurell, J., Westerhold, T., Pujalte, V., Röhl, U., Kroon, D., 2014. Astronomical calibration of the danian stage (early Paleocene) revisited: Settling chronologies of sedimentary records across the atlantic and Pacific oceans. Earth Planet. Sci. Lett. 405, 119–131. https://doi.org/10.1016/j.epsl.2014.08.027.

Dinarès-Turell, J., Pujalte, V., Stoykova, K., Elorza, J., 2013. Detailed correlation and astronomical forcing within the Upper Maastrichtian succession in the Basque Basin. Bol. Geol. y Min. 124 (2), 253–282.

Eldrett, J.S., Ma, C., Bergman, S.C., Lutz, B., Gregory, F.J., Dodsworth, P., Phipps, M., Hardas, P., Minisini, D., Ozkan, A., Ramezani, J., Bowring, S., Kamo, S.L., Ferguson, K., Macalulay, C., Kelly, A.E., 2015. An astronomically calibrated stratigraphy of the cenomanian, turonian and earliest coniacian from the cretaceous western interior Seaway, USA: implications for global chronostratigraphy. Cretac. Res. 56, 316–344. https://doi.org/10.1016/j.cretres.2015.04.010.

Ellwood, B.B., El Hassani, A., Tomkin, J.H., Bultynck, P., 2015. A climate-driven model using time-series analysis of magnetic susceptibility (x) datasets to represent a floating-point high-resolution geological timescale for the Middle Devonian Eifelian stage. In: Da Silva, A.C., Whalen, M.T., Chadimova, J., Chen, D., Spassov, S., Boulvain, F., Devleeschouwer, X. (Eds.), Magnetic Susceptibility Application: A Window onto Ancient Environments and Climatic Variations, vol. 414. Geological Society, London, pp. 209–223. https://doi.org/10.1144/SP414.4. Special Publications.

Ellwood, B.B., Brett, C.E., Tomkin, J.H., Macdonald, W.D., 2013. Visual identification and quantification of Milankovitch climate cycles in outcrop: an example from the upper ordovician Kope Formation, northern Kentucky. In: Jovane, L., Herreo-Bervara, E., Hinnov, L.A., Housen, B.A. (Eds.), Magnetic Methods and the Timing of Geological Processes, vol. 373. Geological Society, London, pp. 341–353. https://doi.org/10.1144/SP373.2. Special Publications.

Elrick, M., Reardon, D., Labor, W., Martin, J., Desrochers, A., Pope, M., 2013. Orbital-scale climate change and glacioeustasy during the early Late Ordovician (pre-Hirnantian) determined from $\delta^{18}O$ values in marine apatite. Geology 41, 775–788. https://doi.org/10.1130/G34363.1.

Edvardsson, S., Karlsson, K.G., Englholm, M., 2002. Accurate spin axes and solar system dynamics: climatic variations for the Earth and Mars. Astronomy Astrophys. 384, 689–701. https://doi.org/10.1051/0004-6361:20020029.

Eros, J.M., Montanez, I.P., Osleger, D.A., Davydov, V.I., Nemyrovska, T.I., Poletaev, V.I., Zhykalyak, M.V., 2012. Sequence stratigraphy and onlap history of the Donets Basin, Ukraine: insight into Carboniferous icehouse dynamics. Palaeogeogr. Palaeoclimatol. Palaeoecol. 313–314, 1–25. https://doi.org/10.1016/j.palaeo.2011.08.019.

Fang, Q., Wu, H., Wang, X., Yang, T., Li, H., Zhang, S., 2018. Astronomical cycles in the Serpukhovian-Moscovian (Carboniferous) marine sequence, South China and their implications for geochronology and icehouse dynamics. J. Asian Earth Sci. 156, 302–315. https://doi.org/10.1016/j.jseaes.2018.02.001.

Fang, Q., Wu, H.C., Hinnov, L.A., Jing, X.C., Wang, X.L., Yang, T.S., Li, H.Y., Zhang, S.H., 2017. Astronomical cycles of Middle Permian Maokou Formation in South China and their implications for sequence stratigraphy and paleoclimate. Palaeogeogr. Palaeoclimatol. Palaeoecol. 474, 130–139. https://doi.org/10.1016/j.palaeo.2016.07.037.

Fang, Q., Wu, H., Hinnov, L.A., Wang, X., Yang, T., Li, H., Zhang, S., 2016. A record of astronomically forced climate change in a late Ordovician (Sandbian) deep marine sequence, Ordos Basin, North China. Sediment. Geol. 341, 163–174. https://doi.org/10.1016/j.sedgeo.2016.06.002.

Fang, Q., Wu, H., Hinnov, L.A., Jing, X., Wang, X., Jiang, Q., 2015. Geological evidence for the chaotic behavior of the planets and its constraints on the third order eustatic sequences at the end of the Late Paleozoic Ice Age. Palaeogeogr. Palaeoclimatol. Palaeoecol. 440, 848–859. https://doi.org/10.1016/j.palaeo.2015.10.014.

Feng, Z.Z., Yang, Y.Q., Bao, Y.Q., Yu, Z.D., Jing, Z.K., 1998. Carboniferous Lithofacies Paleogeography in South China. Geological Press, Beijing, pp. 1–119 (in Chinese with English abstract).

Fiet, N., Gorin, G., 2000. Lithologic expression of Milankovitch cyclicity in carbonate-dominated, pelagic, Barremian deposits in central Italy. Cretac. Res. 21, 457–467. https://doi.org/10.1006/cres.2000.0220.

Fielding, C.R., Frank, T.D., Birgenheier, L.P., Rygel, M.C., Jones, A.T., Roberts, J., 2008. Stratigraphic imprint of the Late Palaeozoic Ice Age in eastern Australia: a record of alternating glacial and nonglacial climate regime. J. Geol. Soc. 165, 129–140.

Fischer, A.G., Herbert, T.D., Napoleone, G., Premoli Silva, I., Ripepe, M., 1991. Albian pelagic rhythms (Piobbico core). J. Sediment. Petrol. 61 (7), 1164–1172. https://doi.org/10.1306/D426785C-2B26-11D7-8648000102C1865D.

Fischer, A.G., 1986. Climatic rhythms recorded in strata. Annu. Rev. Earth Planet. Sci. 14, 351–376. https://doi.org/10.1146/annurev.ea.14.050186.002031.

Fischer, A.G., De Boer, P.L., Premoli Silva, I., 1988. Cyclostratigraphy. In: Beaudoin, B., Ginsburg, R.N. (Eds.), Global Sedimentary Geology Program: Cretaceous Resources, Events, and Rhythms. – NATO ASI Series. Kluwer, Dordrecht, pp. 139–172.

Gambacorta, G., Menichetti, E., Trincianti, E., Torricelli, S., 2018. Orbital control on cyclical primary productivity and benthic anoxia: astronomical tuning of the Telychian Stage (Early Silurian). Palaeogeogr. Palaeoclimatol. Palaeoecol. 495, 152–162. https://doi.org/10.1016/j.palaeo.2018.01.003.

Goldhammer, R.K., Lehmann, P.J., Dunn, P.A., 1993. The origin of high-frequency platform carbonate cycles and third order sequences (Lower Ordovician El Paso Group, west Texas): constraints from outcrop data and stratigraphic modeling. J. Sediment. Petrol. 63, 318–359. https://doi.org/10.1306/D4267AFA-2B26-11D7-8648000102C1865D.

Gong, Y., Droser, M.L., 2001. Periodic anoxic shelf in the Early-Middle Ordovician transition: ichnosedimentologic evidence from west-central Utah, USA. Sci. China, Ser. D 44, 979–989. https://doi.org/10.1007/BF02875391.

Gong, Z., Kodama, K.P., Li, Y.-X., 2017. Rock magnetic cyclostratigraphy of the Doushantuo Formation, South China and its implications for the duration of the Shuram carbon isotope excursion. Precambrian Res. 289, 62–74. https://doi.org/10.1016/j.precamres.2016.12.002.

Grader, G.W., Issacson, P.E., Díaz-Martínez, E., Pope, M.C., 2008. Pennsylvanian and permian sequences in Bolivia: direct responses to Gondwana glaciation. In: Fielding, C.R., Frank, T.D., Isbell, J.L. (Eds.), Resolving the Late Paleozoic Ice Age in Time and Space, vol. 441, pp. 143—159. https://doi.org/10.1130/2008.2441(10. Geological Society of America Special Paper.

Green, J.A.M., Huber, M., Waltham, D., Buzan, J., Wells, M., 2017. Explicitly modelled deep-time tidal dissipation and its implication for Lunar history. Earth Planet. Sci. Lett. 461, 46—53. https://doi.org/10.1016/j.epsl.2016.12.038.

Green, J.A.M., Huber, M., 2013. Tidal dissipation in the early Eocene and implications for ocean mixing. Geophys. Res. Lett. 40, 2707—2713. https://doi.org/10.1002/grl.50510.

Grippo, A., Fischer, A.G., Hinnov, L.A., Herbert, T.D., Premoli Silva, I., 2004. Cyclostratigraphy and chronology of the Albian stage (Piobbico core, Italy). In: D'Argenio, B., Fischer, A.G., Premoli Silva, I., Weissert, H., Ferreri, V. (Eds.), Cyclostratigraphy: Approaches and Case Histories, pp. 57—81. https://doi.org/10.2110/pec.04.81.0057. SEPM Special Publication No. 81.

Grotzinger, J.P., 1986. Cyclicity and paleoenvironmental dynamics, Rocknest platform, northwest Canada. Geol. Soc. Am. Bull. 97, 1208—1231. https://doi.org/10.1130/0016-7606(1986)97<1208:CAPDRP>2.0.CO;2.

Grotzinger, J.P., Fike, D.A., Fischer, W.W., 2011. Enigmatic origin of the largest-known carbon isotope excursion in Earth's history. Nat. Geosci. 4, 285—292. https://doi.org/10.1038/ngeo1138.

Hays, J.D., Imbrie, J., Shackleton, N.J., 1976. Variations in the Earth's orbit: pacemaker of the ice ages. Science 194 (4270), 1121—1132. https://doi.org/10.1126/science.194.4270.1121.

Hennebert, M., 2014. The Cretaceous-Paleogene boundary and its 405-kyr eccentricity cycle phase: a new constraint on radiometric dating and astrochronology. Carnets Géologie 14 (9), 173—189. https://doi.org/10.1029/2012GC004096.

Herbert, T.D., 1999. Toward a composite orbital chronology for the late Cretaceous and early Paleocene GPTS. Phil. Trans. R. Soc. Lond. 357, 1891—1905. https://doi.org/10.1098/rsta.1999.0406.

Herbert, T.D., Stallard, R.F., Fischer, A.G., 1986. Anoxic events, productivity rhythms and the orbital signature in a mid-Cretaceous deep-sea sequence from Central Italy. Paleoceanography 1, 495—506. https://doi.org/10.1029/PA001i004p00495.

Herterich, K., Sarnthein, M., 1984. Brunhes time scale: tuning by rates of calcium-carbonate dissolution and cross spectral analyses with solar insolation. In: Berger, A., Imbrie, J., Hays, J., Kukla, G., Salzman, B. (Eds.), Milankovitch and Climate, Part I, D. Reidel Publishing Company, Dordrecht, pp. 447—466.

Hilgen, F.J., 2010. Astronomical tuning in the 19th century. Earth Sci. Rev. 98, 65—80. https://doi.org/10.1016/j.earscirev.2009.10.004.

Hilgen, F.J., Abels, H.A., Kuiper, K.F., Lourens, L.J., Wolthers, M., 2015. Towards a stable astronomical time scale for the Paleocene: aligning Shatsky rise with the Zumaia — Walvis Ridge ODP site 1262 composite. Newslett. Stratigr. 48 (1), 91—110. https://doi.org/10.1127/nos/2014/0054.

Hilgen, F.J., Hinnov, L.A., Abdul Aziz, H., Abels, H.A., de Boer, B., Bosmans, J.H.C., Hüsing, S.K., Kuiper, K., Lourens, L.J., Tuenter, E., Van de Val, R.S.W., Zeeden, C., 2014. Stratigraphic continuity and fragmentary sedimentation: the success of cyclostratigraphy as part of integrated stratigraphy. Special Publication. In: Strata and Time: Probing the Gaps in Our Understanding, vol. 404. Geological Society of London. https://doi.org/10.1144/SP404.12.

Hilgen, F.J., Lourens, L.J., Van Dam, J.A., 2012. The Neogene period. In: Gradstein, F.M., Ogg, J.G., Schmitz, M., Ogg, G. (Eds.), The Geologic Time Scale 2012. Elsevier, pp. 923—978. https://doi.org/10.1016/B978-0-444-59425-9.00029-9 (Chapter 29).

Hilgen, F.J., Kuiper, K.F., Lourens, L.J., 2010. Evaluation of the astronomical time scale for the Paleocene and earliest Eocene. Earth Planet. Sci. Lett. 300, 139—151. https://doi.org/10.1016/j.epsl.2010.09.044.

Hilgen, F.J., Abdul Aziz, H., Krijgsman, W., Langereis, C.G., Lourens, L.J., Meulenkamp, J.E., Riffi, I., Steenbrink, J., Turco, E., van Vugt, N., Wijbrans, J.R., Zachariasse, W.J., 1999. Present status of the astronomical (polarity) time-scale for the Mediterranean Late Neogene. Phil. Trans. R. Soc. Lond. Ser. A 357, 1931—1947. https://doi.org/10.1098/rsta.1999.0408.

Hilgen, F.J., Lourens, L.J., Berger, A., Loutre, M.F., 1993. Evaluation of the astronomicallay liberated time scale for the late Pliocene and earliest Pleistocene. Paleoceanography 8, 549—565. https://doi.org/10.1029/93PA01248.

Hinnov, L.A., 2018. Astronomical metronome of geological consequence. Proc. Natl. Acad. Sci. 115 (24), 6104—6106. https://doi.org/10.1073/pnas.1807020115.

Hinnov, L.A., 2013. Cyclostratigraphy and its revolutionizing applications in the earth and planetary Sciences. Geol. Soc. Am. Bull. 125, 1703—1734. https://doi.org/10.1130/B30934.1.

Hinnov, L.A., Diecchio, R.J., 2015. Milankovitch cycles in the Juniata Formation, late ordovician central appalachian basin, USA. Stratigraphy 12, 287—296.

Hinnov, L.A., Mei, M., Wu, H., Zhang, S., 2015. The Precambrian Wumishan cyclothems of China: 60+ million year long shallow marine carbonate record of multiple-scale, hierarchical sea level oscillations, an aragonite-calcite sea transition and cyanobacteria calcification. Geol. Soc. Am. Annual Meet. Abstr. Programs 47 (7), 649. Baltimore, Maryland, USA 1-4 November.

Hinnov, L.A., Locklair, R., Ogg, J., 2008. Recent developments in the geologic timescale, construction of the astronomical time scale—Part 1. In: Early Cretaceous, Abstract, 33nd International Geological Congress, Oslo, Norway, 6-14 August.

Hofmann, A., Dirks, P.H.G.M., Jelsma, H.A., 2004. Shallowing upward carbonate cycles in the Belingwe Greenstone Belt, Zimbabwe: a record of Archean sealevel oscillations. J. Sediment. Res. 4, 64—81. https://doi.org/10.1306/052903740064.

Hüsing, S., Benist, A., van der Boon, A., Abels, H.A., Deenen, M.H.L., Ruhl, M., Krijgsman, W., 2014. Astronomically-calibrated magnetostratigraphy of the lower jurassic marine successions at St. Audrie's Bay and East Quantoxhead (Hettangian—Sinemurian; Somerset, UK). Palaeogeogr. Palaeoclimatol. Palaeoecol. 403, 43—56. https://doi.org/10.1016/j.palaeo.2014.03.022.

Huang, C., 2018. Astronomical time scale for the Mesozoic (this volume). Stratigr. Timescales 3 (Chapter 4).

Huang, C., Hinnov, L.A., Fischer, A.G., Grippo, A., Herbert, T., 2010a. Astronomical tuning of the Aptian stage from Italian reference sections. Geology 238, 899—903. https://doi.org/10.1130/G31177.1.

Huang, C., Hinnov, L.A., Swientek, O., Smelnor, M., 2010b. Astronomical tuning of late jurassic-early cretaceous sediments (Volgian-Ryazanian stages), Greenland-Norwegian Seaway, abstract. In: AAPG Annual Convention, New Orleans, LA, 11-14 April.

Huang, C., Hesselbo, S.P., Hinnov, L.A., 2010c. Astrochronology of the late jurassic Kimmeridge Clay (Dorset, England) and implications for earth system processes. Earth Planet. Sci. Lett. 289, 242—255. https://doi.org/10.1016/j.epsl.2009.11.013.

Huret, E., Hinnov, L.A., Galbrun, B., Boulila, S., Collin, P.-Y., 2008a. High-resolution cyclostratigraphy of Upper Jurassic (Callovian to Oxfordian) marly formations (Paris Basin): astronomical calibration and implications for regional correlation, Abstract. In: 33nd International Geological Congress, Oslo, Norway, 6-14 August.

Huret, E., Hinnov, L.A., Galbrun, B., Collin, P.-Y., Gardin, S., Rouget, I., 2008b. Astronomical calibration and correlation of the lower jurassic, paris and lombard basins (tethys), abstract. In: 33nd International Geological Congress, Oslo, Norway, 6-14 August.

Husson, D., Galbrun, B., Laskar, J., Hinnov, L.A., Thibault, N., Gardin, S., Locklair, R.E., 2011a. Astronomical calibration of the maastrichtian (late cretaceous). Earth Planet. Sci. Lett. 305, 328—340. https://doi.org/10.1016/j.epsl.2011.03.008.

Huybers, P., 2007. Glacial variability over the last two million years: an extended depth-derived age model, continuous obliquity pacing, and the Pleistocene progression. Quat. Sci. Rev. 26, 37—55. https://doi.org/10.1016/j.quascirev.2006.07.013.

Huybers, P., 2006. Early Pleistocene glacial cycles and the integrated summer insolation forcing. Science 313, 508—511. https://doi.org/10.1126/science.1125249.

Huybers, P., Wunsch, C., 2004. A depth-derived Pleistocene age model: uncertainty estimates, sedimentation variability, and nonlinear climate change. Paleoceanography 19, PA1028. https://doi.org/10.1029/2002PA000857.

Ikeda, M., Tada, R., Ozaki, K., 2017. Astronomical pacing of the global silica cycle recorded in Mesozoic bedded cherts. Nat. Commun. 8, 15532. https://doi.org/10.1038/ncomms15532.

Ikeda, M., Tada, R., 2014. A 70 million year astronomical time scale for the deep-sea bedded chert sequence (Inuyama, Japan): implications for Triassic-Jurassic geochronology. Earth Planet. Sci. Lett. 399, 30—43. https://doi.org/10.1016/j.epsl.2014.04.031.

Ikeda, M., Tada, R., 2013. Long period astronomical cycles from the Triassic to Jurassic bedded chert sequence (Inuyama, Japan); Geologic evidences for the chaotic behavior of solar planets. Earth Planets Space 65, 351—360. https://doi.org/10.5047/eps.2012.09.004.

Ikeda, M., Tada, R., Sakuma, H., 2010. Astronomical cycle origin of bedded chert; middle Triassic bedded chert sequence, Inuyama, Japan. Earth Planet. Sci. Lett. 297, 369—378. https://doi.org/10.1016/j.epsl.2010.06.027.

Imbrie, J., Hays, J.D., Martinson, D.G., McIntiyre, A., Mix, A.C., Morley, J.C., Pisias, N.G., Prell, W.L., Shackleton, N.J., 1984. The orbital theory of Pleistocene climate: support from a revised chronology of the marine $\delta^{18}O$ record. In: Berger, A., Imbrie, J., Hays, J., Kukla, G., Salzman, B. (Eds.), Milankovitch and Climate, Part I. D. Reidel Publishing Company, Dordrecht, pp. 269—305.

Imbrie, J., Imbrie, J.Z., 1980. Modeling the climatic response to orbital variations. Science 207, 943—953. https://doi.org/10.1126/science.207.4434.943.

Isaacson, P.E., Diaz-Martinez, E., Grader, G.W., Kalvoda, J., Babek, O., Devuyst, F.X., 2008. Late Devonian—earliest Mississippian glaciation in Gondwanaland and its biogeographic consequences. Palaeogeogr. Palaeoclimatol. Palaeoecol. 268 (3—4), 126—142. https://doi.org/10.1016/j.palaeo.2008.03.047.

Ito, T., Tanikawa, K., 2007. Trends in 20th Century Celestial Mechanics, vol. 9. Publications of the National Astronomy Observatory of Japan, pp. 55—112.

Kietzmann, D.A., Palma, R.M., Iglesia Llanos, M.P., 2015. Cyclostratigraphy of an orbitally-driven tithonian—valanginian carbonate ramp succession, southern mendoza, Argentina: implications for the jurassic—cretaceous boundary in the Neuquén basin. Sediment. Geol. 315, 29—46. https://doi.org/10.1016/j.sedgeo.2014.10.002.

Kent, D.V., Olsen, P.E., Rasmussen, C., Lepre, C., Mundil, R., Irmis, R., Gehrels, G., Giesler, D., Geissman, J., Parker, W., 2018. Empirical evidence for stability of the 405 kyr Jupiter-Venus eccentricity cycle over hundreds of millions of years. Proc. Natl. Acad. Sci. USA. https://doi.org/10.1073/pnas.1800891115.

Kent, D.V., Olsen, P.E., Muttoni, G., 2017. Astrochronostratigraphic polarity time scale (APTS) for the Late Triassic and Early Jurassic from continental sediments and correlation with standard marine stages. Earth Sci. Rev. 166, 153—180. https://doi.org/10.1016/j.earscirev.2016.12.014.

Kent, D.V., Olsen, P.E., Witte, W.K., 1995. Late Triassic-earliest Jurassic geomagnetic polarity sequence and paleolatitudes from drill cores in the Newark rift basin, eastern North America. J. Geophys. Res. 100, 14965—14998. https://doi.org/10.1029/95JB01054.

Kim, J.C., Lee, Y.I., 1998. Cyclostratigraphy of the lower ordovician Dumugol Formation, Korea: meter-scale cyclicity and sequence stratigraphic interpretation. Geosci. J. 2, 134–147. https://doi.org/10.1007/BF02910257.

Klein, C., 2005. Some Precambrian banded iron-formations (BIFs) from around the world: their age, geologic setting, mineralogy, metamorphism, geochemistry, and origin. Am. Mineral. 90, 1473–1499. https://doi.org/10.2138/am.2005.1871.

Kodama, K.P., Hinnov, L.A., 2015. Rock Magnetic Cyclostratigraphy. John Wiley and Sons, Chichester, UK. https://doi.org/10.1002/9781118561294, 164 p.

Kodama, K.P., Anastasio, D.J., Pares, J., Hinnov, L.A., 2010. High-resolution rock magnetic cyclostratigraphy in an Eocene flysch, Spanish Pyrenees. G-cubed 11, Q0AA07. https://doi.org/10.1029/2010GC003069.

Kominz, M.A., Beavan, J., Bond, G.C., McManus, J., 1991. Are cyclic sediments periodic? Gamma analysis and spectral analysis of Newark Supergroup lacustrine strata. Kansas Geological Survey Bulletin 233, 320–334.

Kominz, M.A., Bond, G.C., 1990. A new method of testing periodicity in cyclic sediment: application to the Newark Supergroup. Earth Planet. Sci. Lett. 98, 233–244.

Kominz, M.A., Heath, G.R., Ku, T.-L., Pisias, N.G., 1979. Brunhes time scales and the interpretation of climatic change. Earth Planet. Sci. Lett. 45, 394–410.

Kostadinov, T.S., Gilb, R., 2014. Earth Orbit v2.1: a 3-D visualization and analysis model of Earth's orbit, Milankovitch cycles and insolation. Geosci. Model Dev. 7, 1051–2014. https://doi.org/10.5194/gmd-7-1051-2014.

Kotov, S., Pälike, H., 2017. MyDTW - dynamic Time Warping program for stratigraphical time series. In: 19th EGU General Assembly, EGU2017, 23-28 April, 2017, Vienna, Austria., 2157.

Kuiper, K.F., Deino, A., Hilgen, F.J., Krijgsman, W., Renne, P.R., Wijbrans, J.R., 2008. Synchronizing rock clocks of Earth history. Science 320, 500–504. https://doi.org/10.1126/science.1154339.

Lambeck, K., 1980. The Earth's Variable Rotation: Geophysical Causes and Consequences. Cambridge University Press, New York, 449 p.

Laskar, J., 2015. Stability of the Solar System. http://www.scholarpedia.org/article/Stability_of_the_solar_system.

Laskar, J., 2013. Is the solar system stable? Progr. Math. Phys. 66, 239–270.

Laskar, J., 1993. The stability of the solar system. In: Pfenniger, D. (Ed.), Ergodic Concepts and Stellar Dynamics, Lecture Notes in Physics.

Laskar, J., 1990. The chaotic motion of the Solar System: a numerical estimate of the size of the chaotic zones. Icarus 88, 266–291. https://doi.org/10.1016/0019-1035(90)90084-M.

Laskar, J., 1989. A numerical experiment on the chaotic behavior of the Solar System. Nature 338, 237–238. https://doi.org/10.1038/338237a0.

Laskar, J., Farrés, A., Gastineau, M., 2012. Stability of the long period terms in the La2011 astronomical solution. Geophys. Res. Abstr. 14. EGU2012-E2481.

Laskar, J., Gastineau, M., Delisle, J.-B., Farrés, A., Fienga, A., 2011b. Strong chaos induced by close encounters with Ceres and Vesta. Astronomy Astrophys. 532, L4.

Laskar, J., Fienga, A., Gastineau, M., Manche, H., 2011a. La2010: a new orbital solution for the long term motion of the Earth. Astronomy Astrophys. 532, A89. https://doi.org/10.1051/0004-6361/201116836.

Laskar, J., Robutel, P., Joutel, J., Gastineau, M., Correia, A.C.M., Levrard, B., 2004. A numerical solution for the insolation quantities of the Earth. Astronomy Astrophys. 428, 261–285. https://doi.org/10.1051/0004-6361:20041335.

Laskar, J., Joutel, J., Boudin, F., 1993. Orbital, precessional and insolation quantities for the Earth from -20 Myr to +10 Myr. Astronomy Astrophys. 270, 522–533.

Le Hir, G., Donnadieu, Y., Goddéris, Y., Meyer-Berthaud, B., Ramstein, G., Blakey, R.C., 2011. The climate change caused by the land plant invasion in the Devonian. Earth Planet. Sci. Lett. 310 (3—4), 203—212. https://doi.org/10.1016/j.epsl.2011.08.042.

Le Verrier, U.J., 1855. Recherches astronomiques. Annales Observatoire Impérial de Paris 1, 73—383.

Li, M., Kump, L.R., Hinnov, L.A., Mann, M.E., 2018b. Tracking variable sedimentation rates and astronomical forcing in Phanerozoic paleoclimate proxy series with evolutionary correlation coefficients and hypothesis testing. Earth Planet. Sci. Lett. (in press).

Li, M., Huang, C., Hinnov, L.A., Chen, W., Ogg, J.G., Tian, W., 2018a. Astrochronology of the anisian stage (middle triassic) at the Guandao reference section, south China. Earth Planet. Sci. Lett. 482, 591—606. https://doi.org/10.1016/j.epsl.2017.11.042.

Li, M., Huang, C., Hinnov, L.A., Ogg, J., Chen, Z.-Q., Zhang, Y., 2016. Obliquity-forced climate during the early triassic hothouse in China. Geology 44, 623—626. https://doi.org/10.1130/G37970.1.

Lin, L., Khider, D., Lisiecki, L.E., Lawrence, C.E., 2014. Probabilistic sequence alignment of stratigraphic records. Paleoceanography 29, 976—989. https://doi.org/10.1002/2014PA002713.

Lisiecki, L.E., Raymo, M.E., 2005. A Pliocene- Pleistocene stack of 57 globally distributed benthic δ18O records. Paleoceanography 20, PA1003. https://doi.org/10.1029/2004PA001071.

Lisiecki, L.E., Lisiecki, P.A., 2002. Application of dynamic programming to the correlation of paleoclimate records. Paleoceanography 17 (4), 1049. https://doi.org/10.1029/2001PA000733.

Liu, J.-C., Capitaine, N., 2017. Evaluation of a possible upgrade of the IAU2006 precession. Astronomy Astrophys. 597, A83. https://doi.org/10.1051/0004-6361/201628717.

Liu, W., Wu, H., Hinnov, L.A., Ma, C., Li, M., Pas, D., 2017b. Paleoclimate evolution driven by astronomical forcing in the early cretaceous Songliao Basin, northeast China. In: 10th International Symposium on the Cretaceous, 21-26 August, Vienna, AT.

Liu, Z., Huang, C., Algeo, T.J., Liu, J., Hao, Y., Du, X., Lu, Y., Chen, P., Guo, L., Peng, L., 2017a. High-resolution astrochronological record for the paleocene-oligocene (66—23 Ma) from the rapidly subsiding Bohai Bay Basin, northeastern China. Palaeogeogr. Palaeoclimatol. Palaeoecol. https://doi.org/10.1016/j.palaeo.2017.10.030.

Locklair, R.E., Sageman, B.B., 2008. Cyclostratigraphy of the Upper Cretaceous Niobara Formation, Western Interior, U.S.A.: A Coniacian-Santonian orbital timescale. Earth Planet. Sci. Lett. 269, 540—553. https://doi.org/10.1016/j.epsl.2008.03.021.

Loi, A., Ghienne, J.-F., Dabard, M.P., Paris, F., Botquelen, A., Christ, N., Elaouad-Debbaj, Z., Gorini, A., Vidal, M., Videt, B., Destombes, J., 2010. The Late Ordovician glacio-eustatic record from a high-latitude storm-dominated shelf succession: The Bou Ingarf section (Anti-Atlas, Southern Morocco). Palaeogeogr. Palaeoclimatol. Palaeoecol. 296, 332—358. https://doi.org/10.1016/j.palaeo.2010.01.018.

Lorenz, H., 2010. The Swedish Deep Drilling Program: For Science and Society. GFF-Uppsala 132 (1), 25—27. https://doi.org/10.1080/11035891003763354.

Lourens, L.J., Hilgen, F., Shackleton, N.J., Laskar, J., Wilson, D., 2004. The Neogene Period. In: Gradstein, F., Ogg, J., Smith, A. (Eds.), A Geologic Time Scale 2004. Cambridge University Press, Cambridge, UK, pp. 400—440.

Lourens, L.J., Wehausen, R., Brumsack, H.J., 2001. Geological constraints on tidal dissipation and dynamical ellipticity of the Earth over the past three million years. Nature 409, 1029—1033. https://doi.org/10.1038/35059062.

Lourens, L.J., Antonarakou, A., Hilgen, F.J., Van Hoof, A.A.M., Vergnaud-Grazzini, C., Zachariasse, W.J., 1996. Evaluation of the Plio-Pleistocene astronomical timescale. Paleoceanography 11 (4), 391—413. https://doi.org/10.1029/96PA01125.

Loutre, M.-F., 2003. Ice Ages (Milankovitch Theory). In: Curry, J.A., Pyle, J.A. (Eds.), Encyclopedia of Atmospheric Sciences. Elsevier, pp. 995–1003.

Loutre, M.-F., Dehant, V., Berger, A., 1990. Astronomical frequencies in paleoclimatic data and the dynamical ellipticity of the Earth. In: Brosche, P., Sündermann, J. (Eds.), Earth's Rotation from Eons to Days. Springer-Verlag, Berlin, pp. 188–200. https://doi.org/10.1007/978-3-642-75587-3_20.

Lowrie, W., 2007. Fundamentals of Geophysics, second ed. Cambridge University Press, Cambridge. 381 p.

Ma, C., Hinnov, L.A., Carvajal, C.P., Soreghan, G.L., 2018. Earth's Late Bashkirian Precessional Frequency from the Copacabana Formation, Bolivia, South America. Nat. Geosci. (in review).

Ma, C., Meyers, S.R., Sageman, B.B., 2017. Theory of chaotic orbital variations confirmed by Cretaceous geological evidence. Nature 542, 468–470. https://doi.org/10.1038/nature21402.

Malinverno, A., Erba, E., Herbert, T.D., 2010. Orbital tuning as an inverse problem: Chronology of the early Aptian oceanic anoxic event 1a (Selli Level) in the Cismon APTICORE. Paleoceanography 25, PA2203. https://doi.org/10.1029/2009PA001769.

Martinez, M., Deconinck, J.-F., Pellenard, P., Riquier, L., Company, M., Reboulet, S., Moiroud, M., 2015. Astrochronology of the Valanginian–Hauterivian stages (Early Cretaceous): Chronological relationships between the Paraná–Etendeka large igneous province and the Weissert and the Faraoni events. Glob. Planet. Change 131, 158–173. https://doi.org/10.1016/j.gloplacha.2015.06.001.

Martinson, D.G., Menke, W., Stoffa, P., 1982. An inverse approach to signal correlation. J. Geophys. Res. 87, 4807–4818. https://doi.org/10.1029/JB087iB06p04807.

Maurer, F., Hinnov, L.A., Schlager, W., 2004. Statistical time series analysis and sedimentological tuning of bedding rhythms in a Triassic basinal succession (S. Alps, Italy). In: D'Argenio, B., Fischer, A.G., Premoli Silva, I., Weissert, H., Ferreri, V. (Eds.), Cyclostratigraphy: Approaches and Case Histories, vol. 81. Society for Sedimentary Geology Special Publication, pp. 83–99. https://doi.org/10.2110/pec.04.81.0083.

Mazzullo, S.J., 1971. Length of the year during the Silurian and Devonian Periods. Geol. Soc. Am. Bull. 82, 1085–1086.

McGhee Jr., G.R., 2013. When the Invasion of Land Failed: The Legacy of the Devonian Extinctions. Columbia University Press, New York. https://doi.org/10.7312/mcgh16056, 317 p.

Mclemore, Y. and Koonce, L., 2016. Precession, Rigid Body and Basic Concepts of Classical Mechanics, Academic Studio, New York: http://ebooks.wtbooks.com/static/wtbooks/ebooks/9781280131134/9781280131134.pdf. http://www.thefullwiki.org/Axial_precession_(astronomy). https://wikivisually.com/wiki/Axial_precession.

Meech, L.W., 1856. On the relative intensity of the heat and light of the Sun. Smithson. Contribut. Knowl. 9, 1–58.

Mei, M., Tucker, M.E., 2013. Milankovitch-driven cycles in the Precambrian of China: The Wumishan Formation. J. Palaeogeogr. 2 (4), 369–389. https://doi.org/10.3724/SP.J.1261.2013.00037.

Mei, M., Ma, Y., Zhou, H., Du, M., Luo, Z., Guo, Q., 2001. Fischer plots of Wumishan cyclothems, Precambrian records of third-order sea-level changes. J. China Univ. Geosci. 12, 1–10.

Meyers, S.R., 2015. The evaluation of eccentricity-related amplitude modulation and bundling in paleoclimate data: An inverse approach for astrochronologic testing and time scale optimization. Paleoceanography 30, 1625–1640. https://doi.org/10.1002/2015PA002850.

Meyers, S.R., 2014. Astrochron: An R Package for Astrochronology. https://cran.r-project.org/web/packages/astrochron/index.html.

Meyers, S.R., Malinverno, A., 2018. Proterozoic Milankovitch cycles and the history of the solar system. Proc. Natl. Acad. Sci. https://doi.org/10.1073/pnas.1717689115.

Meyers, S.R., Sageman, B.B., 2007. Quantification of deep-time orbital forcing by average spectral misfit. Am. J. Sci. 307, 773—792. https://doi.org/10.2475/05.2007.01.

Milankovitch, M., 1941. Kanon der Erdbestrahlung und seine Anwendung auf das Eiszeitenproblem. Royal Serbian Academy, Section of Mathematical and Natural Sciences, Belgrade, 633 p. (and 1998 reissue in English: Canon of Insolation and the Ice-Age Problem: Belgrade, Serbian Academy of Sciences and Arts, Section of Mathematical and Natural Sciences, 634 p.).

Milankovitch, M., 1920. Théorie Mathématique des Phénomènes Thermiques Produits par la Radiation Solaire. Académie Yougoslave des Sciences et des Arts de Zagreb. Gauthier Villars, Paris, pp. 27—52.

Miller, K.G., Baluyot, R., Wright, J.D., Kopp, R.E., Browning, J.V., 2017. Closing an early Miocene astronomical gap with Southern Ocean $\delta^{18}O$ and $\delta^{13}C$ records: Implications for sea level change. Paleoceanography 32, 600—621. https://doi.org/10.1002/2016PA003074.

Minguez, D., Kodama, K.P., Hillhouse, J.W., 2015. Paleomagnetic and cyclostratigraphic constraints on the synchroneity and duration of the Shuram carbon isotope excursion, Johnnie Formation, Death Valley Region, CA. Precambrian Res. 266, 395—408. https://doi.org/10.1016/j.precamres.2015.05.033.

Miskovitch, V.V., 1931. Variations séculaires des elements astronomiques de l'orbite terrestre. Glas Srpske Kraljevske Akademije 143 (70). Première Classe, Belgrade.

Morgans-Bell, H.S., Coe, A.L., Hesselbo, S.P., Jenkyns, H.C., Weedon, G.P., Marshall, J.E.A., Tyson, R.V., Williams, C.J., 2001. Integrated stratigraphy of the Kimmeridge Clay Formation (Upper Jurassic) based on exposures and boreholes in south Dorset, UK. Geol. Mag. 138 (5), 511—539. https://doi.org/10.1017/S0016756801005738.

Muller, R.A., MacDonald, G.J., 2000. Ice Ages and Astronomical Causes: Data, Spectral Analysis, and Mechanisms. Springer-Praxis, London, 318 p.

Nestor, H., Einasto, R., Nestor, V., Marss, T., Viira, V., 2001. Description of the type section, cyclicity and correlation of the Riksu Formation (Wenlock, Estonia). Proc. Est. Acad. Sci. Geol. 50, 149—173.

Nie, J., 2017. The Plio-Pleistocene 405-kyr climate cycles. Palaeogeogr. Palaeoclimatol. Palaeoecol. https://doi.org/10.1016/j.palaeo.2017.07.022.

Olea, R.A., 2004. CORRELATOR 5.2 — a program for interactive lithostratigraphic correlation of wireline logs. Comput. Geosci. 30, 561—567. https://doi.org/10.1016/j.cageo.2004.01.006.

Olea, R.A., 1994. Expert systems for automated correlation and interpretation of wireline logs. Math. Geol. 26, 879—897. https://doi.org/10.1007/BF02083420 2.

Olsen, P.E., 2010. Fossil Great Lakes of the Newark Supergroup   30 Years Later. In: Benimoff, A.I. (Ed.), Field Trip Guidebook, New York State Geological Association, 83nd Annual Meeting, College of Staten Island, pp. 101—162.

Olsen, P.E., 1997. Stratigraphic record of the early Mesozoic breakup of Pangea in the Laurasia-Gondwana rift system. Annu. Rev. Earth Planet Sci. 25, 337—401. https://doi.org/10.1146/annurev.earth.25.1.337.

Olsen, P.E., Kent, D.V., 1999. Long period Milankovitch cycles from the Late Triassic and Early Jurassic of eastern North America and their implications for the calibration of the early Mesozoic timescale and the long-term behavior of the planets. Phil. Trans. R. Soc. Lond. Ser. A 357, 1761—1786. https://doi.org/10.1098/rsta.1999.0400.

Olsen, P.E., Kent, D.V., 1996. Milankovitch climate forcing in the tropics of Pangaea during the Late Triassic. Palaeogeogr. Palaeoclimatol. Palaeoecol. 122, 1—26. https://doi.org/10.1016/0031-0182(95)00171-9.

Olseger, D.A., Read, J.F., 1991. Relation of eustasy to stacking patterns of meter-scale carbonate cycles, Late Cambrian, USA. J. Sediment. Petrol. 61 (7), 1225–1252. https://doi.org/10.1306/D426786B-2B26-11D7-8648000102C1865D.

Pälike, H., Shackleton, N.J., 2000. Constraints on astronomical parameters from the geological record for the last 25 Myr. Earth Planet. Sci. Lett. 182, 1–14. https://doi.org/10.1016/S0012-821X(00)00229-6.

Pälike, H., Norris, R.D., Herrle, J.O., Wilson, P.A., Coxall, H.K., Lear, C.H., Shackleton, N.J., Tripati, A.K., Wade, B., 2006. The heartbeat of the Oligocene climate system. Science 314, 1894–1898. https://doi.org/10.1126/science.1133822.

Pälike, H., Laskar, J., Shackleton, N.J., 2004. Geological constraints on the chaotic diffusion of the solar system. Geology 32, 929–932. https://doi.org/10.1130/G20750.1.

Paillard, D., Labeyrie, L., Yiou, P., 1996. MacIntosh program performs time series analysis: Eos Transactions. Am. Geophys. Union 77, 379. https://doi.org/10.1029/96EO00259.

Pannella, G., 1972. Paleontological evidence on the Earth's rotational history since early Precambrian. Astrophys. Space Sci. 16, 212–237. https://doi.org/10.1007/BF00642735.

Park, J., Herbert, T.D., 1987. Hunting for paleoclimatic periodicities in a geological time series with an uncertain time scale. J. Geophys. Res. 92, 14027–14040. https://doi.org/10.1029/JB092iB13p14027.

Pas, D., Hinnov, L.A., Day, J., Kodama, K., Sinnesael, M., Liu, W., 2018. Cyclostratigraphic calibration of the Famennian Stage (Late Devonian, Illinois Basin, USA). Earth Planet. Sci. Lett. 488, 102–114. https://doi.org/10.1016/j.epsl.2018.02.010.

Pilgrim, L., 1904. Versuch einer rechnerischen Behandlung des Eiszeitproblems. Jahreshefte des Vereins für Vaterländische Naturkunde Württemberg, Stuttgart 60, 26–117.

Poliakow, E., 2005. Numerical modelling of the paleotidal evolution of the Earth-Moon System. In: Knezevíc, Z., Milani, A. (Eds.), Dynamics of Populations of Planetary Systems, Proceedings IAU Colloquium No. 197, pp. 445–452. https://doi.org/10.1017/S174392130400897X.

Preto, N., Hinnov, L.A., Hardie, L.A., De Zanche, V., 2001. A Middle Triassic orbital signal recorded in the shallow marine Latemar carbonate buildup (Dolomites, Italy). Geology 29, 1123–1128. https://doi.org/10.1130/0091-7613(2001)029<1123:MTOSRI>2.0.CO;2.

Radzevičius, S., Tumakovaitė, B., Spiridonov, A., 2017. Upper Homerian (Silurian) high-resolution correlation using cyclostratigraphy: an example from western Lithuania. Acta Geol. Pol. 67 (2), 307–322. https://doi.org/10.1515/agp-2017-0011. Warszawa.

Rameil, N., 2005. Carbonate Sedimentology, Sequence Stratigraphy, and Cyclostratigraphy of the Tithonian in the Swiss and French Jura Mountains. PhD Thesis. University of Fribourg, Switzerland, 244 p.

Ramezani, J., Bowring, S.A., 2018. Advances in Numerical Calibration of the Permian Timescale Based on Radioisotopic Geochronology, vol. 450. Geological Society, London, pp. 51–60. https://doi.org/10.1144/SP450.17. Special Publications.

Rasmussen, J., Hinnov, L.A., Thibault, N., 2018. Astronomical Time Calibration of the Great Ordovician Biodiversification Event (in preparation).

Renne, P.R., Deino, A.L., Hilgen, F.J., Kuiper, K.F., Mark, D.F., Michell III, W.S., Morgan, L.E., Mundil, R., Smit, J., 2013. Time Scales of Critical Events Around the Cretaceous-Paleogene Boundary. Science 339, 684–687. https://doi.org/10.1126/science.1230492.

Ripepe, M., Fischer, A.G., 1991. Stratigraphic rhythms synthesized from orbital variations. In: Franseen, E.K., Watney, W.L., Kendall, C.G.S., Ross, W. (Eds.), Sedimentary Modeling: Computer Simulations and Methods for Improved Parameter Definition, Kansas Geological Survey Bulletin, 233, pp. 335–344. http://www.kgs.ku.edu/Publications/Bulletins/233/index.html.

Rodrigues, P.O.C., Hinnov, L.A., Franco, D., 2018. A new appraisal of depositional cyclicity in the Neoarchean-Paleoproterozoic Dales Gorge Member (Brockman Iron Formation, Hamersley Basin, Australia). Precambrian Res. (in review).

Rodionov, V.P., Dekkers, M.J., Khramov, A.N., Gurevich, E.L., Krijgsman, W., Duermeijer, C.E., Heslop, D., 2003. Paleomagnetism and cyclostratigraphy of the Middle Ordovician Kriolutsky Suite, Krivaya Luka section, southern Siberian Platform: Record of nonsynchronous NRM-components or a non-axial geomagnetic field? Stud. Geophys. Geodesy. 47, 255−274.

Röhl, U., Westerhold, T., Bralower, T.J., Zachos, J.C., 2007. On the duration of the Paleocene−Eocene thermal maximum (PETM). G-cubed 8. https://doi.org/10.1029/2007GC001784.

Ruhl, M., Hesselbo, S.P., Hinnov, L.A., Jenkyns, H.C., Xu, W., Storm, M., Riding, J., Minisini, D., Ullmann, C.U., Leng, M.J., 2016. Astronomical constraints on the duration of the Early Jurassic Pliensbachian stage and global climate fluctuations. Earth Planet. Sci. Lett. 455, 149−165. https://doi.org/10.1016/j.epsl.2016.08.038.

Sageman, B.B., Singer, B.S., Meyers, S.R., Siewert, S.E., Walaszczyk, I., Condond, D.J., Jicha, B.R., Obradovich, J.D., Sawyer, D.A., 2014. Integrating ^{40}Ar/^{39}Ar, U-Pb, and astronomical clocks in the Cretaceous Niobrara Formation, Western Interior Basin, USA. GSA Bull. 126 (7−8), 956−973. https://doi.org/10.1130/B30929.1.

Schiffelbein, P., Dorman, L., 1986. Spectral effects of time-depth nonlinearities in deep sea sediment cores: a demodulation technique for realigning time and depth scales. J. Geophys. Res. 91, 3821−3835.

Schwarzacher, W., 1993. Cyclostratigraphy and the Milankovitch Theory. In: Developments in Sedimentology, vol. 52. Elsevier, Amsterdam, 224 p.

Scrutton, C.T., 1978. Periodic growth features in fossil organisms and the length of the day and month. In: Brosche, P., Sündermann, J. (Eds.), Tidal Friction and the Earth's Rotation. Springer-Verlag, Berlin, pp. 154−196. https://doi.org/10.1007/978-3-642-67097-8_12.

Schmitz, M.D., Davydov, V.I., 2012. Quantitative radiometric and biostratigraphic calibration of the Pennsylvanian-Early Permian (Cisuralian) time scale and pan-Euramerican chronostratigraphic correlation. Geol. Soc. Am. Bull. 124, 549−577. https://doi.org/10.1130/B30385.1.

Shackleton, N.J., Crowhurst, S.J., Weedon, G.P., Laskar, J., 1999. Astronomical calibration of Oligocene-Miocene time. Phil. Trans. R. Soc. Lond. Ser. A 357, 1907−1929. https://doi.org/10.1098/rsta.1999.0407.

Sharaf, S.G., Budnikova, N.A., 1969. Secular changes in the Earth's orbital elements and the astronomical theory of climate variations. Trudy Instituta Teoreticheskoy Astronomii, Leningrad 14, 48−84.

Sharaf, S.G., Budnikova, N.A., 1967. Secular variations of elements of the Earth's orbit which influence the climate of the geological past. Trudy Instituta Teoreticheskoy Astronomii, Leningrad 11, 231−261.

Sinnesael, M., Mauviel, A., Desrochers, A., McLaughlin, P.I., De Weirdt, J., Vandenbroucke, T.R.A., Claeys, P., 2017b. Cyclostratigraphy of the Upper Ordovician Vaureal Formation, Anticosti Island, eastern Canada. Geol. Soc. Am. Annual Meet. Seattle, WA, Abstracts with Programs 49 (6). https://doi.org/10.1130/abs/2017AM-302947.

Sinnesael, M., Loi, A., Dabard, M.-P., Vandenbroucke, T., Claeys, P., 2017a. Cyclostratigraphic analysis of the Middle to lower Upper Ordovician Postolonnec Formation in the Armorican Massif (France): integrating XRF, gamma-ray and lithological data. Geophys. Res. Abstr. 19. EGU2017-9565, General Assembly, Vienna, Austria.

Sprovieri, M., Sabatino, N., Pelosi, N., Batenburg, S.J., Coccioni, C., Iavarone, M., Mazzola, S., 2013. Late Cretaceous orbitally-paced carbon isotope stratigraphy from the Bottaccione Gorge (Italy). Palaeogeogr. Palaeoclimatol. Palaeoecol. 379-380, 81—94. https://doi.org/10.1016/j.palaeo.2013.04.006.

Stockwell, J.N., 1873. Memoir on the secular variations of the elements of the eight principal planets. Smithsonian Contributions to Knowledge, Washington 18 (3), 199 p.

Sucheras-Marx, B., Giraud, F., Fernandez, V., Pittet, B., Lecuyer, C., Olivero, D., Mattioli, E., 2013. Duration of the Early Bajocian and the associated $\delta^{13}C$ positive excursion based on cyclostratigraphy. J. Geol. Soc. Lond. 170, 107—118. https://doi.org/10.1144/jgs2011-133.

Svensen, H.H., Hammer, Ø., Corfu, F., 2015. Astronomically forced cyclicity in the Upper Ordovician and U—Pb ages of interlayered tephra, Oslo Region, Norway. Palaeogeogr. Palaeoclimatol. Palaeoecol. 418, 150—159. https://doi.org/10.1016/j.palaeo.2014.11.001.

Thibault, N., Jarvis, I., Voight, S., Gale, A.S., Attree, K., Jenkyns, H.C., 2016. Astronomical calibration and global correlation of the Santonian (Cretaceous) based on the marine carbon isotope record. Paleoceanography 31, 847—865. https://doi.org/10.1002/2016PA002941.

Thomson, D.J., 1990. Quadratic-Inverse Spectrum Estimates: Applications to Palaeoclimatology. Phil. Trans. Phys. Sci. Eng. 332 (1627), 539—597. https://doi.org/10.1098/rsta.1990.0130.

Thomson, D.J., 2009. Time-series analysis of paleoclimate data. In: Gornitz, V. (Ed.), Encyclopedia of Paleoclimatology and Ancient Environments, pp. 949—959. https://doi.org/10.1007/978-1-4020-4411-3_222.

Thomson, D.J., 1982. Spectrum estimation and harmonic analysis. Proc. IEEE 70, 1055—1096. https://doi.org/10.1109/PROC.1982.12433.

Trendall, A.F., 1973. Varve cycles in the Weeli Wolli Formation of the Precambrian Hamersley Group, Western Australia. Econ. Geol. 68, 1089—1097. https://doi.org/10.2113/gsecongeo.68.7.1089.

Trendall, A.F., 1966. Second progress report on the Brockman Iron Formation in the Wittenoom-Yampire area. In: Annual Report for the Year 1965, Geological Survey of Western Australia, pp. 75—87.

Trendall, A.F., 1965. Progress report on the Brockman Iron Formation in the Wittenoom-Yampire area. In: Annual Report for the Year 1964, Geological Survey of Western Australia, pp. 55—64.

Vanyo, J.P., Awramik, S.M., 1985. Stromatolites and Earth-Sun-Moon dynamics. Precambrian Res. 29, 121—142. https://doi.org/10.1016/0301-9268(85)90064-6.

Vernekar, A.D., 1972. Long-period global variations of incoming solar radiation. Meteorol. Monogr. 12 (34), 130 p.

Wahr, J.M., 1988. The Earth's rotation. Annu. Rev. Earth Planet Sci. 16, 231—249. https://doi.org/10.1146/annurev.ea.16.050188.001311.

Wang, C., Feng, Z., Zhang, L., Huang, Y., Cao, K., Wang, P., Zhao, B., 2013. Cretaceous paleogeography and paleoclimate and the setting of SKI borehole sites in Songliao Basin, northeast China. Palaeogeogr. Palaeoclimatol. Palaeoecol. 385, 17—30. https://doi.org/10.1016/j.palaeo.2012.01.030.

Walker, J.C.G., Zahnle, K.J., 1986. Lunar nodal tide and distance to the Moon during the Precambrian. Nature 320, 600—602. https://doi.org/10.1038/320600a0.

Waltham, D., 2015. Milankovitch period uncertainties and their impact on cyclostratigraphy. J. Sediment. Res. 85, 990—998. https://doi.org/10.2110/jsr.2015.66.

Weedon, G.P., Coe, A.L., Gallois, R.W., 2004. Cyclostratigraphy, orbital tuning and inferred productivity for the type Kimmeridge Clay (Late Jurassic), Southern England. J. Geol. Soc. 161, 655—666. https://doi.org/10.1144/0016-764903-073.

Wells, J.W., 1970. Problems of annual and daily growth-rings in corals. In: Runcorn, S.K. (Ed.), Palaeogeophysics. Academic, San Diego, CA, pp. 3—9.

Westerhold, T., Röhl, U., Frederichs, T., Bohaty, S.M., Zachos, J.C., 2015. Astronomical calibration of the geological timescale: closing the middle Eocene gap. Clim. Past 11, 1181—1195. https://doi.org/10.5194/cp-11-1181-2015.

Westerhold, T., Röhl, U., Pälike, H., Wilkens, R., Wilson, P.A., Acton, G., 2014. Orbitally tuned timescale and astronomical forcing in the middle Eocene to early Oligocene. Clim. Past 10, 955—973. https://doi.org/10.5194/cp-10-955-2014.

Westerhold, T., Röhl, U., Laskar, J., 2012. Time scale controversy: accurate orbital calibration of the early Paleogene. G-cubed 13, Q06015. https://doi.org/10.1029/2012gc004096.

Westerhold, T., Röhl, U., Raffi, I., Fornaciari, E., Monechi, S., Reale, V., Bowles, J., Evans, H.F., 2008. Astronomical calibration of the Paleocene time. Palaeogeogr. Palaeoclimatol. Palaeoecol. 257, 377—403. https://doi.org/10.1016/j.palaeo.2007.09.016.

Westerhold, T., Röhl, U., Laskar, J., Bowles, J., Raffi, I., Lourens, L.J., Zachos, J.C., 2007. On the duration of magnetochrons C24r and C25n and the timing of early Eocene global warming events: implications from the Ocean Drilling Program Leg 208 Walvis Ridge depth transect. Paleoceanography 22, PA2201. https://doi.org/10.1029/2006PA001322.

Williams, G.E., 2000. Geological constraints on the Precambrian history of Earth's rotation. Rev. Geophys. 38 (1), 37—59. https://doi.org/10.1029/1999RG900016.

Williams, G.E., 1991. Milankovitch-band cyclicity in bedded halite deposits contemporaneous with Late Ordovician—Early Silurian glaciation, Canning Basin, western Australia. Earth Planet. Sci. Lett. 103, 143—155. https://doi.org/10.1016/0012—821X(91)90156—C.

Williams, G.E., 1989. Tidal rhythmites: Geochronometers for the ancient Earth-Moon system. Episodes 12, 162—171.

Williams, J.G., 1994. Contributions to the Earth's obliquity rate, precession and nutation. Astronomical J. 108 (2), 711—724. https://doi.org/10.1086/117108.

Wu, H., Hinnov, L.A., Chu, R., Zhang, S., Jiang, G., Yang, R., Li, H., Xi, D., Wang, C., 2018b. Terrestrial evidence for two Solar System chaotic events in the Late Cretaceous geological record. Earth Planet. Sci. Lett. (in review).

Wu, H., Fang, Q., Wang, X., Hinnov, L.A., Qi, Y., Shen, S., Yang, T., Li, H., Chen, J., Zhang, S., 2018a. A ~34 m.y. astronomical time scale for the uppermost Mississippian through Pennsylvanian of the Carboniferous System. Geology (in review).

Wu, H., Zhang, S., Hinnov, L.A., Jiang, G., Yang, T., Li, H., Wan, X., Wang, C., 2014. Cyclostratigraphy and orbital tuning of the terrestrial upper Santonian—Lower Danian in Songliao Basin, northeastern China. Earth Planet. Sci. Lett. 407, 82—95. https://doi.org/10.1016/j.epsl.2014.09.038.

Wu, H., Zhang, S., Hinnov, L.A., Feng, Q., Jiang, G., Li, H., Yang, T., 2013. Late Permian Milankovitch cycles. Nat. Commun. https://doi.org/10.1038/ncomms3452.

Yao, X., Zhou, Y., Hinnov, L.A., 2015. Astronomical forcing of Middle Permian chert in the Lower Yangtze area, South China. Earth Planet. Sci. Lett. 422, 206—221. https://doi.org/10.1016/j.epsl.2015.04.017.

Yu, Z.W., Ding, Z.L., 1998. An automatic orbital tuning method for paleoclimate records. Geophys. Res. Lett. 25, 4525—4528. https://doi.org/10.1029/1998GL900197.

Zachos, J.C., Pagani, M., Sloan, L., Thomas, E., Billups, K., 2001. Trends, rhythms, and aberrations in global climate 65 Ma to present. Science 292, 686—693. https://doi.org/10.1126/science.1059412.

Zachos, J.C., Dickens, G.R., Zeebe, R.E., 2008. An early Cenozoic perspective on greenhouse warming and carbon-cycle dynamics. Nature 451, 279—283. https://doi.org/10.1038/nature06588.

Zeebe, R.E., 2017. Numerical solutions for the orbital motion of the Solar System over the past 100 Myr: limits and new results. Astronomical J. 154, 193. https://doi.org/10.3847/1538-3881/aa8cce.

Zeeden, C., Kaboth, S., Hilgen, F.J., Laskar, J., 2018. Taner filter settings and automatic correlation optimisation for cyclostratigraphic studies. Comput. Geosci. 119, 18—28. https://doi.org/10.1016/j.cageo.2018.06.005.

Zeeden, C., Meyers, S.R., Louren, L.J., Hilgen, F.J., 2015. Testing astronomically tuned age models. Paleoceanography 30. https://doi.org/10.1002/2014PA002762.

Zeeden, C., Hilgen, F., Hüsing, S., Lourens, L., 2014. The Miocene astronomical time scale 9—12 Ma: new constraints on tidal dissipation and their implications for paleoclimatic investigations. Paleoceanography 29 (4), 296—307. https://doi.org/10.1002/2014PA002615.

Zhang, S., Wang, X., Hammarlund, E.U., Wang, H., Mafalda Costa, M., Bjerrum, C.J., Connelly, J.N., Zhang, B., Bian, L., Canfield, D.E., 2015. Orbital forcing of climate 1.4 billion years ago. Proc. Natl. Acad. Sci. 112 (12), E1406—E1413. https://doi.org/10.1073/pnas.1502239112.

CHAPTER TWO

Astronomical Time Scale for the Mesozoic

Chunju Huang

State Key Laboratory of Biogeology and Environmental Geology, School of Earth Sciences, China University of Geosciences, Wuhan, China
E-mail: huangcj@cug.edu.cn

Contents

Abstract

The astronomical theory and its application to sediment cyclostratigraphy have been widely studied over the past thirty years. Especially during the past decade, these have been successfully applied for a continuous high-resolution calibration of portions of the geologic time scale. The method of astronomical tuning enhances the traditional geological dating methods, such as paleontology, paleomagnetism and radioisotope dating. In *The Geologic Time Scale 2012* (GTS2012), most of the Cenozoic Era was directly calibrated to the absolute astronomical time scale

Stratigraphy & Timescales, Volume 3
ISSN 2468-5178
https://doi.org/10.1016/bs.sats.2018.08.005

(ATS); however, only a few intervals of the Mesozoic-Paleozoic were scaled by cyclo-stratigraphy in discontinuous floating segments. Since then, other cyclostratigraphy studies have been published using the most stable 405-kyr long-eccentricity cycle to calibrate portions of the Mesozoic geological time scale. Other than some gaps within the mid-Triassic and Middle Jurassic, it is possible to compile a nearly complete ATS status for the Mesozoic. This chapter compiles and recalibrates an extensive suite of selected paleoclimate proxies series and applies a tuning to the master astronomical scale based on the stable 405-kyr long-eccentricity period. This enables the synthesis of a nearly complete Mesozoic ATS to the base of the Triassic.

1. INTRODUCTION

An accurate geological time scale is the essential key for understanding and deciphering the Earth's evolutionary history and geologic processes. The age model for the geologic time scale with integrated biostratigraphy, chemostratigraphy and magnetostratigraphy that we currently use was constructed largely based on only a few age calibrations from radioisotope dating (Gradstein, 2012). However, a high-resolution age model is not always available or only vaguely defined for many stratigraphic sections (Hinnov and Hilgen, 2012). For example, until recently, due to a lack of precise dating, it was debated whether the major end-Permian mass extinction event was instantaneous in different ecosystems or how it was associated with different stages of the Siberian large igneous province (e.g., Benton and Twitchett, 2003; Shen and Bowring, 2014; Burgess et al., 2014; 2017; Shen et al., 2018). In addition, most of the isotopic dates are not located at the geological stage boundaries. The compilation of *The Geologic Time Scale* (GTS2012) is one of the most important reference books for the geoscience community that provided the current status of an integrated time scale; however, its main framework for the associated age model for the set of geologic stages and primary biozonations of chronostratigraphic units was largely constructed by different interpolation methods of uncertain precision (Gradstein, 2012). Astronomical tuning of the orbital forced stratigraphic records to construct the high-resolution Astronomical Time Scale (ATS) can be developed to solve this problem. In GTS2012, this was applied for a full astronomical calibration of the Quaternary, Neogene and most of Paleogene periods (Hinnov and Hilgen, 2012).

For pre-Cenozoic intervals, the astronomical signals in the cyclic stratigraphic records could not be calibrated directly to an absolute ATS due to model limitations and uncertainties on the full orbital solution and the lack of a continuous cyclic sediment record connecting to the base of the Cenozoic. Nonetheless, it has been possible in numerous cyclostratigraphic studies to develop "floating" time scales for portions of the Mesozoic and Paleozoic eras based on recognition and tuning of the sediment records to different Milankovitch orbital periods (e.g., short- or long-eccentricity, obliquity and/or precession). Applying shorter orbital periods for tuning has an inherent uncertainty in that minor hiatuses or ambiguities in the sediment record can easily result in omission or non-recognition of one or more cycles. Indeed, any cyclostratigraphy is generally considered to be a minimum estimate of the actual duration.

However, the 405-kyr long-eccentricity period is often present in the deep time sedimentary records and this term is much more stable than the ~100-kyr short-eccentricity or other orbital terms (Laskar et al., 2011), therefore it has been proposed as the primary "metronome" for establishment of geological timescales (Laskar et al., 2004; Hinnov and Hilgen, 2012). In order to construct the high-precision ATS for the Mesozoic Era based on the 405-kyr long-eccentricity tuning, we need apply the uniform standards for the astronomically calibration of continuous cyclostratigraphic records that span major intervals of geologic time and contain records of global biostratigraphic, chemostratigraphic or magnetostratigraphic zones and events. Even though this chapter mainly presents the estimates of ages and durations of international geologic stages, the main goal of cyclostratigraphy and the ATS is determining the actual rates and precise relative and absolute timing of the succession of evolutionary, geomagnetic, geochemical and other events through Earth's history.

2. ASTRONOMICAL THEORY AND THE 405-KYR ASTRONOMICAL TUNING

2.1 Astronomical Theory

A major paleoclimate factor is the difference between summer warmth and winter cooling at a given latitude, which in turn are largely governed by the difference in the amount of solar radiation (insolation) received through the year at that latitude. For example, a continental ice sheet will grow and advance if the amount of summer heat is inadequate to completely melt the snow that accumulated during the winter. The mean seasonal insolation received at different latitudes on the Earth's surface are gradually changing due to slow cyclic variations in the magnitude and in the direction of the Earth's rotational tilt relative to the Sun (obliquity and precession) and in the magnitude of Earth's elliptical orbit around the Sun (eccentricity). Milutin Milanković, a Serbian mathematician and atmospheric scientist, was the first to compute Earth's insolation parameters for a given season and latitude and proposed that long-climate responded to these "Milankovitch" orbital forcings (Milankovitch, 1941). His computation of insolation quantities on Earth has been enhanced by Laskar et al. (1993, 2004, 2011) to provide us an astronomical solution or Astronomical Time Scale (ATS) spanning from 250 million years (Myr) in the past and 250 million years in the future. However, Laskar's current solution is valid for calibrations of paleoclimatic data from all the contributions only for the past/future 50 million years. Beyond this time, the solution's accuracy decreases for some components, especially the precision for the exact ages of maximum precession and obliquity, due to chaotic behavior in the gravitational interaction of the planets and minor planets and tidal-induced exchange of energy between the Earth and Moon (Laskar et al., 2011).

The Earth's main orbital parameters, also known as Milankovitch cycles, are the precession, obliquity and eccentricity cycles and their long-term modulation. The most important for causing seasons on the Earth is its tilt relative to the Sun, because if the Earth was not tilted, then there not be any significant seasonal change at any latitude. The Earth's orbital obliquity or axial tilt (ε) is the angle between the Earth's equatorial plane and its orbital plane, or the angle between an Earth's rotational axis and its orbital axis (Berger and Loutre, 2004). The current obliquity ε_0 is 23.44°, and the obliquity oscillates between 22.5° and 24.5° and will be progressively decreasing during the next 10 ka. The principal period of obliquity cycles are 41 kyr, with lesser cycles at 54, 39.5 and 29 kyr for the time interval 0–5 Ma. Due to the tidal dissipation interactions of the Earth and Moon, the periods for obliquity oscillation were more rapid in the past. During the early Triassic (244–249 Ma), the principal period was 34.5 kyr, and less ones were 33.6, 43 and 26 kyr (Fig. 1B).

Precession of the axial tilt relative the Earth's elliptical orbit around the Sun magnifies the seasonal contrast for a hemisphere. Today, the northern hemisphere is closer to the Sun during its winter season than during its summer season. The definition of the Earth's orbital precession parameter is $e*\sin(\varpi)$, where ϖ is the Earth's spin rate and e is the eccentricity. Due to the dissipative effects of the Earth-Moon system, the Earth rotation

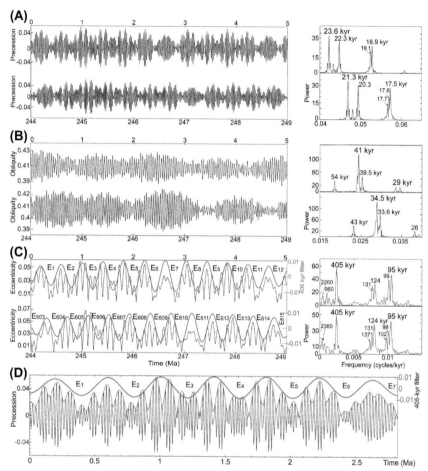

Figure 1 The La2010d nominal solution of Earth's orbital parameters for the past 0—5 Ma and for the Early Triassic (244—249 Ma) (Laskar et al., 2011). (A) Precession index time series for 0—5 and 244—249 Ma with its power spectrum on the right. (B) Obliquity time series for 0—5 and 244—249 Ma with its power spectrum on the right. (C) Eccentricity time series with 405 kyr filter output for 0—5 and 244—249 Ma with its power spectrum on the right. (D) precession index (black curve) for 0—2.82 Ma modulated by eccentricity time series (green curve (gray in print version)) with 405 kyr filtered output. Labels on spectral peaks indicate periodicity in kyr. The 405 kyr filtered curve used a Gaussian filter centered at 0.002469 with passband ±0.0003 in the AnalySeries 2.0.8 software package (Paillard et al., 1996).

rate is slowing down and the Earth-Moon distance is increasing. This causes both a significant change in the length of day (e.g., it was only 22 h during the early Triassic, 250 Ma) and a slowing of the precession cycle. At present, the main periods of precession cycles are 23.6, 22.3, 19.1 and 18.9 kyr, however, during the early Triassic the main periods were 21.3, 20.3, 17.7, 17.6 and 17.5 kyr (Fig. 1A).

The amplitude of precession signal on climate is strongly modulated by the magnitude of eccentricity of Earth's elliptical orbit; if the Earth's orbit was circular around the Sun, then its precession cycle would not change the seasonal contrasts through that cycle. The Earth's orbital eccentricity e (a measure of the degree of ellipticity of the Earth's orbit around the Sun, in which a value of "0.0" is circular) is currently 0.016 (Laskar et al., 2004). The interaction of the Earth with the gravitational pull of the other planets causes the eccentricity to vary between 0.00021318 and 0.066957 with the main periods being 95, 99, 124, 131, 405, ~1 and ~2.3 Myr (Fig. 1C). The amplitude variations of the precession cycles are modulated by the ~100-kyr short-eccentricity cycles, and the amplitude of ~100 kyr cycles are modulated by the 405-kyr long-eccentricity cycles (Fig. 1D). The ~100 and 405 kyr periods remain quite stable through geologic time, especially 405-kyr long-eccentricity cycle. Indeed, due to the huge mass of Jupiter and the gravitational interactions between the orbital perihelia of Venus and Jupiter, the 405-kyr long-eccentricity cycle has remained stable in the past 250 Myr and is less influenced by the Solar system's chaotic diffusion than the frequency for the ~100-kyr short-eccentricity cycle (Laskar et al., 2004; 2011).

These cyclic variations of the Earth's orbit around the Sun govern the seasonal insolation received at a latitude on the Earth's surface, thereby driving cyclical fluctuations of the climate system. The record of these periodic fluctuations of climate can be preserved by proxies of temperature, runoff, productivity, current intensity or other climatic-sensitive features within the sedimentary deposits to produce rhythmic cycles superimposed on longer-term sediment trends. Extracting these cyclic variations in preserved climatic proxies and determining if they were induced by a suite of Milankovitch orbital cycles is the goal of cyclostratigraphy. Once the main climatic cycle signals are extracted and proven to be associated with certain Milankovitch cycles, then the period of those cycles can be applied to the sediment record to convert the meter-record into a time-record. This astronomical tuning from the depth domain into a time domain often enables the resolution of additional climatic cycles that correspond to additional major Milankovitch cycles.

2.2 The 405-kyr Astronomical Tuning

The astronomical tuning method to assign a certain time scale for the recognized astronomical cycles recorded in a sedimentary succession leads to the construction of a tentative Astronomical Time Scale (ATS) for that sedimentary succession (Hilgen, 2010; Hinnov and Hilgen, 2012), which under ideal conditions can be directly matched to the astronomical solution for absolute age assignments. This process of progressive tuning, development of ATS segments and calibration to the full orbital solution has enabled the construction of a precise Neogene through Quaternary time scale. Calibration to the eccentricity components of the orbital solution underpins the majority of the current Paleogene time scale.

This discipline of cyclostratigraphy to study the cyclic stratigraphic records to search for Milankovitch cycles uses climate proxies such as carbon and oxygen isotopes, depositional facies, magnetic susceptibility, carbonate content, total organic carbon (TOC), grayscale, well logs, etc. The most stable cycles are those of eccentricity, especially 405-kyr long-eccentricity cycle, which are commonly signals superimposed on the geological sedimentary succession. Examples of different proxies and different Mesozoic sedimentary successions include a grayscale series from Cretaceous pelagic sediments in the Piobbico core of central Italy (Fig. 2A), a TOC content series from the Jurassic hemipelagic sediment in a borehole from Dorset, England (Fig. 2C), and relative-depth rank series from Triassic lacustrine deposits cored in the Newark Basin (USA) (Fig. 2E). In each of these depositional settings, the 405-kyr long-eccentricity cycles are well preserved even they have different thickness and facies characterisics (Fig. 2BDF).

The 405-kyr long-eccentricity cycle is the most stable period among the Earth's orbital parameters and has been called the 405-kyr metronome for geologic time. Therefore, we chose this 405 kyr term as a chrono-cycle for calibrating the Mesozoic geologic time scale and designated the chrono-cycle scale from E163 at the Cretaceous/Paleogene boundary (66 Ma) to E622 at the Permian/Triassic boundary (251.9 Ma) (Fig. 1). The goal of the Cyclostratigraphy Research Working Group of International Stratigraphic Commission is trying to identify each of these 405-kyr long-eccentricity cycles in the sedimentary geologic records throughout Phanerozoic. The long-term goal is to assign the appropriate 405-kyr chrono-cycles to biozones, magnetic chrons and other stratigraphic intervals (Hinnov and Hilgen, 2012).

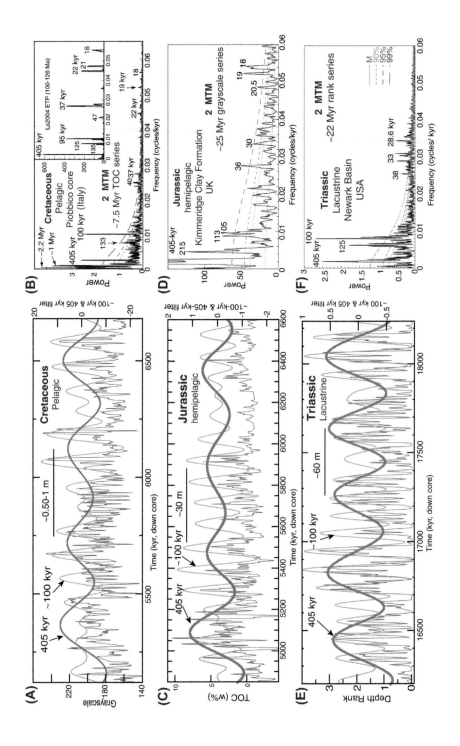

Figure 2 The dominant 405-kyr long-eccentricity cycles recorded in different facies and different paleoclimate proxies in the Mesozoic. (A) An example interval of the grayscale series of the pelagic facies in the Piobbico core (central Italy) of the Aptian Stage of Early Cretaceous with ∼ 100 ky and 405-kyr short and long-eccentricity filter output. (B) 2π multitaper (MTM) power spectrum of the 405 kyr tuned grayscale 25.85 Myr-long ATS for the Piobbico core. (C) An example interval from the TOC content series from the hemipelagic Kimmeridge Clay Formation of Dorset (England) of Late Jurassic. (D) 2π multitaper (MTM) power spectrum of the 405 kyr tuned 7.5 Myr-long TOC ATS for the Kimmeridge Clay Formation. (E) An example interval for the lacustrine depth rank series of Newark Basin (USA) of Late Triassic. (F) 2π multitaper (MTM) power spectrum of the 405 kyr tuned ∼ 22 Myr-long depth rank ATS for the Newark composite cores. ∼ 100 and 405 kyr filter output using the Gaussian filter, passband at 0.01 ± 0.003 and 0.002469 ± 0.0006 cycles/kyr, respectively.

These 405-kyr long-eccentricity chrono-cycles have an estimated cumulative uncertainty of ~500 kyr at 250 Ma for calibrating the geologic time scale (Hinnov and Hilgen, 2012). As noted above, the 405-kyr long-eccentricity cycles are very often present in many Mesozoic stratigraphic records, even where the short periods signals, such as obliquity or precession cycles and even ~100-kyr short-eccentricity cycle, may not be clearly visible or resolved in the ancient stratigraphic sequences. Therefore, the stable 405-kyr long-eccentricity cycle should be used to calibrate the Mesozoic and Paleozoic time scale (Laskar et al., 2004; Hinnov and Hilgen, 2012). Several intervals with astronomically calibrated "floating time scales" have been used for partial calibration of events and zone durations within intervals in the GTS2012 (Gradstein, 2012). In this chapter, we assemble and standardize the analysis of large portions of the available data from those and later cyclostratigraphy studies in order to apply 405-kyr long-eccentricity tuning. These 405-kyr-tuned intervals are assigned to potential 405-kyr chrono-cycles in Mesozoic Era.

3. THE CURRENT STATUS OF THE TRIASSIC ATS

The Triassic period is bounded by end-Permian and end-Triassic mass extinctions. There are U–Pb dates of 251.902 ± 0.024 Ma at the Permian/Triassic boundary (PTB) (Burgess et al., 2014) of 201.36 ± 0.17 Ma at the Triassic/Jurassic boundary (revised by Wotzlaw et al., 2014 from 201.31 ± 0.18 Ma in Schoene et al., 2010); therefore, the Triassic spans a total duration of about 50.5 Myr. The bases of the 7 international Triassic stages as assigned at GSSPs or at candidate definitions are at levels corresponding to the lowest occurrences of ammonoid or conodont taxa within exposures in Alpine, Mediterranean or Himalayan exposures (e.g., reviews in GTS2012 and in Ogg et al., 2016). Over 30 cyclostratigraphic studies have been published for the Triassic in the past three decades (Fig. 3).

3.1 The Early Triassic

The Early Triassic includes **Induan** and **Olenekian stages** and spans about 5 Myr from ~252 Ma at Permian/Triassic boundary (PTB) to ~247 Ma at Early/Middle Triassic boundary.

Astronomically-forced stratigraphic records through the PTB interval have been proposed by several researchers with differing conclusions

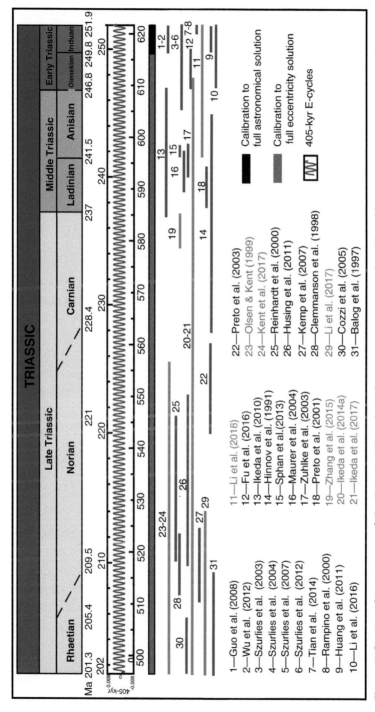

Figure 3 Stratigraphic coverage of the Triassic ATS and 405-kyr metronome. Intervals with *red lines* (gray in print version) indicate the collected data coverage subjected to a 405 kyr tuning to compile the ATS in this study, and those with *blue lines* (dark gray in print version) indicate additional supporting cyclostratigraphy studies.

(Algeo et al., 2010; Hansen et al., 2000; Huang et al., 2011; Rampino et al., 2000; Tian et al., 2014; Yin et al., 2001).

The Induan Stage has been studied in Europe and in South China. In the Carnic Alps in Austria, Rampino et al. (2000) interpreted Milankovitch cycles for the PTB and the basal Griesbachian substage of the Induan from the 331 m Gartnerkofel-1 core that mainly consisted of cyclic dolomitized limestones and interbedded thin marls/shales of shallow-marine facies. Power spectrum of the stacked normalized gamma-ray and density logging data shows $\sim$36, $\sim$11, $\sim$4.7 and $\sim$2.2 m cycles with a ratio of $\sim$40:10:4.7:2.2 that is approximately correlative to the ratio $\sim$400:100: 37:19 for the Late Permian astronomical parameters (Berger., 1988). Using a count of the dominant $\sim$100-kyr short-eccentricity cycles, the estimated total duration for the Griesbachian substage was about $\sim$1.6 Myr (Rampino et al., 2000). In contrast, a much shorter duration was interpreted at the West Pingdingshan section of Chaohu, Anhui province of South China by Guo et al. (2008), who collected 2184 samples for the magnetic susceptibility (MS) data measurement from the 44 m succession of limestone cyclicly interbedded with mudstone. His interpretation of 56 precession and 12 short-eccentricity cycles implied that the entire Induan Stage spanned only 1.1 Myr. From the Daxiakou section of Hubei province in South China, Wu et al. (2012) interpreted a similar Induan duration of 1.16 Myr based on tuning the 55 m MS and anhysteretic remanent magnetization (ARM) series (2440 samples) to 405, 100 and 20 kyr cycles.

For the Olenekian Stage, Fu et al. (2016) collected high-resolution carbonate carbon-isotope $\delta^{13}C_{carb}$ data from the slope-facies succession in the 200 m Majiashan section at Chaohu in South China. They interpreted a 3.2 Myr-long ATS for the late Smithian through Spathian of Olenekian stage, of which the Spathian substage spanned an estimated 2.89 Myr duration based on assigning a 115.3-kyr short-eccentricity tuning to the observed main cycles. Their age model assumed that $\delta^{13}C_{carb}$ minima correspond to long-eccentricity maxima and an anchor at a horizon of 162.62 m to the eccentricity curve of La2010d at 247.95 Ma. Their interpretation was partly influenced by the estimated duration of 3.2 ± 0.6 Myr from a pair of earlier U–Pb radioisotopic dates of 250.55 ± 0.51 Ma for the earliest Spathian (Ovtcharova et al., 2006) and 247.32 ± 0.08 Ma for the end-Spathian (Lehrmann et al., 2006). The end-Spathian U–Pb date has been partly independently verified in another section (Ovtcharova et al., 2015), but the interpreted U-Pb date for earliest Spathian has yet to be verified.

Those estimates of a short-duration Induan Stage and a long-duration Olenekian Stage (especially for the Spathian substage) are very different than those derived from extensive cycle—magnetostratigraphy correlation in Europe and to South China. Szurlies et al. (2003, 2012) and Szurlies (2004) interpreted ~100-kyr short-eccentricity cycles from outcrops and boreholes (especially gamma-ray well logs) to scale his geomagnetic polarity zones of the lower Buntsandstein continental facies of the Central Germany of latest Permian through Early Triassic. This resulted in an estimated duration of Induan Stage of about 1.8—2.2 Myr and duration for the Olenekian of about 3 Myr depending on the placements of the PTB, the Induan/Olenekian boundary and the top of the Olenekian within the lower Buntsandstein.

The cycle-scaled magnetostratigraphy of the continental Germanic Basin correlates well with the 5.1 Myr astronomical-calibrated time scale constructed by Li et al. (2016) using a 405-kyr long-eccentricity cycles tuning of spectral gamma-ray logs from the marine sections of Chaohu, Daxiakou, Guangdao and Meishan in South China. That study yielded durations for Induan and Olenekian stages of 2.0 and 3.1 Myr, respectively, and durations of 1.4, 0.6, 1.7 and 1.4 Myr for the Griesbachian, Dienerian, Smithian and Spathian substages, respectively. This estimated duration of 1.4 Myr for the Griesbachian substage based on 405 kyr tuning is also consistent with a duration of 1.6 Myr estimated from the Carnic Alps in Austria based on the ~100 kyr counting by Rampino et al. (2000). A similar cyclostratigraphy-derived duration of 1.7 ± 0.1 Myr for the Induan Stage was obtained from the conodont-dated distal Montney Formation of British Columbia that yielded a full suite of long-eccentricity, short-eccentricity, obliquity and precession cycles (Shen et al., 2017).

One contributing reason why the earlier studies by Guo et al. (2008) and Wu et al. (2012) had interpreted a short duration of only ~1.1 Myr for the Induan stage that is because they assumed a fairly constant sedimentation accumulation rate for the spectral analysis of the entire section. However, the sedimentation accumulation rate appears to significantly change from the lower Yinkeng Formation (Griesbachian part) with thin mud/claystone interbedded with limestones into the middle Yinkeng Formation (Dienerian part) with the increasing thickness of mudstone interbedded with limestones. They had also assumed that the predominant cycles of ~0.8 m are ca. 20-kyr precession cycles for the whole series; however, these predominant ∼0.8 m cycles should be the ca. 33-kyr obliquity cycles from the middle Griesbachian to the end of Dienerian (10−74 m interval in Chaohu section) according to the ratio of the wavelengths of ∼10:2.3:0.9:0.5 that is similar with the ratios of Milankovitch cycles of 405:100:33:20.

Therefore, based on an anchoring age of the U–Pb date of 251.9 Ma for the PTB (Burgess et al., 2014), an absolute ATS for the Early Triassic spanning 5.1 Myr from 251.9 to 246.8 Ma has been constructed based on the composite marine sections of Chaohu, Daxiakou, Guangdao and Meishan in South China (Li et al., 2016) (Fig. 4A). The calculated Induan/Olenekian boundary (I/OB) and Olenekian/Anisian boundary (O/AB) ages are 249.9 and 246.8 Ma from this ATS. The astronomical-calibrated O/AB age of 246.8 Ma is slightly younger than the U–Pb dates for the FAD of the proposed base-Anisian marker conodont *Chiosella timorensis* of 247.28 ± 0.12 Ma at the Guandao section (Lehrmann et al., 2015) and an interpreted 247.31 ± 0.06 Ma at the Monggan/Wantou section (Ovtcharova et al., 2015; although they recognized anomalies in the succession of U–Pb dates in the suite of ash beds). Therefore, when combined with the ammonoid and magnetostratigraphy global correlation, the estimated age for Early/Middle Triassic boundary, which has yet to be assigned a formal GSSP definition, is placed at a rounded 247 Ma (Ogg et al., 2016).

The 5.1 Myr duration for the Early Triassic composite ATS could be assigned as the appropriate 405-kyr chrono-cycles of E622 to E609.4 (Fig. 4A).

3.2 The Middle Triassic

The Middle Triassic includes **Anisian** and **Ladinian** stages and spanned about 10 Myr from the Early/Middle Triassic boundary at ∼247 Ma to the Middle/Late Triassic boundary at 237 Ma. There are many cyclostratigraphy studies for the Middle Triassic Latemar limestone from 1980s (Goldhammer et al., 1987, 1990; Hardie et al., 1997; Hinnov and Goldhammer, 1991; Kent et al., 2004; Maurer et al., 2004; Mundil et al., 2003; Preto et al., 2001, 2004; Zühlke et al., 2003; Zühlke, 2004) (Fig. 3).

The Latemar carbonate platform in the Dolomites of Northern Italy was formed in the western tropical Tethys Ocean. The exposed platform facies of laminated limestones, wackestone, packstone and grainstone has about 500 meter-scale cycles (Goldhammer et al., 1987) or 600−700 shallowing-upward cycles or microcycles (Preto et al., 2004; Zühlke, 2004). The

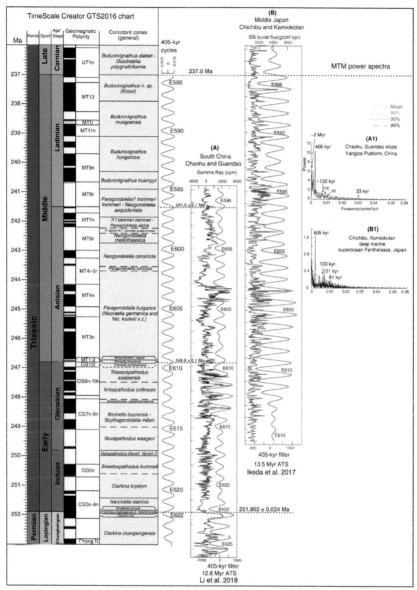

Figure 4 Our reanalysis and 405-kyr tuning of selected series for the Early-Middle Triassic ATS. Chronostratigraphy columns on the left were generated using TimeScale Creator 7.0 (*engineering. purdue.edu/Stratigraphy/tscreator*) with the GTS2016 integrated stratigraphy (Ogg et al., 2016). (A) Composite section of GR series (the original data from Li et al., 2018, using the GR residuals series after removing 10% weighted average) from Chaohu and Guandao section (South China) with 405 kyr filter output (red curve, passband: 0.00247 ± 0.0005 cycles/kyr). (B) BSi burial fluxes of chert series (the original data from Ikeda et al., 2017, using the BSi residuals series after removing 25% weighted average) with 405 kyr filter output (red curve, passband: 0.00247 ± 0.0005 cycles/kyr). (A1) and (B1) are 2π MTM power spectra of series in A and B, respectively.

estimated duration for the Latemar platform based on the orbital interpretations of these cycles is about 9—14 Myr (Goldhammer et al., 1990;
Preto et al., 2004), which is ca. 4 times longer than its span indicated by
the U-Pb zircon ages (Mundil et al., 1996, 2003; Brack et al., 1996,
2007). This "Middle Triassic Latemar controversy" mismatch between
standard cyclostratigraphic analyses and the other stratigraphic interpretations was difficult to resolve. Kent et al. (2004) suggested a total duration
of only ~1 Myr for the ~470 m thick Latemar carbonate platform in the
Dolomites based on the lack of magnetic reversals, thereby implying a
mean duration of ~1.7 kyr for the observed meter-scale cycles. A similar
interpretation of high-frequency (1—2 kyr) meter-scale rhythmic beds in a
basinal facies at the Rio Sacuz section of northern Italy was derived from
the magnetostratigraphy and cyclostratigraphy of magnetic susceptibility
proxies (Spahn et al., 2013). In this case, the total duration of the Latemar
platform deposition, coupled with its uncertain biostratigraphic age span,
is inadequate for calibrating the geological time scale.

In contrast, the Guandao marine section in the Guizhou Province of
South China is an important reference section for integrated Anisian
stratigraphy (Lehrmann et al., 2015). Li et al. (2018) resolved long-
eccentricity cycles in over 300 m of these marine slope deposits of
pelagic micrite-rich limestones with some carbonate packstone—
grainstone turbidites and debris flow breccia beds. The analyses were based
on 2061 magnetic susceptibility data points and 1071 spectral gamma-ray
measurements at 5—20 cm sample intervals for a ~260 m series after
removing the redeposited breccia beds. The predominant ~23 m cycles
were assumed to be the stable 405-kyr long-eccentricity cycle based on
the biostratigraphy and magnetostratigraphy constraints (Lehrmann et al.,
2015), and tuning to this 405-kyr period produced a ca. 6 Myr ATS for
all the Anisian conodont and magnetic polarity zones. The Anisian has a
5.3 Myr duration at this Guandao reference section; therefore, based on
the 246.8 Ma age for the Olenekian/Anisian boundary based on the Early
Triassic ATS from Chaohu and Daxiakou sections of South China,
then the Anisian/Ladinian boundary is at 241.5 Ma (Li et al., 2016)
(Fig. 4A). We reanalyzed the ~11 Myr composite GR series from South
China marine sections spanning the entire Early Triassic and Anisian
(Fig. 4A1, Table 1).

Volcanic ash beds in the Buchenstein Formation at Passo Feudo and
Seceda in the Dolomites of Italy provides the U-Pb CA-ID-TIMS
dates that bracket the Anisian/Ladinian boundary to be between

Table 1 Reanalysis of Selected Early-Middle Triassic Reference Sections Using Standardized Tuning to 405-kyr Long-Eccentricity Cycle and MTM Power-Spectra Analysis. Span (Myr) and Age Limits (Ma) are Based on the Interpreted Tuning Results

Location	Geologic-age	Data source	Proxy	Span (Myr)	Significant spectral cycles (kyr)	Assigned E-cycles	Age limits (Ma)
Chaohu and Guandao (South China)	Uppermost Lopingian–lowest Ladinian	Li et al. (2018)	Gamma Ray (GR)	~12.6	E: *405 (tuned)* e: 130, 110, 95 O: 33	E626–E595	253.4–240.8
Inuyama, Japan	Lower Olenekian–basal Carnian	Ikeda et al. (2017)	Biologic silica flux	13.5	*405* e: 131, **121**, 81	E616–E583	249.5–236

241.705 ± 0.045 Ma and 240.576 ± 0.042 Ma; for an estimate on the GSSP Anisian/Ladinian boundary age of 241.43 ± 0.15/0.17/0.31 Ma. A boundary age of 241.464 ± 0.064/0.097/0.28 Ma was achieved by integrating high-resolution cyclostratigraphy (Wotzlaw et al., 2018), which is identical to the 241.5 Ma based on the ATS by Li et al. (2018).

The middle Triassic bedded chert sequence in the Inuyama area of central Japan consists of centimeter-scale rhythmic alternations of chert and shale beds with rich radiolarian fossils that is considered to have been deposited in a pelagic deep-sea environment (Ikeda et al., 2010). The 720 chert—shale couplets with an average 21 mm thickness recognized in the 33-m bedded chert succession appear to be productivity oscillations induced by the ∼20 kyr precession cycle, and the bundles of about 20 chert—shale couplets represent modulation by the 405-kyr long-eccentricity cycle. However, that initial deep-sea 15 Myr-long floating ATS (Ikeda et al., 2010) had only limited biostratigraphy and lacked any magnetostratigraphy or isotopic dates. Therefore, Ikeda and Tada (2014) extended their work to compile 3346 chert—shale couplets through a 110-m thick deep sea bedded chert sequence in the Inuyama area that spans the Lower Triassic to Lower Jurassic and applied 405 kyr tuning. This ∼70 Myr long ATS was anchored at the end-Triassic radiolarian extinction level at chert bed-number 2525 as 201.4 ± 0.2 Ma (Fig. 4B). Ikeda et al. (2017) presented the biogenic silica flux (BSi) series from radiolarian chert. We collected the BSi data for this ∼70 Myr-long record from this author and performed the spectral analysis for the Early-Middle Triassic part (Fig. 4B, Table 1). Combined with radiolarian biostratigraphy events and U-Pb dates, Ikeda and Tada (2014) projected an age of 235.2 ± 0.9 Ma for a radiolarian-based Ladinian/Carnian boundary, but the lack of radiolarian calibration to Middle Triassic stages precludes obtaining accurate durations for the Anisian and Ladinian stages based on this deep sea bedded chert ATS (Fig. 4B). But when combining the estimated 235.2 ± 0.9 Ma for the Ladinian/Carnian boundary by Ikeda and Tada (2014) with the 241.5 Ma base-Ladinian age from Guandao ATS, then estimated duration for the Ladinian stage from these ATS scales is about 6.3 ± 0.9 Myr. However, in contrast to the radiolarian-based placement in the deep-sea cherts, Mietto et al. (2012) reported a U-Pb zircon date of 237.77 ± 0.14 Ma for an ash layer near the top of the Ladinian Stage from the Rio Nigra section, therefore suggested an older age of ∼237 Ma for the Ladinian/Carnian boundary, therefore implying that the Ladinian Stage spans only 4.5 Myr.

In summary, the 5.3 Myr duration for the Anisian stage based on the ATS corresponds to the 405-kyr chrono-cycles from E609.4 (at 246.8 Ma) to E596 (at 241.5 Ma). The 11.6 Myr duration for the entire Middle Triassic from merging the south China and central Japan ATS appears to correspond to the 405-kyr chrono-cycles from E608.5 to E581 (Fig. 4, Table 1). However, for the Ladinian and Carnian stages of middle Triassic, more cyclostratigraphy studies is crucial and need to be developed and correlated and integrated with biostratigraphy, chemostratigraphy, magnetostratigraphy and radioisotopic dating to verify of the main 405-kyr tuned ATS.

3.3 The Late Triassic

The Late Triassic includes **Carnian, Norian** and **Rhaetian** stages, and spans about 36 Myr from ~237 Ma at Middle/Late Triassic boundary to 201.4 Ma at Triassic/Jurassic boundary. There are about 10 cyclostratigraphic studies for the Late Triassic published in the past three decades (Balog et al., 1997; Clemmanson et al., 1998; Cozzi et al., 2005; Hüsing et al., 2011; Ikeda and Tada, 2014; Ikeda et al., 2017; Kemp and Coe, 2007; Kent et al., 2017; Li et al., 2017; Olsen and Kent, 1996; 1999; Olsen et al., 2011; Preto and Hinnov, 2003; Vollmer et al., 2008; Zhang et al., 2015) (Fig. 3).

The Carnian Stage does not yet have a continuous cyclostratigraphy in a non–deep-sea chert facies. Zhang et al. (2015) analyzed the GR series from South China carbonate-rich facies with a transition to clastic-rich facies ~2.4 Myr cycle-calibrated magnetostratigraphy that spans the upper part of the Julian substage and lower part of the Tuvalian substage of the Carnian Stage (Fig. 5E).

A superb cyclostratigraphy study of terrestrial strata was enabled by a ~3500-m thick succession in the Newark Basin derived from 7 overlapping boreholes that cores a total of 6770 m. This unparalleled continuous record spans the Late Carnian to Early Hettangian stages (Olsen and Kent, 1996; Olsen et al., 2011; Kent et al., 2017). The succession in the lacustrine to fluvial facies is characterized by 3–6 m thick cycles that correspond to lake level variations responding to ~20 kyr precession periodicity in relative monsoonal rainfall. These basic cycles are grouped into ~12–18 m and ~60 m bundles that represent strongly modulation by ~100-kyr short and 405-kyr long-eccentricity cycles, respectively. This ~3500 m composite depth rank series presents about 22 Myr duration for Late Triassic (Olsen and Kent, 1996). We restudied the composited depth rank series (relative

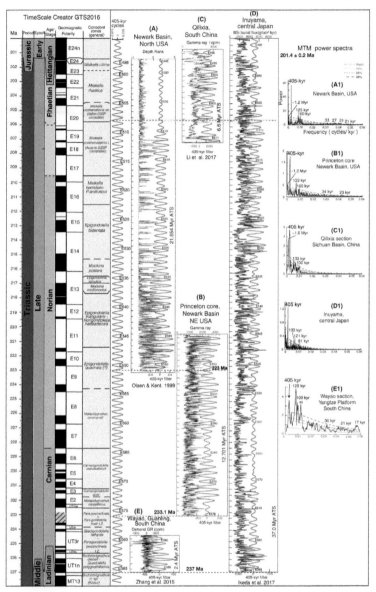

Figure 5 Our restudy and 405-kyr tuning of selected series for the Late Triassic ATS. Chronostratigraphy columns on the left were generated using TimeScale Creator 7.0 (*engineering.purdue.edu/Stratigraphy/tscreator*) with the GTS2016 integrated stratigraphy (Ogg et al., 2016). (A) The composite section depth rank series (the original data compiled by Olsen and Kent (1999) was downloaded from https://www.ldeo.columbia.edu/~polsen/nbcp/data.html, and retuned the composite depth rank residuals series using 405-kyr period in this study) from the Newark Basin (USA).

lake-depth or exposure intensity by Olsen and Kent, 1996) to recognized and tune the 53 predominant ~66 m wavelength cycles (405-kyr) in a large portion of their record. We constructed a ~21.65 Myr–long ATS based on tuning these ~66 m cycles to the 405-kyr long-eccentricity period (Fig. 5A, Table 2). In addition to this depth-rank proxy, we also analyzed the GR data from the Princeton drilled core in that Newark Basin set. A ~12.7 Myr–long ATS has been constructed based on the recognition of 31 predominant ~40 m cycles that were tuned to the 405-kyr long-eccentricity period (Fig. 5B, Table 2). According the correlation provided by Kent et al. (2017; their Fig. 1), the 294 m level in the Princeton core could correlate with the base of composite depth-rank series at 223 Ma or chrono-cycle E551; thereby projecting the base of the Princeton GR ATS at 233.1 Ma or chrono-cycle E576 (Fig. 5B).

Applying the ATS to obtain cycle-durations of Late Triassic stages is hindered by their lack of standard GSSP definitions. Olsen et al. (2011) used their astronomically tuned geomagnetic polarity timescale (called Newark-Hartford APTS) to estimate approximate durations of ~7 ± 4 Myr duration for the recovered portion of the Carnian Stage, a 6.2 Myr or 9.7 Myr for the Rhaetian Stage (depending whether a short or long option was used for its definition) and a 17.6–20.1 Myr duration for the Norian Stage depending upon the selected Rhaetian option. The main Newark lacustrine succession is capped by the massive basalt flows of the Central Atlantic Magmatic Province (CAMP) large igneous event, which are considered to be a major contributor to the end-Triassic global extinctions, and the base of the Jurassic corresponds to the first appearance of *Psiloceras spelae* ammonite. These have been dated by U-Pb methods, thereby enabling an absolute age scale for this Late Triassic ATS. The Triassic/Jurassic boundary is dated as 201.36 ± 0.14 Ma from bracketing ash beds on the ammonite occurrence in Peru by Wotzlaw et al. (2014; revising

----

(B) GR series of the Princeton core of the Newark Basin (USA) (we downloaded the original data from https://www.ldeo.columbia.edu/~polsen/nbcp/prince.core.html, after removing some spikes to tune it to 405-kyr period). (C) GR series of the Qilixia Section from Sichuan Basin (South China) (the original data from Li et al., 2017). (D) BSi burial fluxes of chert series (data from Ikeda et al. (2017), the BSi residuals series after removing 25% weighted average). (E) GR series of the Wayao composite section (South China) (the original data from Zhang et al., 2015). All the 405-kyr filter output (red curve, passband: 0.00247 ± 0.0003–0.0008 cycles/kyr). A1-E1 are 2π MTM power spectrum of A-E series.

Table 2 Reanalysis of Selected Late Triassic Reference Sections Using Standardized Tuning to 405-kyr Long-Eccentricity Cycle and MTM Power-Spectra Analysis. Span (Myr) and Age Limits (Ma) are Based on the Interpreted Tuning Results

Location	Geologic-age	Data source	Proxy	Span (Myr)	Significant spectral cycles (kyr)	Assigned E-cycles	Age limits (Ma)
Composite core, Newark Basin (North USA)	Low Norian–uppermost Rhaetia	Oslen &Kent (1999), https://www.ldeo.columbia.edu/~polsen/nbcp/data.html	Depth rank series	21.65	E: **405 (*tuned*)** e: **125, 100** O: 33, 27 P: 23, 21	E551-E498	223.1–201.4
Princeton core, Newark Basin (North USA)	Middle Carnian–middle Norian	https://www.ldeo.columbia.edu/~polsen/nbcp/prince.core.html	Gamma Ray (GR)	12.7	**405** e: **133, 100** O: 34 P: 23	E576-E545	233.1–220.6
Qilixia section, Sichuan Basin (China)	Upper Norian-Lowermost Hettangian	Li et al. (2017)	Gamma Ray	6.4	**405** e: **133, 100**	E512-E496	207.2–200.8
Gayao, Guanling (South China)	Lower Carnian	Zhang et al. (2015)	Gamma Ray	2.4	**405** e: **125, 100**, 85 O: 30 P: 21, 17	E586-E579	234.6–237
Inuyama, Japan	Basal Carnian-lower Hettangian	Ikeda et al. (2017)	Biologic silica flux	37.0	**405** e: **131, 121**, 81	E585-E494	237–200

the previous 201.3 ± 0.18 Ma by Schoene et al., 2010), and the basalts at the top of the Newark ATS have a suggested age of 201.4 Ma (Sha et al., 2015).

Applying the Newark-Harford APTS to the magnetostratigraphy correlation from the Tethyan marine sections of Pizzo Mondello (Italy) (Muttoni et al., 2004) and Silicka Brezova (Slovakia) (Channell et al., 2003), combined with conodont biostratigraphy (Mazza and Rigo, 2012) and a U-Pb date of 230.91 ± 0.33 Ma within the uppermost Carnian at the Pignola 2 marine section in Italy (Furin et al., 2006), Kent et al. (2017) estimated an age of about 227 Ma for the Carnian/Norian boundary and 205.5 Ma for the Norian/Rhaetian boundary. Therefore, including the base-Carnian date of 237 Ma (Mietto et al., 2012), then the durations of Carnian, Norian and Rhaetian (short-option) could be assigned as 10, 21.5 and 4.1 Myr, respectively based on the Newark-Hartford APTS. In contrast, based on radiolarians in their deep-sea chert-derived 70-Myr-long Inuyama-ATS, Ikeda and Tada (2014) proposed stage boundary ages of 235.2 ± 0.9, 225.0 ± 1.1 and 208.5 ± 0.3 Ma for the Ladinian/Carnian, Carnian/Norian and Norian/Rhaetian boundaries, respectively; implying durations for Carnian, Norian and Rhaetian of 10.2, 16.5 and 7.1 Myr, respectively. Hüsing et al. (2011) also suggested a cycle-calibrated magnetostratigraphy that implied both a Long-Norian (~17 Myr) and a Long Rhaetian (~9 Myr), which is consistent with the Inuyama-ATS model. The GTS2016 (Ogg et al., 2016) assigned ages of 237 and 228.5 Ma for the Ladinian/Carnian and Carnian/Norian boundaries, and either 209.5 or 205.7 Ma for the Norian/Rhaetian boundary age based on the Austrian GSSP candidate (long-Rhaetian option) and the Italian GSSP candidate (short-Rhaetian option). It remains an embarrassment that we have a rather well-developed ATS for the ages of the entire Late Triassic magnetic polarity time scale, dinosaur evolution, atmospheric carbon dioxide history (Schaller et al., 2015) and other events; but have not yet decided on the definitions for the Norian and Rhaetian stages or their potential substages.

If the "short" Rhaetian option is used, which is based on a more restricted taxonomic definition for the basal-Rhaetian conodont marker, then the current best estimate age for the resulting Norian/Rhaetian boundary would be 205.50 ± 0.35 Ma based on bracketing U-Pb dates of 205.70 ± 0.15 and 205.30 ± 0.14 Ma (Wotzlaw et al., 2014) combined with the magnetostratigraphic (reversed-polarity Chron E20r.2r) cyclostratigraphy correlation from Newark APTS to a potential Norian/Rhaetian boundary candidate GSSP section at Pignola-Abriola in southern Italy (Maron et al., 2015; Kent et al., 2017). The combined cyclo-magnetic

ATS can also be applied to non-fluvial terrestrial deposits (e.g., Xujiahe and Qilixia sections) in Sichuan, China (Li et al., 2017a), and demonstrates a common negative carbon isotope excursion in organic matter (Li et al., 2017b,c). This short version of the Rhaetian spans 4.3 Myr in the ATS (Fig. 5AC).

Based on the Triassic-Jurassic boundary age of 201.4 Ma, the ~21.65 Myr-long ATS from Newark Basin could be assigned to the 405-kyr chrono-cycles from E 551 to E498. Even though the Inuyama-ATS could provide a 35.6 Myr scale for the Late Triassic spanning potentially 405-kyr chrono-cycles from E 585 to E497 (Fig. 5D), it does not yet have potential use for a reference standard due to lack of a more globally useful biostratigraphy, magnetostratigraphy or isotopic dating constraints. In contrast, when combining with the 405 kyr tuned GR series in this study from the Princeton drilled core, the Newark Basin provides nearly a 32 Myr-long ATS that could be assigned to the 405-kyr chrono-cycles from E576 to E498 (Fig. 5AB). This ATS with its magnetic chron and other calibrations could serve as a standard reference for the international geological time scale.

4. THE CURRENT STATUS OF THE JURASSIC ATS

The Jurassic Period begins with the lowest occurrence of ammonite *P. spelae* after the end-Triassic extinction is U-Pb dated as 201.4 ± 0.17 Ma (Wotzlaw et al., 2014). The top of Jurassic (Jurassic/Cretaceous boundary) has not been defined, but the candidates currently include the base of micro-fossil *Calpionella alpina* or the base of magnetic polarity Chron M18r at 145.5 Ma in GTS2012 (Ogg et al., 2012a; 2016). The bases of the 11 international Jurassic stages as assigned at GSSPs or at candidate definitions are at levels corresponding to the lowest occurrences of ammonoid taxa. Over 30 cyclostratigraphic studies have been published for the Jurassic in the past three decades (Fig. 6).

4.1 The Early Jurassic

The Early Jurassic includes **Hettangian**, **Sinemurian**, **Pliensbachian** and **Toarcian stages** and spans about 27 Myr from the Triassic/Jurassic boundary at 201.4 Ma to the Early/Middle Jurassic boundary at 174.2 Ma. Nearly 20 Early Jurassic cyclostratigraphic studies have been published over the past twenty years (Boulila et al., 2014; Hinnov and

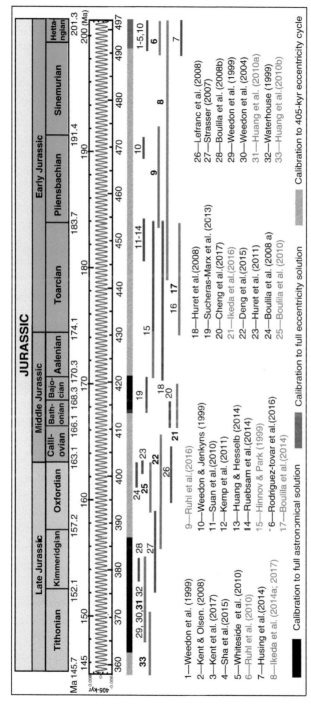

Figure 6 Stratigraphic coverage of the Jurassic ATS and 405-kyr metronome. Intervals with *red lines* (gray in print version) indicate the collected data coverage subjected to a 405-kyr tuning to compile the ATS in this study, and those with *blue lines* (dark gray in print version) indicate additional supporting cydostratigraphy studies.

Park, 1999; Huang and Hesselbo, 2014; Hüsing et al., 2014; Ikeda and Hori, 2014; Ikeda et al., 2017; Kemp et al., 2011; Kent and Olsen, 2008; Kent et al., 2017; Rodriguez-tovar et al., 2016; Ruebsam et al., 2014; Ruhl et al., 2010; 2016; Sha et al., 2015; Suan et al., 2010; Weedon et al., 1999; Weedon and Jenkyns, 1999; Whiteside et al., 2010). Milankovitch signals of precession, obliquity and eccentricity have been recognized from outcrop sections in Britain and Southern Alps (Weedon et al., 1999; Hinnov and Park, 1999) (Fig. 6).

The Hettangian Stage has a duration of ca. 1.8 Myr according to the cyclostratigraphy at the St Audrie's Bay and East Quantoxhead (UK) outcrop section of the Bristol Channel Basin (Ruhl et al., 2010) and its correlation with the astronomically calibrated magnetostratigraphy sequence from the Hartford Basin (USA) of Kent and Olsen (2008). At the St Audrie's Bay, the 120 m thick succession of black-shale and limestone display 3.5–6 m thick cycles in $\delta^{13}C_{org}$, TOC, $CaCO_3$ and magnetic susceptibility data that were interpreted as ~100-kyr short-eccentricity cycles (Ruhl et al., 2010). We applied 405-kyr long-eccentricity cycle tuning for the $\delta^{13}C_{org}$ series based on the recognition of ~4.5 m (~100 kyr) and ~16 m (405 kyr) cycles to construct a 3.55-Myr-long ATS (Fig. 7A, Table 3). The estimate duration of 1.97 Myr for the Hettangian Stage implies that the Hettangian/Sinemurian boundary age is 199.43 Ma (Fig. 7A), which is the same as the 2 Myr duration used in GTS2012 (Ogg et al., 2012a). Kent and Olsen (2008) constructed a 2.4 Myr-long APTS from the 2500 m succession in the Hardford Basin and estimated the duration for Hettangian Stage was also about 1.9 Myr; and, relative to the 201.4 Ma age for the Triassic/Jurassic boundary, this implied an age of 199.5 Ma for the Hettangian/Sinemurian boundary (Kent et al., 2017). Sha et al. (2015) obtained a similar result from the 300-m thick fluvial-lacustrine sequences in the Badaowan Formation of the Junggar Basin (China) based on the 405-kyr long-eccentricity correlation between the Newark–Hartford basins (USA), the Bristol Channel Basin (UK) and the Pucara Basin (Peru) and La2010d 405-kyr eccentricity theory curve.

Weedon and Jenkyns (1999) estimated a minimum duration of 4.82 Myr for the Pliensbachian Stage based on the cycle counting for the *T. ibex* (0.34 Myr) and *U. jamesoni* (1.42 Myr) ammonite zones from the assumed precession cycles of the light to dark marl bedding couplets sequence in

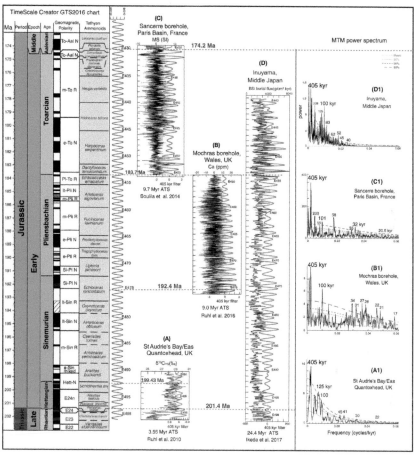

Figure 7 Our restudy and 405-kyr tuning of selected series for the Early Jurassic ATS. Chronostratigraphy columns on the left were generated using TimeScale Creator 7.0 (*engineering.purdue.edu/Stratigraphy/tscreator*) with the GTS2016 integrated stratigraphy (Ogg et al., 2016). (A) $\delta^{13}C_{org}$ series of the St Audrie's Bay (UK) (the original data from Ruhl et al. (2010, we retuned the $\delta^{13}C_{org}$ residuals series (after removing 25% weighted average) using 405 kyr period). (B) Ca content series of the Mochras Farm (Llanbedr) Borehole (UK) (data from Ruhl et al., 2016). (C) BSi burial fluxes of cherts residuals series (data from Ikeda et al., 2017), after removing 25% weighted average. (D) MS residuals series (after removing 10% weighted average) of the Sancerre-Couy drill-core in the southern Paris Basin (data download from https://doi.pangaea.de/10.1594/PANGAEA.821258, Boulila et al. (2014)). All the 405 kyr filter output (red curve, passband: 0.00247 ± 0.0003−0.0008 cycles/kyr). A1-D1 are 2π MTM power spectrum of A-D series.

Table 3 Reanalysis of Selected Early Jurassic Reference Sections Using Standardized Tuning to 405-kyr Long-Eccentricity Cycle and MTM Power-Spectra Analysis. Span (Myr) and Age Limits (Ma) are Based on the Interpreted Tuning Results

Location	Geologic-age	Data source	Proxy	Span (Myr)	Significant spectral cycles (kyr)	Assigned E-cycles	Age limits (Ma)
St. Audrie's Bay (UK)	Uppermost Rhaetian–lowest Sinemurian	Rohl et al. (2010, online Suppl.)	$\delta^{13}C_{org}$	3.55	E: **405** *(tuned)* e: **125, 100** O: 45, **41**, 30 P: **22**	E499- E491	202.2–198.65
Mochras Borehole (UK)	Uppermost Sinemurian–end Pliensbachian	Rohl et al. (2016)	Ca-series	9	**405** e: **100** O: 34, 31, 27, 26 P: 22, 21, 17	E476- E454	192.7–183.7
Sancerre-Couy Borehole, Paris Basin (France)	Base-Toarcian–basal Aalenian	Boulila et al. (2014)	Magnetic suscepti-bility	9.7	**405** e: 120, **101** O: **58**, 35, **32** P: 21	E454- E430	183.7–174.0
Inuyama, Japan	Upper Rhaetian–mid-Toarcian	Ikeda et al. (2017)	Biologic silica flux	24.4	**405** e: **125, 100**, 83 O: 62, 52, 45, 40	E499- E440	202.3–177.9

the epicontinental sea of the Belemnite Marls (Dorset, southern England). Based on scaling the rate-of-change of marine Sr-isotope ratios, they projected minimum durations 2.86, 7.6 and 6.67 Myr for Hettangian, Sinemurian and Pliensbachian stages respectively. However, their estimated durations of the *T. ibex* (0.34 Myr) and *U. jamesoni* (1.42 Myr) ammonite zones from the Belemnite Marls is much shorter than the duration of ~1.8 and ~2.7 Myr for these two zones estimated from the more expanded records in the Mochras Borehole by Ruhl et al. (2016).

For the Pliensbachian Stage, Ruhl et al. (2016) present an over 9-Myr-long ATS based on the 405-kyr tuned Ca series from the marl and clayey limestone record of the Mochras Borehole drilled in the 1968–70 at Cardigan Bay Basin on the west coast of Wales. This ATS record includes well-defined ammonite zones, indicating durations of ~2.7, ~1.8, ~0.4, ~2.4 and ~1.4 Myr for the *U. jamesoni*, *T. ibex*, *Psiloceras davoei*, *A. margaritatus* and *Psiloceras spinatum* zones, respectively (Fig. 7B). We applied 405 kyr tuning to their series to assign the interval to potential "E" cycles (Fig. 7B1, Table 3). Ruhl et al. (2016) obtained a ~8.7 Myr duration for the Pliensbachian stage and suggested that the Sinemurian/Pliensbachian boundary had an age of 192.5 ± 0.4 Ma relative to a radiometric date of 183.8 ± 0.4 Ma for the base of the Toarcian.

There is not yet a verified cyclostratigraphy scaling for the intervening Sinemurian Stage. However, the constraints of the Hettangian/Sinemurian boundary age of 199.5 Ma from the Newark-Hartford APTS proposed by Kent et al. (2017) and the 192.6 Ma age for the Sinemurian/Pliensbachian boundary age imply that the Sinemurian Stage spans ~7 Myr (Ruhl et al., 2016).

The Toarcian Stage astrochronology became a hot topic when trying to determine the rates and duration for the Early Toarcian Oceanic Anoxic Event (T-OAE). Depending on the interpretation of the dominant and superimposed cycles, the interpreted durations for the T-OAE and its associated negative carbon-isotope excursion (CIE) have ranged from 200 kyr to 1.0 Myr (Ait-Itto et al., 2018; Boulila et al., 2014; Huang and Hesselbo, 2014; Hüsing et al., 2014; Ikeda and Hori, 2014; Kemp et al., 2011; Martinez et al., 2017; Müller et al., 2017; Ruebsam et al., 2014; Ruhl et al., 2010; Suan et al., 2010). However, these cyclostratigraphy studies mainly focused on the brief T-OAE interval. The main current reference ATS for the entire Toarcian stage is from the gray marine marls in the Sancerre-Couy drill-core (194.55–360 m) in the southern Paris Basin that has a well-defined ammonite biostratigraphy. Boulila et al. (2014)

measured a high-resolution magnetic susceptibility data series for the Sancerre-Couy drill-core and interpreted ~8 m (405 kyr) and ~32 m (1.6 Myr) cycles characterized the entire Toarcian Stage. Based on the 405-kyr long-eccentricity cycle tuning, they estimated a ~300 to 500 kyr duration for the Toarcian CIE and a minimum duration of ~8.3 Myr for the Toarcian Stage. We reanalyzed this MS series and recognized 24 predominant ~8 m cycles (405-kyr) to construc a ~9.7 Myr-long ATS based on a similar 405-kyr tuning (Fig. 7C, Table 3). Our estimate duration for the Toarcian Stage is ~9.5 Myr, which is similar to 9.5 Myr duration assigned for this stage in GTS2016 (Ogg et al., 2016).

Toarcian cyclostratigraphy has also been studied in the Italian and French Alps. Hinnov and Park (1999) studied the rank series of the bedded basinal carbonate succession of the Colle di Sogno section in the Lombard Basin of the Southern Alps. Based on counting the obliquity cycles, they estimated a minimum duration of 11.37 ± 0.05 Myr for the combined Toarcian-Aalenian stages. Huret et al. (2008) restudied the rank series of the Aalenian part between 83–130 m from this Sogno section and esti-mated a duration of 3.85 Myr for the Aalenian Stage. Their duration for the Toarcian Stage from the Sogno Section of 7.52 Myr is significantly less than the 8.3 Myr duration estimated from the Sancerre core by Boulila et al. (2014).

On the broad scale for nearly the entire Triassic through Early Jurassic, Ikeda et al. (2017) constructed ~70 Myr-long BSi ATS from the deep-sea bedded-chert sequence exposed in the Inuyama area of the central Japan based on the 405-kyr long-eccentricity cycle tuning and anchored at 201.4 ± 0.2 Ma for the end Triassic radiolarian extinction event (Fig. 7D). Their super long ATS spans 249.5 Ma to ca. 177.9 Ma. Their proposed ATS age of 183.4 ± 0.5 Ma for their placement of the Pliensba-chian/Toarcian boundary is similar to the 183.7 Ma assigned age in the GTS2016 (Ogg et al., 2016). Our analysis (Fig. 7D, Table 3) suggested that the Jurassic portion of this Inuyama series could be the 405-kyr chrono-cycles from E497 to E440. However, as with the Triassic, these cyclostratigraphy studies could not provide us an accurate time scale for the Early Jurassic there has been better calibration of the deep-sea stratigraphy.

The Hettangian Stage could be assigned to chrono-cycles from E497.3 to E493 based on the age of 201.4 Ma for the Triassic/Jurassic boundary and the 1.97 Myr duration from the 405 kyr tuned $\delta^{13}C_{org}$ series from the Bristol Channel Basin (UK). Combining the ~9.3 Myr-long ATS for the Late Sinemurian and the entire Pliensbachian Stage from the 405 kyr tuned Ca series from the Mochras core (Ruhl et al., 2016) and the GTS2016 assigned age of 183.7 Ma for the base of the Toarcian (Ogg et al., 2016) implies a possible coverage of 405-kyr chrono-cycles from E476 to E454. The 9.5 Myr-long ATS for the Toarcian Stage could be assigned the chrono-cycles from E454 to E430.

4.2 The Middle Jurassic

The Middle Jurassic includes **Aalenian, Bajocian, Bathonian** and **Callovian stages** that spanned about 11 Myr from the Early/Middle Jurassic boundary at ~174 Ma to the Middle/Late Jurassic boundary at ~163 Ma. There are only a few cyclostratigraphic studies published for the Middle Jurassic over the past twenty years (Cheng et al., 2017; Hinnov and Park, 1999; Huret et al., 2008; Ikeda et al., 2016; Sucheras-Marx et al., 2013; Weedon et al., 1999) (Fig. 6).

The Aalenian Stage has been analyzed for cyclostratigraphy mainly using a lithostratigraphy rank series of the Sogno Section in the Lombardy Pre-Alps, Italy. Huret et al. (2008) reanalyzed the rank-series data collected by Hinnov and Park (1999) to propose a minimum duration of 3.85 Myr, and this result was incorporated as part of the calibration of the Aalenian time scale in GTS2012 (Ogg et al., 2012a). However, we re-analyzed this data in this study and derived a 4.44 Myr-long ATS based on the 405 kyr tuning the envelope of rank series (Fig. 8A, Table 4). The resulting estimated duration for the Aalenian stage of ~4 Myr is slightly longer than the 3.9 Myr duration assigned in the GTS2016, although that estimate had relatively high uncertainties of ±1.0 to ±1.4 Myr on the bounding stage boundaries (Ogg et al., 2016). Based on the projected age of 174.2 Ma for the Toarcian/Aalenian boundary, the 4 Myr Sogno ATS for the Aalenian stage could be assigned as the 405-kyr chrono-cycles from E430 to E420 (Fig. 8A).

Bajocian through Callovian chert abundances and color at the Torre De Busi and Corre Di Sogno sections in the Lombardian Basin (Northern Italy) was analyzed by Ikeda et al. (2016) to establish a ~4 Myr-long ATS. Their scale was based on the assumption that the observed cycles of 8, 16, 40 and 160 cm corresponded to ~20, 40, 100 and 405 kyr cycles based on the biostratigraphic age constraints. We used their chert abundance series from Ikeda et al. (2016) supplementary data to construct a ~4.2 Myr-long ATS and based on tuning the predominant ~1.6 m thick

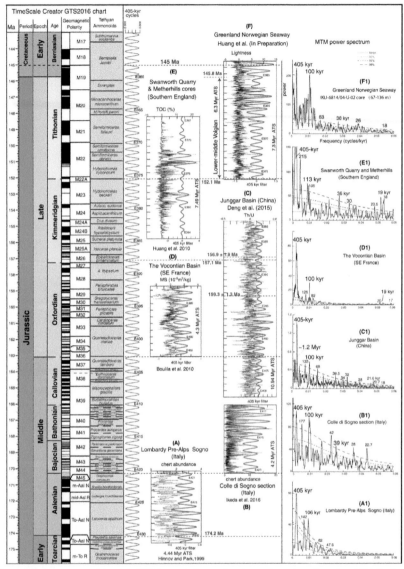

Figure 8 Our restudy and 405-kyr tuning of selected series for the Middle-Late Jurassic ATS. Chronostratigraphy columns on the left were generated using TimeScale Creator 7.0 (*engineering.purdue.edu/Stratigraphy/tscreator*) with the GTS2016 integrated stratigraphy (Ogg et al., 2016). (A) Chert abundance series (the original data from Hinnov and Park, 1999). (B) Chert abundance series of Chichibu and Kamiokotan areas (Japan) (the original data from Ikeda et al. (2016) supplementary data online at https://doi.org/10.1016/j.palaeo.2016.06.009). (C) Th/U residuals series (the original data from Deng et al., 2015, after removing 15% weighted average). (D) MS residuals series from the composited three Terres Noires sections of the Vocontian Basin

cycles to the 405-kyr long-eccentricity period (Fig. 8B, Table 4). This ~4 Myr-long Corre Di Sogno ATS which could be considered as the extension of the Toarcian-Aalenian Sogno ATS from the Lombardy Pre-Alps had been interpreted to span much of the Bajocian and Bathonian stages and could be assigned as the 405-kyr chrono-cycles from E420 to E410.

In contrast, a much longer duration for the lower Bajocian was estimated by Sucheras-Marx et al. (2013) based on a 169 m thick $CaCO_3$ wt% series from the Chaudon—Norante section of the French Subalpine Basin in France. They estimated a duration of 4.082 ± 0.144 Myr for the Early Bajocian based on the recognition of the 0.48—0.84, 0.98—1.43, 2.8—3.84 and 12—16 m cycles that assumed to correspond to precession, obliquity, short and long-eccentricity cycles. However, a 4.08 Myr duration for only the Early Bajocian is difficult to reconcile with the GTS2012 assigned duration of 2 Myr for the entire Bajocian stage without greatly shortening at least one other geologic stage within the Middle Jurassic through Early Cretaceous.

For the Bathonian Stage, Cheng et al. (2017) constructed a 2.09 Myr ATS based on the recognition of ~100 kyr cycles (~26 m) in the 557 m thick GR well log series of the drilled core from the marine carbonate Buqu Formation in the Qiangtang Basin (Tibet, China) and proposed a 2.09 Myr-long ATS for the Bathonian Stage. This result is compare closely to the 2.2 ± 1.3 Myr duration assigned in the GTS 2012 and 2016 that was partly constrained by radiometric dates from Pacific ODP Site 801 near the Bajocian/Bathonian biostratigraphic boundary (Ogg et al., 2012a; 2016). Due to the lack of consistent results for the Bajocian and Bathonian stage, and the lack of a calibrated ATS for the Callovian Stage, we could not compile the majority of the Middle Jurassic into the composite cyclo-chron figure.

(SE France) (the original data from Boulila et al., 2010, after removing 25% weighted average). (E) TOC residuals series (the original data from Huang et al., 2010, after removing 10% weighted average). (F) Lightness residuals series of the IKU-6814/04-U-02 core in the Greenland-Norwegian Seaway (the original data download from http://doi.pangaea.de/10.1594/PANGAEA.141090 posted by Swientek (2004), after removing 15% weighted average). All the 405 kyr filter output (red curve, passband: 0.00247 ± 0.0003—0.0008 cycles/kyr). A1-F1 are 2π MTM power spectrum of A-F series.

Table 4 Reanalysis of Selected Middle–Late Jurassic Reference Sections Using Standardized Tuning to 405-kyr Long-Eccentricity Cycle and MTM Power-Spectra Analysis. Span (Myr) and Age Limits (Ma) are Based on the Interpreted Tuning Results

Location	Geologic-age	Data source	Proxy	Span (Myr)	Significant spectral cycles (kyr)	Assigned E-cycles	Age limits (Ma)
Lombardy Pre-Alps Sogno (Italy)	Uppermost Toarcian–Basal Bajocian	Hinnov & Park (1999)	Chert abundance	4.44	**E: 405 (tuned)** e: 142, **106** O: 62, 47.6	E431-E420	174.6–170.16
Colle di Sogno section (Italy)	Basal Bajocian-basal Callovian	Ikeda et al. (2016, online suppl.)	Chert abundance	4.2	***405*** e: **100** O: **42**, **39**, 28 P: **22.7**	E420.5-E410	170.3–166.1
Junggar Basin (China)	Lower Callovian-middle Kimmeridgian	Deng et al. (2015)	Th/U	10.9	***405*** e: 133, **100** O: **39.5**, 34, **32** P: **21.6**, 20.7, 18	E409-E382	165.6–154.7
Vocontian Basin (SE France)	Basal-upper Oxfordian	Boulila et al. (2010)	Magnetic suscepti-bility	4.3	***405*** e: **126**,**100**, 82 P: **20**, 19, 17	E403-E392.5	163.1–158.8
Swanworth Quarry & Metherhills cores (England)	Lower Kimmeridgian-upper Tithonian	Huang et al. (2010)	TOC (%)	7.49	***405*** e: **113**, **105** O: **36**, 30 P: 20.5, **19**, **18**	E385-E366.5	155.74–148.25
Greenland Norwegian Seaway	Basal Tithonian-basal Berriasian	Huang et al. (preparation)	Lightness	7.3	***405*** e: **100** O: **36**, 26 P: **18**	E376-E358	152.1–144.8

4.3 The Late Jurassic

The Late Jurassic includes **Oxfordian, Kimmeridgian** and **Tithonian stages** and spanned about 18 Myr from ~163 Ma at Middle/Late Jurassic boundary to ~145 Ma at Jurassic/Cretaceous boundary. There are nearly a dozen cyclostratigraphic papers have been published for the Late Jurassic over the past twenty years (Boulila et al., 2008a, b, 2010; Deng et al., 2015; Huang et al., 2010a,b; Strasser, 2007; Swientek, 2002; Waterhouse, 1999; Weedon et al., 1999, 2004) (Fig. 6).

An interval of lower Callovian through lower Kimmeridgian of Junggar Basin of northwest China was analyzed by Deng et al. (2015) to propose a ~10.9 Myr-long synthesized time scale. They used the Th/U series from an inner lacustrine succession that mainly consists of red mudstone-sandstone interbedded with gray shale, gypsum and volcanic ash beds. The geologic age assignments to the cyclostratigraphy were based on comparing to biostratigraphy, magnetostratigraphy and two SHRIMP U-Pb dates of 159.3 ± 1.3 Ma and 156.9 ± 1.9 Ma, thereby interpreting this series as spanning ~165.6 Ma (Early Callovian) to ~154.7 Ma (Early Kimmeridgian) and with projected ages of 161.5 Ma for the Callovian/Oxfordian boundary and 157.1 Ma for the Oxfordian/Kimmeridgian boundary. Their proposed Oxfordian/Kimmeridgian boundary age of 157.1 Ma is similar to the GTS2012 assigned age of 157.3 ± 1.0 Ma, but their estimated base-Oxfordian age is about 1.5 Myr younger. We reanalyzed this Th/U series and recognized 27 dominant ~50 m thick cycles to construct a 10.93 Myr-long ATS based on tuning these ~50 m cycles to the 405-kyr long-eccentricity period (Fig. 8C, Table 4). This 10.93 Myr-long Junggar lacustrine ATS could be assigned as 405-kyr chrono-cycles from E409 to E382 (Fig. 8C).

The cyclostratigraphic durations of Oxfordian ammonite zones are well constrained from the cyclic clayey and marly sediments of the Terres Noires Formation in the Vocontian Basin of southeast France and coeval deposits in the Paris Basin. Boulila et al. (2008a) collected 667 samples for the MS data measurement from the ~333-m thick outcrop in the Aspres-sur-Buëch section and measured the MS series from the EST342 drill-core in the eastern Paris Basin (France). Based on the recognition of clear visible ~55 and ~6 m cycles in the two series and assuming these cycles response to the 405-kyr long-eccentricity cycles and tuned these two MS series, they constructed an over 2 Myr-long ATS. Boulila et al. (2010) composited three Terres Noires sections and recognized 10 major cycles that were assigned as 405-kyr long-eccentricity cycles to construct a ~4.3 Myr-long ATS for the Early-Middle Oxfordian stage (Fig. 8D). Even though their estimated duration for the entire Early-Middle Oxfordian of about 4.07 Myr from the 405 kyr tuned composite MS ATS, was similar with assigned duration of 3.8 ± 1.4 Myr in the GTS2004 (Gradstein et al., 2004), the relative durations of the component ammonite zones were significantly different. For example, they estimated a ~2.2 Myr duration for the *Q. mariae* ammonite zone in the Early Oxfordian stage, which was much longer than the 0.6 Myr duration assigned in GTS2004 based on applying equal ammonite subzonal durations (Gradstein et al., 2004), but was also shorter than the estimated duration of 2.7 Myr based on French borehole cyclostratigraphy by Lefranc et al. (2008). The composite Terres Noires ATS by Boulila et al. (2010) was used as the reference standard for the early portion of the Oxfordian time scale in GTS2012 (Ogg et al., 2012a). This interval could be assigned as the 405-kyr chrono-cycles from E403 to E392.5 (Fig. 8D, Table 4).

The Kimmeridgian cyclostratigraphy has been studied in both the Tethyan and Subboreal biogeographic realms. In the Tethyan realm, Strasser (2007) proposed a 3.2 Myr duration for the early Middle Oxfordian through earliest Kimmeridgian and ~3.2–3.3 Myr duration for the Kimmeridgian from the shallow carbonate platform of the Swiss and French Jura Mountains record based on the counting the small-scale sequences considering as ~100-kyr short-eccentricity cycles. Boulila et al. (2008b) constructed a ~1.6 Myr ATS for the Early Kimmeridgian from the ~43-m thick alternating marl–limestone succession of pelagic outcrop at La Méouge section of the Vocontian Basin (southeastern France) that shows high amplitude precession cycles and strongly 405-kyr long-eccentricity cycles.

In the Subboreal realm, Weedon et al. (2004) measured a 542-m succession for MS, PEF, total gamma-ray (GR) and total organic carbon (TOC) from a composite of three boreholes into the type Kimmeridge Clay Formation in Dorset (Southern England). Spectral analysis results showed long–wavelength cycles of 1.87–4.05 m; and they assumed these long cycles were ~38 kyr obliquity cycles to construct a 7.5 Myr-long ATS based on tuned the series using obliquity cycle. The estimate durations were 3.6 Myr (95 obliquity cycles) for the "early" Kimmeridgian Stage (from *Psiloceras baylei* to *A. autissiodorensis* Subboreal ammonite zones) in Subboreal usage that is equal to the international Kimmeridgian Stage as defined by Tethyan ammonites, and 3.9 Myr (103 cycles) for the "late" Kimmeridgian (from *Psiloceras elegans* to *V. fittoni* zones) in Subboreal usage that is equal to the early Tithonian Stage in international usage.

These initial outcrop-based Kimmeridgian Clay studies were enhanced by detailed borehole-based analyses. Huang et al. (2010a) analyzed the TOC series from the composite core section for the Kimmeridge Clay Formation and found hierarchy of cycles of ~1.6, ~3.8, ~9.1 and ~40 m wavelengths that corresponding to ~20 kyr precession, ~36 kyr obliquity, ~100 and 405-kyr short and long-eccentricity cycles respectively. They constructed a ~6.8 Myr ATS based on progressively tuning the TOC series to 405 kyr, ~100 and ~36 kyr cycles. However, we reanalyzed the TOC series and added two more possible 405 kyr cycles compared with the previous study to construct a 7.49 Myr-long ATS based on the 405 kyr cycle tuning (Fig. 8E, Table 4). The estimated duration for the "early" Kimmeridgian (except its lower *R. cymodoce* and *P. baylei* zones) and "late" Kimmeridgian are 3.64 Myr and 3.85 Myr, respectively, which is similar to the ~7.5 Myr ATS based only on 38 kyr obliquity tuning by Weedon et al. (2004). Assigning the base of the Subboreal *P. elegans* ammonite zone as the Kimmeridgian/Tithonian boundary using its assigned age of 152.1 Ma based on scaling Pacific marine magnetic anomalies (GTS2012; GTS2016), then the TOC ATS spans 155.74 and 148.25 Ma, respectively. This ~7.49 Myr-long Kimmeridge Clay ATS could be assigned as the 405-kyr chrono-cycles from E384.5 to E366.

An ATS for the combined Tithonian and Berriasian stages from cyclo-magnetostratigraphy studies in Argentina (Kietzmann et al., 2018 this volume) is briefly summarized in the next section on Early Cretaceous.

The Boreal Volgian Stage, which has a base that might be equivalent to the base of the Tethyan (international) Tithonian Stage, has been analyzed in boreholes from the Norwegian Sea. Huang et al. (2010b) analyzed the ~184-m Lightness-data series from the core IKU-6814/04-U-02 that had been drilled in the Greenland-Norwegian Seaway (Swientek, 2002). The facies of laminated organic rich shales with in meter-scale intercalations of carbonate beds had been deposited in relatively deep water (Swientek, 2002). The power spectrum of this high-resolution lightness series (65–191 m) reveals sedimentary cycles with wavelengths at ~4, ~1, ~0.34 and ~0.2 m. We construct a ~13.66 Myr-long ATS based on the calibrating ~4 m cycles to the 405-kyr long-eccentricity cycles and estimated a 6.08 Myr duration for the combined Early and Middle Volgian substage (135–191 m). In this study, we reanalyzed the Lightness series of the 8.39—191.29 m interval for the entire core record and recognized totally 63 cycles with ∼2- to 5-m wavelength. A ∼25.6 Myr-long high-resolution continuous ATS is based on tuning the ∼2—5 m cycles to 405-kyr long-eccentricity cycles (Fig. 8F, Table 4). The resulting ∼6.3 Myr duration for the combined Early and Middle Volgian substages (135—191 m) is close to the span of the Tithonian Stage if a lower Jurassic/Cretaceous boundary definition is used (Fig. 8F). This ∼6.3 Myr-long Greenland-Norwegian ATS could be assigned as the 405-kyr chrono-cycles from E376 to E360 (Fig. 8F).

However, for the Jurassic, one or more cyclostratigraphy studies is crucial and need to be developed to cover the Sinemurian of the Early Jurassic, the most of Middle Jurassic includes **Aalenian, Bajocian, Bathonian** and **Callovian stages**. For the Late Jurassic, a duplicate study sections need to correlated and integrated with biostratigraphy, chemostratigraphy, magnetostratigraphy and radioisotopic dating to verify of the main 405-kyr tuned ATS for these three stages (**Oxfordian, Kimmeridgian** and **Tithonian stages**).

5. THE CURRENT STATUS OF THE CRETACEOUS ATS

5.1 The Early Cretaceous

The Early Cretaceous includes **Berriasian, Valanginian, Hauterivian, Barremian, Aptian** and **Albian stages** and spanned about 45 Myr from ∼145 Ma at Jurassic/Cretaceous boundary to ∼100 Ma at Early/Late Cretaceous boundary. There are about 16 cyclostratigraphic studies published for the Early Cretaceous over the past twenty years (Amodio et al., 2013; Charbonnier et al., 2013; Fiet, 2000; Fiet et al., 2001; Fiet and Gorin, 2000; Gale et al., 2011; Giraud et al., 1995; Grippo et al., 2004; Huang et al., 1993; Huang et al., 2010b,c; Liu et al., 2017; Martinez et al., 2012; 2013; 2015; Sprenger and ten Kate, 1993; Sprovieri et al., 2006) (Fig. 9).

The Berriasian Stage currently has a partial ATS that is mainly derived from magnetostratigraphy sections in Argentina (Kietzmann et al., 2015, and in press this volume). Sprenger and ten Kate (1993) analyzed the 77.5 m long CaCO3 content series collected from the rhythmic hemipelagic limestone-marl sequence in the Caravaca region of southeast Spain and recognized ∼30, ∼6.67 and ∼1 m cycles. They assigned the ∼6.67 m cycles as ∼100-kyr short-eccentricity cycle to tune this series,

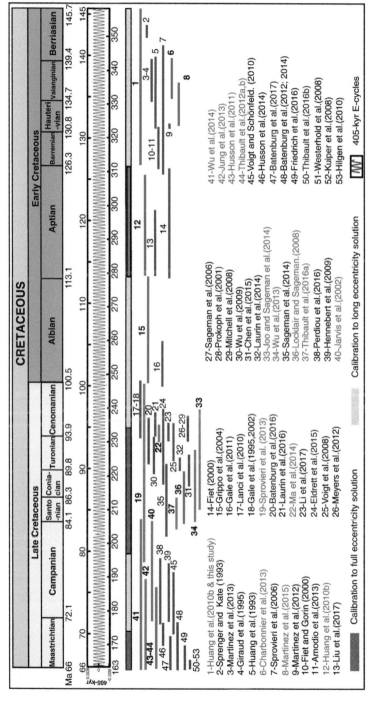

Figure 9 Stratigraphic coverage of the Cretaceous ATS and 405-kyr metronome. Intervals with *red lines* (gray in print version) indicate the collected data coverage subjected to a 405 kyr tuning to compile the ATS in this study, and those with *blue lines* (dark gray in print version) indicate additional supporting cyclostratigraphy studies.

and estimated a duration of 1.1—1.2 Myr for the *Calpionellopsis simplex* (D1) and *Calpionellopsis oblonga* (D2) subzones of upper Berriasian. A nearly complete Berriasian ATS was derived from alterations of marl and lime-stone in multiple sections of the Vaca Muerta Formation in the Neuquén Basin of Argentina (Kietzmann et al., 2015, and in press this volume). Even though the ammonite zonation of the Neuquén Basin is mainly endemic, the resolution of the complete suite of cycle-scaled magnetic polarity Chron M22r through Chron M15r enable a detailed correlation to the magnetic polarity time scale that has been calibrated to most of the Tethyan ammonite zones. The composite suite of sections spans the lowest Tithonian to nearly the top of the Berriasian. The bundling of the ca. 20- to 40-cm basic cycles is interpreted as long- and short-eccentricity modulation of precession cycles. They conclude that the Tithonian spanned a minimum of 5.67 Myr, and the Berriasian spans at least 5.27 Myr (Kietzmann et al., 2018 this volume).

The Valanginian Stage has been extensively analyzed in limestone-dominant to marl-dominant alternations in the Vocontian Basin of Southeastern France. Giraud et al. (1995) recognized 91 precession cycles and 137 obliquity cycles based on spectral analysis of the CaCO3 content series, and proposed a duration of 7.04 Myr for the Valanginian Stage. However, Martinez et al. (2013) provides a shorter duration of 5.08 Myr for the Valanginian Stage based on the 405-kyr long-eccentricity cycle tuning the Gamma-Ray Spectrometry (GRS) series from the Vergol-Morénas, La Charce, Angles and Reynier sections with intercalibration be-tween ammonite and calcareous nannofossil biozones and carbon-isotope stratigraphy in these well-defined biostratigraphy sections. Charbonnier et al. (2013) analyzed the 250-m thick high-resolution magnetic suscepti-bility data series from the hemipelagic marl-limestone alternations at the Orpierre section spanning from the Upper Berriasian to Valanginian. The prominent ∼21, ∼6.46, ∼1.1—1.44 and ∼0.71—1.02 m cycles could be correspond to 405-kyr, ∼100 kyr eccentricity cycles, ∼34 kyr obliquity and ∼20 kyr precession cycles. A ∼5.08 Myr-long high-resolution ATS was been constructed based on tuning the ∼21 m cycles to the 405-kyr long-eccentricity cycles, and they proposed a 4.695 Myr duration for the Valanginian stage, which was used in GTS2016 (Ogg et al., 2016) rather than 5.5 Myr duration assigned in GTS2012 (Ogg et al., 2012b).

The Hauterivian and Barremian stages in the Vocontian Basin has similar alternations of limestone and marl as in the underlying Valanginian.

Martinez et al. (2015) analyzed the GRS series of the La Charce—Pommerol (240 m) sections spanning Upper Valanginian to Hauterivian, plus compared to the Río Argos (160 m) sections in southern Spain They derived a 5.93 ± 0.41 Myr duration for the Hauterivian stage based on the identification of more than fourteen $\sim 11-28$ m wavelengths cycles that were assumed to correspond to the 405 kyr eccentricity. We combined the three sections of Reynier, La Charce—Pommerol and Río Argos as a composite 442.93 m long GRS series from the Late Berriasian to Early Barremian stages. This enabled a ~ 12.56 Myr-long ATS based on the recognition thirty-one ~ 15 m cycles tuned to the 405-kyr long-eccentricity cycles (Fig. 10A, Table 5). The estimate durations for Valanginian and Hauterivian stages are ~ 5.3 and 5.93 Myr, respectively. When we anchored the 12.3 m level at the Reynier section to a 139.4 Ma age for the Berriasian/Valanginian boundary, then the top of this ATS series reaches 127.4 Ma. In this case, the ~ 12.56 Myr-long ATS could be assigned as the 405-kyr chrono-cycles from E345 to E314 (Fig. 10A).

The Valanginian through Barremian cyclic sediments are also characteristic of the deep-sea successions in the Central Atlantic and uplifted Tethyan sections in Italy. Huang et al. (1993) estimated the minimum durations of 5.9 and 5.3 Myr for the Valanginian and Hauterivian stages based on comparing the cyclic sequences from the Central Atlantic and the Vocontian Basin (Southeast France). Sprovieri et al. (2006) analyzed the composite high-resolution pelagic bulk carbonate stable isotope record in three well-exposed sequences sections (Chiaserna Monte Acuto, Bosso, and Gorgo a Cerbara) in central Italy. They constructed a ~ 19 Myr-long ATS and estimated durations of ~ 6.9, ~ 3.5 and ~ 4.4 Myr for the Valanginian, Hauterivian and Barremian stages, respectively, and proposed that the base of a positive carbon isotope excursion was at ~ 136.34 Ma. However, their $\delta^{13}C$ data series is not yet available for independent analysis.

As previously mentioned, we had constructed a ~ 19.3 Myr-long ATS of the Lightness series of the 8.39—135 m interval from the Late Volgian to Barremian from the IKU-6814/04-U-02 drilled core in the Greenland-Norwegian Seaway (Fig. 10B, Table 5). Based on the Greenland-Norwegian Seaway biostratigraphy, the estimated duration is ~ 6.75 Myr for Berriasian Stage (70.7—135 m), a ~ 7.29 Myr duration for combined Valanginian and Hauterivian stages (35.8—70.7 m), and ~ 5.27 Myr duration for the Barremian Stage (8—35.8 m) (Fig. 10B). A duration of 5.13 ± 0.34 Myr for the Barremian Stage was also estimated by Fiet and Gorin (2000) from their analysis of a cyclic carbonate-dominated

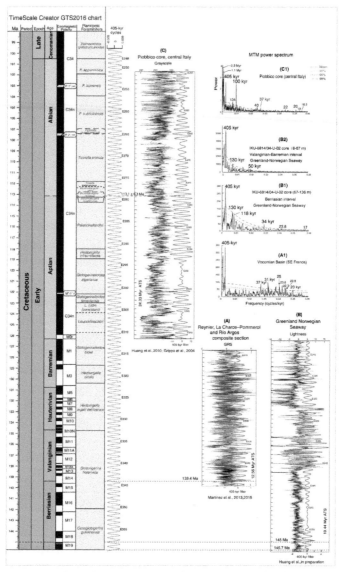

Figure 10 Our restudy and 405-kyr tuning of selected series for the Early Cretaceous ATS. Chronostratigraphy columns on the left were generated using TimeScale Creator 7.0 (*engineering.purdue.edu/Stratigraphy/tscreator*) with the GTS2016 integrated stratigraphy (Ogg et al., 2016). (A) GRS residuals series (after subtracting a 15% weighted average) from the Vocontian Basin (SE France) (the original data from Martinez et al. (2013, 2015), using the 405 kyr period to tune the 442.93 m composite section from the Reynier, La Charce–Pommerol and Río Argos three cores). (B) Lightness residuals series (after subtracting a 15% weighted average) of the IKU-6814/04-U-02 core in

pelagic sequence in the Umbria–Marche Basin (central Italy), and a similar 5.2 Myr minimum duration was estimated by Amodio et al. (2013) from the shallow–marine carbonates of the two drilled cores in the central Apennines (central Italy) and one outcrop section at Monte Faito in the southern Apennines (southern Italy). Although we could not define the exactly position for the Valanginian/Hauterivian boundary in our data set, if we assume an age of 145.8 or 145 Ma for the Jurassic/Cretaceous boundary, then the top of Barremian stage age is at about 126.5 Ma based on the ∼19.3 Myr ATS. Therefore, there is significant agreement on the combined duration for the Berriasian through Barremian stages of the Early Cretaceous, even though the actual stages have not yet been officially defined by GSSPs. This total Berriasian-Barremian cycle-scaled duration was implicitly incorporated into GTS2016 to assign approximate ages for the common Tethyan stage boundary usage relative to an age of 126.3 ± 0.4 Ma for the Barremian/Aptian boundary (Ogg et al., 2012b). This ∼19.3 Myr-long Greenland-Norwegian ATS and the corresponding Berriasian-Barremian stae interval could be assigned as the 405-kyr chrono-cycles from E360 to E312 (Fig. 10B).

The Aptian and Albian stages have been studied for cyclostratigraphy in Italian sections and boreholes. The most important, but also controversial, cyclostratigraphy study was the 12.9 Myr-long ATS interpreted for the Aptian stage in the Piobbico core that was drilled in the Fucoid Marls at Piobbico (central Italy) by Huang et al. (2010), which used as the reference scale for Aptian events in GTS2012 and GTS2016 (Ogg et al., 2012b; 2016). This ATS was based on the 405-kyr long-eccentricity cycle tuning of the high-resolution grayscale series of the 33.7 m of Aptian in that core that was well-dated by calcareous nannofossils, planktonic foraminifers and organic-rich episodes. The Piobbico core did not recover the basal-Aptian interval of magnetic chron M0r, therefore the Aptian ATS was extended by the ∼0.5 Myr duration for the M0r zone based on the correlation with the Cismon core to yield an the entire Aptian stage duration of about 13.4 Myr. Even though this 13.4 Myr-long Piobbico core ATS for the

──────────◄───────────────────────────────────

the Greenland-Norwegian Seaway (the original data download from http://doi. pangaea.de/10.1594/PANGAEA.141090 posted by Swientek (2004)). (C) Grayscale residuals series (after subtracting a 15% weighted average) of Piobbico core (central Italy) (the original data from Huang et al. (2010) and Grippo et al. (2004), for the entire 77-m-long Piobbico core grayscale series). All the 405 kyr filter output (red curve, passband: 0.00247 ± 0.0006 cycles/kyr). A1-C1 are 2π MTM power spectrum of A-C series.

Table 5 Reanalysis of Selected Early Cretaceous Reference Sections Using Standardized Tuning to 405-kyr Long-Eccentricity Cycle and MTM Power-Spectra Analysis. Span (Myr) and Age Limits (Ma) are Based on the Interpreted Tuning Results

Location	Geologic-age	Data source	Proxy	Span (Myr)	Significant spectral cycles (kyr)	Assigned E-cycles	Age limits (Ma)
Reynier, La Charce– Pommerol and Río Argos composite core (Southern Spain)	Uppermost Berriasian–upper Barremian	Martinez et al. (2013), (2015)	GRS	12.56	E: **405 (*tuned*)** O: **37, 31** P: **25, 23.8, 22.7,** 22.5, **20**	E345-E314	139.88–127.32
Greenland Norwegian Seaway	Uppermost Tithonian– upper Barremian	Huang et al. (preparation)	Lightness	19.44	**405** e: **130**, 118 O: **34** P: 23.8, 17	E360-E312	145.94–126.5
Piobbico core (central Italy)	Basal Aptian– uppermost Albian	Huang et al. (2010), Grippo et al. (2004)	Grayscale	24.5	**405** e: 134,**100** O: **40, 37** P: **22**, 20, 19, 18.5	E311.5-E251	126.3–101.8

Aptian stage is much longer than the 6.4 ± 0.2 Myr duration estimated by Fiet (2000) from the nearby outcrop or the 6.8 ± 0.4 Myr duration estimated by Fiet et al. (2006) from the cyclostratigraphic study in the Vocontian basin (SE of France), it is consistent with 13.8 Myr duration estimated by Al-Husseini and Matthews (2010) based on the Arabian sequence stratigraphy scaled to their 405 kyr "straton" chronology.

For the Albian Stage, Grippo et al. (2004) constructed a 12.4 Myr-long ATS (including three 405 kyr missing in the top of the core and therefore spliced from an outcrop study) based on the 405-kyr long-eccentricity cycle tuning of the 43 m high-resolution grayscale series of the same Piobbico core. This Albian Piobbico ATS is longer than the duration of 10.64 Myr estimated from the Col de Palluel section of Rosans in Hautes-Alpes (France) by Gale et al. (2011).

We combined these two cyclostratigraphic studies for the entire 77-m-long Piobbico core plus the Chron M0r part of the Cismon core to derive a 25.8 Myr-long ATS spanning the Aptian and Albian stages (Fig. 10C, Table 5). The Aptian/Albian boundary interval has a U–Pb date of 113.1 ± 0.3 Ma on a basal Albian volcanic ash layer in the northwest Germany (Selby et al., 2009), and the corresponding biostratigraphic boundary is known in the Piobbico core. Therefore, the calculated base of the Aptian age of 126.5 Ma in this restudy does not alter the 126.3 ± 0.4 Ma in GTS2012 and GTS2016, and calculated the Albian/Cenomanian boundary age of about 100.485 Ma is identical to the radiometric dating of 100.5 ± 0.4 Ma used GTS2012 and GTS2016. The entire 24.4 Myr-long Piobbico core ATS reference section could be assigned as the 405-kyr chrono-cycles from E311.5 to E251 (Fig. 10C).

It should be noted that the age model used for the Late Jurassic through Early Cretaceous in GTS2016 is undergoing revision; partly from the apparent discrepancies between its reliance on published radiometric dates on oceanic basalts drilled by the Ocean Drilling Program (ODP) and some significantly younger new U–Pb dates from volcanic ash beds in the Neuquén Basin of Argentina (e.g., Vennari et al., 2014; Aguirre-Urreta et al., 2015) and elsewhere (e.g., Midtkandal et al., 2016). For example, an ID-TIMS U–Pb date of 139.55 ± 0.18 Ma occurring in the middle of the regional *Argentiniceras noduliferum* ammonite zone (Vennari et al., 2014) has been correlated to lowermost Chron M16r (Kietzmann et al., 2018, this volume), at a level that had been assigned an approximate age of 142 Ma in GTS2016. Therefore, even though the cycle-derived durations of most of the Oxfordian through Barremian stages as summarized in this current study are similar to the stage-durations estimated in GTS2016, then either the GTS2016 scale must be shifted nearly 2 Myr younger, which would imply either lengthening one or more of the Middle Jurassic stages which have conflicting cycle-derived durations while shortening the Aptian Stage or with some other method of compensation. At this point, there are not enough reliable and verified cyclostratigraphic studies on Middle Jurassic or on Aptian sections a lack of independent radioisotopic ages to resolve these inconsistencies among published interpretations and dating.

5.2 The Late Cretaceous

The Late Cretaceous includes **Cenomanian, Turonian, Coniacian, Santonian, Campanian** and **Maastrichtian stages** and spans about 34 Myr from ~100 Ma at Early/Late Cretaceous boundary to ~66 Ma at Cretaceous/Paleogene boundary. There are nearly 40 cyclostratigraphic studies published for the Late Cretaceous over the past twenty years (Batenburg et al., 2012; 2014; 2016; 2017; Chen et al., 2015; Eldrett et al., 2015; Friedrich et al., 2016; Gale, 1995; Gale et al., 2002; Hennebert et al., 2009; Hilgen, 2010; Husson et al., 2011; 2014; Jung et al., 2013; Kuiper et al., 2008; Lanci et al., 2010; Laurin et al., 2014, 2016; Li et al., 2017; Locklair and Sageman, 2008; Ma et al., 2014; Meyers et al., 2012; Mitchell et al., 2008; Perdiou et al., 2016; Prokoph et al., 2001; Sageman et al., 2006, 2014; Sprovieri et al., 2013; Thibault et al., 2012a,b, 2016a,b; Voigt et al., 2008; Voigt and Schönfeld., 2010; Wu et al., 2009, 2013; 2014; Westerhold et al., 2008) (Fig. 9).

The important Late Cretaceous reference section spanning ~200-m in the Bottaccione Gorge region in the Umbria-Marche Basin (Gubbio, central Italy) was used by Sprovieri et al. (2013) to construct a ~23 Myr-long high-resolution $\delta^{13}C_{carb}$ (~20 cm sample interval) series. The lower part of section is recorded pelagic sequence with the light gray to light green limestones interbedded bands of black chert in the Scaglia Bianca Formation from the late Albian to early Turonian age. There is ~1 m-thick of black laminated shales interbedded with gray radiolarian sands known as the Bonarelli Level (considered as OAE2) in the late Cenomanian at the uppermost of the Scaglia Bianca Formation. The Scaglia Rossa Formation in the upper part of the section consists of pink clay-rich limestones and cherts variations. Sprovieri et al. (2013) derived durations of 5.40, 4. Myr, 3.46 and 2.94 Myr for the Cenomanian, Turonian, Coniacian and Santonian stage, respectively. We re-analyzed their data and recognized 55 cycles of ~3.44–4.88 m wavelengths to construct a 22.275 Myr-long ATS based on tuning these $\sim$4 m cycles to the stable 405-kyr long-eccentricity cycles (Fig. 11A, Table 6). Anchoring this to the radioisotopic and astrochronologic intercalibration date of 93.9 $\pm$ 0.15 Ma for the Cenomanian/Turonian boundary (Bonarelli Level) (Meyers et al., 2012), then this 22.275 Myr-long Bottaccione $\delta^{13}C_{carb}$ ATS spans 101.2 to 78.94 Ma and could be assigned as the 405-kyr chrono-cycles from E249 to E195 (Fig. 11A).

This cycle-tuning of the Tethyan-based carbon-isotope record can be verified and extended upward using the $\delta^{13}C_{carb}$ series data from Jarvis et al. (2002), who presented a 467.3-m thick $\delta^{13}C_{carb}$ series from the Tethyan pelagic-hemipelagic section near El Kef of the northern Tunisia spanning Campanian to basal Maastrichtian. We re-analyzed the available $\delta^{13}C_{carb}$ from the Table 1 in the Jarvis et al. (2002) paper. The power spectral of this $\delta^{13}C_{carb}$ series display significant peaks at $\sim$45, $\sim$13 and $\sim$5 m wavelengths. We constructed a 6.952 Myr-long ATS based on tuning the $\sim$45 m cycles to the 405-kyr long-eccentricity (Fig. 11G, Table 6). When we anchor the Campanian/Maastrichtian boundary at 476 m to its estimated age of 72.1 Ma, this 6.95 Myr-long ATS spans 78.88 to 71.94 Ma and could be assigned as the 405-kyr chrono-cycles from E195 to E178 (Fig. 11G). Combined with the 22.275 Myr-long $\delta^{13}C$ ATS constructed from the Umbria-Marche Basin (Gubbio, central Italy), a 29 Myr-long $\delta^{13}C_{carb}$ ATS could be compiled spanning most of the Cenomanian through Maastrichtian stages that could be assigned as the 405-kyr chrono-cycles from E250 to E178.

The coeval succession of Upper Cretaceous chalk-marl alternations in the Western Interior Seaway of North America are the main reference sections for intercalibration of radiometric ages to biostratigraphic zones and carbon-isotope excursions. Joo and Sageman (2014) compiled a composite carbon isotope record for the Cenomanian-Campanian from well-defined biostratigraphic cores (Aristocrat Angus (AA), Portland (PO) and CL-1). A $\sim$14 Myr-long $\delta^{13}C_{org}$ time scale has been constructed based on the Meyers et al. (2012) and Sageman et al. (2014) intercalibrated astrochronologic and radioisotopic ($^{40}Ar/^{39}Ar$ and U-Pb) ages of 93.90 $\pm$ 0.15 Ma, 89.75 $\pm$ 0.38 Ma, 86.49 $\pm$ 0.44 Ma and 84.19 $\pm$ 0.38 Ma for the Cenomanian/Turonian, Turonian/Coniacian, Coniacian/Santonian and Santonian/Campanian boundaries, respectively. Spectral analysis of this $\delta^{13}C_{org}$ time series shows strong peaks at 120, $\sim$200, 360, 500, $\sim$900 and $\sim$1500 kyr. The 360 and 500 kyr peaks should be 405 kyr period in the astronomical theory. We used the composite section (2140.8–2417.6 m) from the AA,

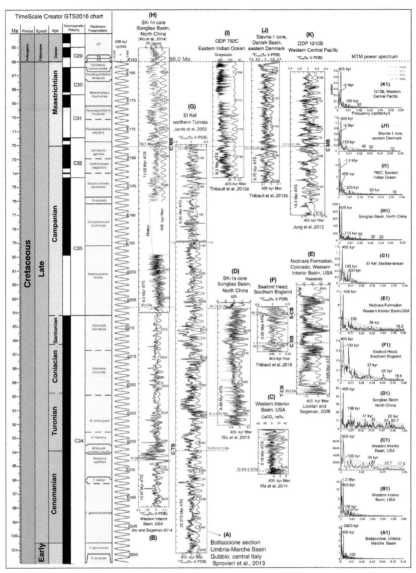

Figure 11 Our restudy and 405-kyr tuning of selected series for the Late Cretaceous ATS. Chronostratigraphy columns on the left were generated using TimeScale Creator 7.0 (*engineering.purdue.edu/Stratigraphy/tscreator*) with the GTS2016 integrated stratigraphy (Ogg et al., 2016). (A) $\delta^{13}C_{carb}$ residuals series (after subtracting a 15% weighted average) of the Bottaccione section in the Umbria-Marche Basin (the original data from Sprovieri et al., 2013). (B) $\delta^{13}C_{org}$ residuals series (after subtracting a 6% weighted average) of the Western Interior Basin (USA) (the original data from Joo and Sageman, 2014). (C) $CaCO_2$ (wt%) residuals series (after subtracting a 10% weighted average) of the Western Interior Basin (USA) (the original data from Ma et al., 2014).

PO and CL-1 cores and recognized 36 cycles with ~7 m thickness that is compares well with the 33 cycles from the $\delta^{13}C_{org}$ time series of 83.8—97.235 Ma obtained by band-passing for 325—524 kyr cycles (filter frequency from 0.00187 to 0.00307). Tuning these 36 cycles of ~7 m as 405-kyr long-eccentricity cycles, then a 14.58 Myr-long ATS has been constructed (Fig. 11B, Table 6). Anchoring with a 94.58 Ma age for the base of the OAE2 (Joo and Sageman, 2014), then this 14.58 Myr-long ATS spans 97.97 to 83.39 Ma. This 13.97 Myr-long $\delta^{13}C_{org}$ ATS from 97.97 to 84.0 Ma could be assigned as the 405-kyr chrono-cycles from E241 to E207 (Fig. 11 B).

The Cenomanian/Turonian boundary interval in this Western Interior Seaway region has been studied at very high resolution by Ma et al. (2014). They obtained high-resolution (5 mm) X-ray fluorescence (XRF) core scanning data including the weight percent $CaCO_3$ series and a $^{40}Ar/^{39}Ar$ age of 95.39 ± 0.18 Ma from bentonites in the *Dunveganoceras pondi* ammonite biozone in the uppermost Lincoln Limestone Member of the Aristocrat Angus core. When constrained by two $^{40}Ar/^{39}Ar$ ages of 93.79 ± 0.26 and 94.20 ± 0.28 Ma for the *Watinoceras devonense* and *Sciponoceras gracile* ammonite biozones (Meyers et al., 2012), the cyclostratigraphy indicates a strong precessional control on the deposition of the organic-carbon—rich strata. We re-tuned the $CaCO_3$ series using the 405-kyr long-eccentricity cycles to construct a ~3.16 Myr ATS. This was anchored to 95.39 Ma at the ~2304 m level in the Angus core

---

(D) GR residuals series (after subtracting a 30% weighted average) of the SK-I south borehole in the Songliao Basin (NE China) (the original data from Wu et al., 2013). (E) Resistivity residuals series (after subtracting a 10% weighted average) of the two wells from the Western Interior basin (Colorado, USA) (the original data from Locklair and Sageman, 2008). (F) $\delta^{13}C_{carb}$ residuals series (after subtracting a 10% weighted average) of the Seaford Head (southern England) and Bottaccione (central Italy) (the original data from Thibault et al., 2016a). (G) $\delta^{13}C_{carb}$ residuals series (after subtracting a 15% weighted average) near El Kef (Mediterranean, northern Tunisia) (the original data from Jarvis et al., 2002). (H) Th series of SK-I south borehole in the Songliao Basin (NE China) (the original data from Wu et al., 2014). (I) Grayscale residuals series (after subtracting a 10% weighted average) of the ODP 762C borehole (the original data from Thibault et al., 2012a). (J) $\delta^{13}C_{carb}$ residuals series (after subtracting a 10% weighted average) of the Stevns-1 core of the Danish Basin (eastern Denmark) (the original data from Thibault et al., 2012b). (K) $\delta^{13}C_{pl}$ residuals series (after subtracting a 10% weighted average) of the ODP 1210B borehole (the original data from Jung et al., 2012). All the 405 kyr filter output (red curve, passband: 0.00247 ± 0.0003—0.0006 cycles/kyr). A1-K1 are 2π MTM power spectrum of A-K series.

Table 6 Reanalysis of Selected Late Cretaceous Reference Sections Using Standardized Tuning to 405-kyr Long-Eccentricity Cycle and MTM Power-Spectra Analysis. Span (Myr) and Age Limits (Ma) are Based on the Interpreted Tuning Results

Location	Geologic-age	Data source	Proxy	Span (Myr)	Significant spectral cycles (kyr)	Assigned E-cycles	Age limits (Ma)
Bottaccione section, Umbria-Marche Basin (central Italy)	Uppermost Albian-lower Campanian	Sprovieri et al. (2013)	$\delta^{13}C_{carb}$ (‰)	22.275	**E: 405 (*tuned*)** **e: 100**	E249-E194.4	101.2-78.94
El Kef (northern Tunisia)	Middle Campanian-basal Maastrichtian	Jarvis et al. (2002)	$\delta^{13}C_{carb}$ (‰)	6.95	**405** e: 134, **100** O: **40, 37** P: **22**, 20, 19, 18.5	E195-E178	78.88-71.94
Western Interior Basin (USA)	Lower Cenomanian-basal Campanian	Joo and Sageman (2014)	$\delta^{13}C_{org}$ (‰)	13.97	**405** e: **138, 102**	E241-E207	97.97-84
Angus core, Western Interior Basin (USA)	Upper Cenomanian-lower Turonian	Ma et al. (2014)	$CaCO_3$ (wt%)	3.16	**405** e: **100** O: **38**, 33 P: 22.7, 17.9	E236-E228	95.6-92.4
SK-1s core, Songliao Basin (north China)	Middle Turonian-basal Campanian	Wu et al. (2013)	GR	8.68	**405** e: 106 O: **41**, 33 P: 23, **22**, 20.7	E227.5-E206	92.08-83.4
Colorado, Western Interior Basin (USA)	Uppermost Turonian-lower Campanian	Locklair and Sageman (2008)	Resistivity	7.695	**405** e: **100** O: **35** P: 18.2	E222-E203	89.9-82.2

Site	Age	Reference	Proxy	Value	Cycles	Elements	Age range
Seaford Head (southern England)	Uppermost Coniacian-lowermost Campanian	Thibault et al. (2016a)	$\delta^{13}C_{carb}$ (‰)	3.2	**405** e: **133** O: **37** P: **25**, 18.6	E214-E207	86.8–83.6
SK-1n core, Songliao Basin (north China)	Basal Campanian-Lower Paleocene	Wu et al. (2014)	Th (ppm)	14 (4 Myr gap)	**405** e: **110** O: **50**, 30 P: 25	E207-E161	83.9–65
ODP hole 762C (Eastern Indian Ocean)	Upper Campanian-Basal Paleocene	Thibault et al. (2012a)	grayscale	8.3	**405** e: **100** O: **33** P: 19	E184-E164	74.3–66
Stevns-1 core (eastern Denmark)	Upper Campanian-Basal Paleocene	Thibault et al. (2012b)	$\delta^{13}C_{carb}$ (‰)	8.75	**405** e: **133** O: 40, 32 P: 19	E185-E163	74.68–65.93
ODP hole 1210B (Western Central Pacific)	Middle Campanian - uppermost Maastrichtian	Jung et al. (2013)	$\delta^{13}C_{pl}$ (‰)	10.35	**405** e: **100** O: **50**	E190-E165	77.01–66.66

(Fig. 11 C, Table 6). This ~3.16 Myr Angus core ATS could be assigned as the 405-kyr chrono-cycles from E236 to E228 (Fig. 11 C).

The Coniacian and Santonian stages have also been analyzed in this Western Interior Seaway region with its well-defined biostratigraphy. Locklair and Sageman (2008) presented a ~85-m thick high-resolution borehole resistivity data series from two wells that record a decimeter-scale rhythmic alternations of chalk and marl beds and decameter-scale oscillations of chalky and marly succession in the Niobrara Formation of Colorado. They constructed a ~6.1—6.7 Myr ATS based on tuning ~1—2 m cycles in the resistivity series to the 95-kyr short-eccentricity and estimated durations of 3.4 ± 0.13 and 2.39 ± 0.15 Myr for the Coniacian and Santonian stages, respectively. However, the power spectral of the 95 kyr tuned series displays an anomalous suite of significant peaks at 696, 370, 236, 140, 120, 95, 71, 44, 38, 31, 21.3, 19.6 and 18.3 kyr periods that is lacking the stable 405 kyr long-eccentricity cycle peak. Therefore, we used the 19 cycles of ~3—5 m wavelength that were recognized from the bandpassed ~85-m resistivity series of Locklair and Sageman (2008) to construct a 7.695 Myr-long ATS based on tuning these ~3—5 m cycles to the 405-kyr long-eccentricity cycles (Fig. 11E, Table 6). This results in estimated durations of 3.66 and 2.6 Myr for Coniacian and Santonian stages, respectively, which is similar to the assigned durations of 3.5 ± 0.4 and 2.6 ± 0.5 Myr for Coniacian and Santonian stages in the GTS2012 that were based on interpolations from radiometric dating (Siewert, 2011; Ogg et al., 2012b). These durations are also consistent with the estimates of 3.46 and 2.94 Myr for the Coniacian and Santonian by Sprovieri et al. (2013) and the calculated durations of 3.26 ± 0.82 and 2.30 ± 0.82 Myr for Coniacian and Santonian stages based on the intercalibration of astrochronologic and $^{40}Ar/^{39}Ar$ and U-Pb ages of 89.75 ± 0.38 Ma for the Turonian/Coniacian, 86.49 ± 0.44 Ma for the Coniacian/Santonian, and 84.19 ± 0.38 Ma for the Santonian/Campanian boundary. When we anchor the 89.8 Ma for the Turonian/Coniacian boundary age, this ~7.695 Myr-long resistivity ATS could be assigned as the 405-kyr chrono-cycles from E222 to E203 (Fig. 11E).

This late Coniacian to early Campanian time interval was also studied by Thibault et al. (2016a) in marine sections at Seaford Head (southern England) and Bottaccione (central Italy). They compiled a ~3.2 Myr ATS based on tuning the high-resolution bulk carbonate carbon isotopes series to the 405-kyr long-eccentricity cycle (Fig. 11F, Table 6). There are five 405 kyr cycles for the Tethyan Santonian based on the ~16 m filtered output of the $\delta^{13}C_{carb}$ and the global correlation between the Niobrara Formation of the Western Interior Basin (USA) and English Chalk of

Seaford Head of England and La2010 solution (Laskar et al., 2011). They indicated that the 405 kyr filtered output minima of the resistivity series in the Niobrara Formation correlate to the solution 405-kyr long-eccentricity minima and the $\delta^{13}C_{carb}$ filtered output maxima in the Seaford Head (Fig. 11E and F). We accepted their age model option 2 and anchoring to the 84.2 Ma age for the Santonian/Campanian boundary, which in turn was based on the 405-kyr long-eccentricity solution curve minima and the cycle-calibrated isotopic-dating at 84.19 ± 0.38 Ma (Sageman et al., 2014). This results in the Coniacian/Santonian boundary is placed at 86.17 Ma. This ~ 3.2 Myr Seaford Head $\delta^{13}C_{carb}$ ATS could be assigned as the 405-kyr chrono-cycles from E214 to E207 (Fig. 11F).

The recent drilling of Upper Cretaceous in the Songliao Basin of northeast China has yielded expanded records in a lacustrine succession. Wu et al. (2013) obtained a ~ 950-m (from 960 to 1910 m depth) thick GR well log from the SK-Is (south) borehole from the dark mudstones intercalated with thin carbonate layers and black shales of the Nenjiang Formation (960 to 1128.17 m depth), brownish-greenish/grayish mudstone and greenish muddy siltstone of the Yaojia Formation (1128.17 to 1285.91 m depth), gray-dark gray-black mudstone inter-bedded with marlstone and shale in the lower and thin bedded gray silt-stone of the upper part of the Qingshankou Formation (1285.91 to 1782.93 m depth), and grayish green or brown mudstone, siltstone and sandstone in the Quantou Formation (1782.93 to 1910.0 m depth). An 8.68 Myr-long 'absolute' ATS was constructed for the early Turonian—early Campanian (92.08 to 83.4 Ma) based on tuning $\sim 36-65$ m wavelengths to the 405-kyr long-eccentricity cycles (Fig. 11D, Table 6). Combined with the four SIMS U—Pb zircon radioisotope ages of by 83.7 ± 0.5 Ma at 1019 m, 90.4 ± 0.4 Ma at 1673 m, 90.1 ± 0.6 Ma at 1705 m, and 91.4 ± 0.5 Ma at 1780 m (He et al., 2012), and the magnetic reversal C33r/C34n boundary at 985.95 m assigned as 83.64 Ma by Ogg et al. (2012). This 8.68 Myr-long SK-Is core GR ATS could be assigned as the 405-kyr chrono-cycles from E227.5 to E206 (Fig. 11D).

In a continuation and enhancement of their previous study, Wu et al. (2014) presented a 1541.66 m Th (thorium) series from the SK1n core (north Songke-1 borehole) of the Songliao Basin. The lithology consists of purple—red-gray mudstone and muddy siltstone in the lower Mingshui Formation; greenish gray-black and purple—red mudstone to greyish green muddy siltstone and sandstone in the upper Mingshui Formation; purple—red and black to gray mudstone and clayey siltstone and sandstone in the Sifangtai Formation; and gray to black mudstone to marl and silty mudstone in the lower part, and black mudstone to greyish black sandy mudstone and gray siltstone interbedded with sandstone in the upper part of the Nenjiang Formation. The significant peaks of power spectra show the cycle wavelengths ratios of $\sim$20:5:2:1 in Th series that approximately corresponds to the Milankovitch cycles of 405 and $\sim$100 kyr eccentricity, 38 kyr obliquity and $\sim$20 kyr precession cycles, respectively. A $\sim$14 Myr-long ATS has been constructed from 83.9 to 65 Ma based on the Th series maxima of 405 kyr sedimentary cycles filter output tuned to the maxima of the 405 kyr eccentricity solution curve of La2010d (Laskar et al., 2011) (Fig. 11H, Table 6). This was anchored to an initial magnetostratigraphy age control of the Chron C30n/C29r boundary at 66.3 Ma and the Cretaceous/Paleogene (K/Pg) boundary at 66.0 Ma in GTS2012 (Ogg et al., 2012b). The placement of the K/Pg, Campanian-Maastrichtian, Santonian-Campanian boundaries at depths of 318 m, 752.8 m and 1751.1 m, imply ATS ages of 66, 72.1 and 83.6 Ma, respectively. However, a ca. 4-Myr gap at the top of Nenjiang Formation separates this Th ATS series into two parts of 65–76.077 and 79.9–83.917 Ma. In this case, the SK1n ATS could be assigned as a pair of 405-kyr chrono-cycle intevals from E207 to E197 and E188 to E161 (Fig. 11 H).

The Upper Campanian through Maastrichtian interval was also studied by Husson et al. (2011) in a suite of ODP sites. They constructed a $\sim$8 Myr-long ATS based on the recognition of the 405-kyr long-eccentricity variation from the high-resolution MS in the Hole 1258A of ODP Leg 207 (Equatorial Atlantic), MS in 1267B of ODP Leg 208 (South Atlantic), grayscale series in the Hole 762C of ODP Leg 122 (Indian Ocean) and in the Hole 525A of DSDP Leg 74 (South Atlantic). By integrating with the magnetostratigraphy and the astronomical eccentricity solution of La2010a, they estimated the cycle-scaled durations of each chron from C32n1n to C29r and proposed an absolute ATS age for the K/Pg boundary as 65.59 ± 0.07 Ma or 66.0 ± 0.07 Ma and for the Campanian/Maastrichtian boundary as 72.34 and 72.75 Ma. The 66.0 Ma age for K/Pg boundary was also the age model in GTS2012 (Ogg et al., 2012b). Thibault et al. (2012a) identified twenty-one $\sim$2–7 m cycles in ODP Hole 762C grayscale series. By assigning these as 405-kyr long-eccentricity cycles, they proposed the age for the Campanian/Maastrichtian boundary as 72.15 ± 0.05 Ma and estimated a duration of 6.15 ± 0.05 Myr for the Maastrichtian Stage. We re-tuned this dataset and similarly recognized twenty-one $\sim$2–7 m cycles that we assigned as 405-kyr long-eccentricity cycles. The a $\sim$8.3 Myr high-resolution grayscale ATS of ODP Hole 762C spans the Late Campanian at 74.3 Ma to the K/Pg boundary of 66 Ma (Fig. 11I, Table 6). This $\sim$8.3 Myr grayscale ATS could be assigned as the 405-kyr chrono-cycles from E184 to E164 (Fig. 11I).

This Upper Campanian through Maastrichtian cyclostratigraphy was further verified by Thibault et al. (2012b), who provided a 456-m thick and 1968 samples to construct the high-resolution carbon isotope series of this interval from the Stevns-1 core drilled in the Danish Basin of eastern Denmark. The core records consist of white chalk with several intercalated marly intervals. The age model and Campanian/Maastrichtian boundary was identified between 320–325 m through the correlation of biostratigraphy and negative $\delta^{13}C_{carb}$ excursion event between Stevns-1 core and the GSSP section at Tercis les Bains (SW France). Our spectral analysis of this high-resolution carbon isotope series shows $\sim$25 m thick cycles predominant through the series. We constructed a $\sim$8.75 Myr-long ATS from 74.677 to 65.925 Ma based on tuning this $\sim$25 m cycles to the 405-kyr long-eccentricity and anchoring the 321 m at 72.1 Ma (Fig. 11J, Table 6). This $\sim$8.75 Myr-long $\delta^{13}C_{carb}$ ATS could be assigned as the 405-kyr chrono-cycles from E185 to E163 (Fig. 11J).

The carbon-isotope record from the same time interval was analyzed by Jung et al. (2013) using a high-resolution planktonic foraminifera ($\delta^{13}C_{pl}$) data series from the Hole 1210B of ODP Leg 198 in the tropical Pacific Ocean (Shatsky Rise). The age model for $\delta^{13}C_{pl}$ data series based on Voigt et al. (2012) enabled a global correlation of high-resolution $\delta^{13}C$ records for the Late Cretaceous spanning 77.01 to 66.66 Ma. We reanalyzed this $\delta^{13}C_{pl}$ series and tuned the dominant $\sim$5 m cycles to the 405-kyr long-eccentricity cycles to construct a $\sim$10.4 Myr-long ATS. This compares well with the 10.35 Myr duration from the global $\delta^{13}C$ age model of 77.01–66.66 Ma (Fig. 11K, Table 6). This 10.4 Myr-long ODP Hole 1210B $\delta^{13}C_{pl}$ ATS from ODP Hole 1210B could be assigned as the 405-kyr chrono-cycles from E190 to E165 (Fig. 11K).

For a full Late Cretaceous ATS, there still some correlation problems between different paleoclimate proxies due to the anchor/constraint ages uncertainties. In addition, similar paleoclimate proxies for the 405-kyr E cycles, such as the carbon isotopes but when measured from different materials (bulk carbonate, organic matter, plankton foraminifers, etc.), often do not correlate very well (Fig. 11J and K). But, in other cases, the bulk carbonate $\delta^{13}C_{carb}$ ATS from the Western Interior Basin (USA) and the organic carbon $\delta^{13}C_{org}$ ATS from the Bottaccione reference section (Gubbio, central Italy) do could correlate well (Fig. 11A and B).

This extensive suite of replicate studies enables coverage and verification of the main 405-kyr tuned ATS for the entire Late Cretaceous (Fig. 11, Table 7). There are at least three ATS sections correlated to each other for the **Turonian, Coniacian, Santonian** and **Maastrichtian** stages,

Table 7 Summary of the Astronomical Calibrated the Stage Boundary Ages and Durations for the Mesozoic and Compare With the GTS2012

	Stage	Base age (Ma) GTS2012/GTS2016	Duration (Myr)	ATS age (Ma)	ATS duration (Myr)	405-Kyr chrono-cycles
PALE-OGENE	Danian	66.0 ± 0.1		66.0		E163
CRETACEOUS	Maastrichtian	72.1 ± 0.2	6.1	72.1	6.1	E178-E163
	Campanian	83.6 ± 0.3/84.19 ± 0.38	11.5/12.09	84.2	12.1	E208-E178
	Santonian	86.3 ± 0.5/86.49 ± 0.44	2.7/2.3	86.2	2	E213-E208
	Coniacian	89.8 ± 0.4/89.75 ± 0.38	3.5/3.26	89.8	3.6	E221.5-E213
	Turonian	93.9 ± 0.2	4.1/4.15	93.9	4.1	E232-E221.5
	Cenomanian	100.5 ± 0.4	6.6	100.485	6.6	E248-E232
	Albian	113.0 ± 0.4/113.14 ± 0.4	12.5/12.65	113.1	12.6	E279-E248
	Aptian	126.3 ± 0.4	13.3/13.16	126.5	13.4	E311.5-E251
	Barremian	130.8 ± 0.5	4.5			E344-E314
	Hauterivian	133.9 ± 0.6/134.7 ± 0.7	3.1/3.9		5.9	
	Valanginian	139.4 ± 0.7	5.5/4.7	139.4	5.3	
	Berriasian	145.0 ± 0.8/145.7 ± 0.8	5.6/6.3	145.8	6.4	E360-E344
JURASSIC	Tithonian	152.1 ± 0.9	7.1/6.4	152.1	6.3	E376-E360
	Kimmeridgian	157.3 ± 1.0	5.2	157.1	5	E388-E376
	Oxfordian	163.5 ± 1.1/163.1 ± 1.1	6.2/5.8	163.1	6	E403-E388

Stage					
Callovian	166.1 ± 1.2	2.6/3.0			*E410-E403*
Bathonian	168.3 ± 1.3	2.2		4	*E420.6-E410*
Bajocian	170.3 ± 1.4	2.0	170.3		E430-E420.6
Aalenian	174.1 ± 1.0/174.2 ± 1.0	3.8/3.9	174.2	4	
Toarcian	182.7 ± 0.7/183.7 ± 0.5	8.6/9.5	183.7	9.5	E453.5-E430
Pliensbachian	190.8 ± 1.0/191.4 ± 0.3	8.1/7.7	192.4	8.7	E475-453.5
Sinemurian	199.3 ± 0.3/199.4 ± 0.3	8.5/8.0	199.430		*E429.5-E475*
Hettangian	201.3 ± 0.2/201.4 ± 0.17	2.0/2.0	201.4	1.97	E497.3-E492.5
TRIASSIC Rhaetian	205.4/209.5	4.1/8.2	205.7	4.3	E508-E497.3
Norian	221/228.4	15.7/18.9			E585.7-508
Carnian	237 ± 1.0	16/8.7	237.0		
Ladinian	241.5 ± 1.0	4.5	241.5	4.5	*E596.5-585.7*
Anisian	247.1 ± 0.1/246.8 ± 0.2	5.6/5.3	246.8	5.3	E609.5-E596.5
Olenekian	250 ± 0.2/249.8 ± 0.2	2.9/3.0	249.9	3.1	E617-E609.5
Induan	252.2 ± 0.2/251.9 ± 0.02	2.2/2.1	251.9	2	E622-E617

and two ATS sections for most of the **Cenomanian** and the **Campanian** stages. A precise duration for each stage could be obtained based on these ATS reference sections.

6. CONCLUSIONS

In order to calibrate the accurate geological time scale and the accurate durations for each stage, the cyclostratigraphy of multiple sections (marine and continental) correlation need to be developed, carefully cross-correlated and integrated with biostratigraphy, chemostratigraphy, magne-tostratigraphy and radioisotopic dating. Also, wherever possible, we need apply the same types of paleoclimate proxies for each section and also compare among different depositional settings (e.g., lacustrine versus marine) and climatic hemispheres. The goal is to recognize the 405-kyr long-eccentricity E cycles and apply tuning to construct the ATS for the Mesozoic with one-to-one correspondence to the predicted astronomical model for these E cycles. Even though this paper mainly focused on the estimates of ages and durations of international geologic stages (Table 1), the main goal of cyclostratigraphy and the ATS is determining the actual rates and precise relative all biologic, geochemical and other events within those stages. Future work needs to focus on closing the main gaps in coverage in the middle Triassic (Ladinian and Carnian stages), the Sinemurian Stage of Early Jurassic, and much of the Middle Jurassic, and verifying the middle Cretaceous (Aptian Stage).

ACKNOWLEDGMENTS

I thank Dr. Mario Sprovieri, Masayuki Ikeda and Huaichun Wu provide their data for this study. I thank Dr. Alessandro Grippo for providing his data from the Aptian-Albian Piobbico core. This work was supported by the National Natural Science Foundation of China (No. 41772029) and Natural Science Foundation for Distinguished Young Scholars of Hubei Province of China (No. 2016CFA051), the 111 Project (No. B14031, B08030), 973 Program (No. 2014CB239101) and the Fundamental Research Funds for the Central Universities, China University of Geosciences (Wuhan) (No. CUGQYZX1705, CUGCJ1703).

REFERENCES

Ait-Itto, F.Z., Martinez, M., Price, G.D., Addi, A.A., 2018. Synchronization of the astronomical time scales in the Early Toarcian: a link between anoxia, carbon-cycle perturbation, mass extinction and volcanism. Earth Planet. Sci. Lett. 493, 1−11.
Al-Husseini, M., Matthews, R.K., 2010. Tuning Late Barremian − Aptian Arabian Plate and global sequences with orbital periods. GeoArabia Special Publication 1, 199−228.

Algeo, T.J., Hinnov, L., Moser, J., Maynard, J.B., Elswick, E., Kuwahara, K., Sano, H., 2010. Changes in productivity and redox conditions in the Panthalassic Ocean during the latest Permian. Geology 38, 187–190.

Amodio, S., Ferreri, V., D'Argenio, B., 2013. Cyclostratigraphic and chronostratigraphic correlations in the Barremian–Aptian shallow-marine carbonates of the central-southern Apennines (Italy). Cretac. Res. 44, 132–156.

Aguirre-Urreta, B., Lescano, M., Schmitz, M.D., Tunik, M., Concheyro, A., Rawson, P.F., Ramos, V.A., 2015. Filling the gap: new precise Early Cretaceous radioisotopic ages from the Andes. Geol. Mag. 152 (3), 557–564. https://doi.org/10.1017/S001675681400082X.

Balog, A., Haas, J., Read, J.F., Coruh, C., 1997. Shallow marine record of orbitally forced cyclicity in a late Triassic carbonate platform, Hungary. J. Sediment. Res. 67, 661–675.

Batenburg, S.J., De Vleeschouwer, D., Sprovieri, M., Hilgen, F.J., Gale, A.S., Singer, B.S., Koeberl, C., Coccioni, R., Claeys, P., Montanari, A., 2016. Orbital control on the timing of oceanic anoxia in the Late Cretaceous. Clim. Past 12, 1995–2009.

Batenburg, S.J., Friedrich, O., Moriya, K., Voigt, S., Cournède, C., Moebius, I., Blum, P., Bornemann, A., Fiebig, J., Hasegawa, T., Hull, P.M., Norris, R.D., Röhl, U., Sexton, P.F., Westerhold, T., Wilson, P.A., Scientists, I.E., 2017. Late Maastrichtian carbon isotope stratigraphy and cyclostratigraphy of the Newfoundland Margin (site U1403, IODP Leg 342). Newsl. Stratigr.

Batenburg, S.J., Gale, A.S., Sprovieri, M., Hilgen, F.J., Thibault, N., Boussaha, M., Orue-Etxebarria, X., 2014. An astronomical time scale for the Maastrichtian based on the Zumaia and Sopelana sections (Basque country, northern Spain). J. Geol. Soc. 171, 165–180.

Batenburg, S.J., Sprovieri, M., Gale, A.S., Hilgen, F.J., Hüsing, S., Laskar, J., Liebrand, D., Lirer, F., Orue-Etxebarria, X., Pelosi, N., Smit, J., 2012. Cyclostratigraphy and astronomical tuning of the Late Maastrichtian at Zumaia (Basque country, northern Spain). Earth Planet. Sci. Lett. 359-360, 264–278.

Benton, M.J., Twitchett, R.J., 2003. How to kill (almost) all life: the end-Permian extinction event. Trends Ecol. Evol. 18, 358–365.

Berger, A., Loutre, M.F., 2004. Astronomical theory of climate change. J. de Physique IV 121, 1–35.

Berger, A.L., 1988. Milankovitch theory and climate. Rev. Geophys. 26, 624–657.

Boulila, S., Galbrun, B., Hinnov, L.A., Collin, P.-Y., 2008a. High-resolution cyclostratigraphic analysis from magnetic susceptibility in a Lower Kimmeridgian (Upper Jurassic) marl–limestone succession (La Méouge, Vocontian Basin, France). Sediment. Geol. 203, 54–63.

Boulila, S., Galbrun, B., Hinnov, L.A., Collin, P.Y., Ogg, J.G., Fortwengler, D., Marchand, D., 2010. Milankovitch and sub-Milankovitch forcing of the Oxfordian (Late Jurassic) Terres Noires Formation (SE France) and global implications. Basin Res. 22, 717–732.

Boulila, S., Galbrun, B., Huret, E., Hinnov, L.A., Rouget, I., Gardin, S., Bartolini, A., 2014. Astronomical calibration of the Toarcian Stage: implications for sequence stratigraphy and duration of the early Toarcian OAE. Earth Planet. Sci. Lett. 386, 98–111.

Boulila, S., Hinnov, L.A., Huret, E., Collin, P.-Y., Galbrun, B., Fortwengler, D., Marchand, D., Thierry, J., 2008b. Astronomical calibration of the early Oxfordian (Vocontian and Paris basins, France): Consequences of revising the Late Jurassic time scale. Earth Planet. Sci. Lett. 276, 40–51.

Brack, P., Mundil, R., Oberli, F., Meier, M., Rieber, H., 1996. Biostratigraphic and radiometric age data question the Milankovitch characteristics of the Latemar cycles (Southern Alps, Italy). Geology 24, 371–375.

Brack, P., Rieber, H., Mundil, R., Blendinger, W., Maurer, F., 2007. Geometry and chronology of growth and drowning of middle Triassic carbonate platforms (Cernera and Bivera/Clapsavon) in the southern Alps (northern Italy). Swiss J. Geosci. 100, 327—348.

Burgess, S.D., Bowring, S., Shen, S.Z., 2014. High-precision timeline for Earth's most severe extinction. Proc. Natl. Acad. Sci. USA 111, 3316—3321.

Burgess, S.D., Muirhead, J.D., Bowring, S.A., 2017. Initial pulse of Siberian Traps sills as the trigger of the end-Permian mass extinction. Nat. Commun. 8, 164—169. https://doi.org/10.1038/s41467-017-00083-9.

Channell, J.E.T., Labs, J., Raymo, M.E., 2003. The Réunion Subchronozone at ODP site 981 (Feni Drift, north Atlantic). Earth Planet. Sci. Lett. 215, 1—12.

Charbonnier, G., Boulila, S., Gardin, S., Duchamp-Alphonse, S., Adatte, T., Spangenberg, J.E., Föllmi, K.B., Colin, C., Galbrun, B., 2013. Astronomical calibration of the Valanginian "Weissert" episode: the Orpierre marl—limestone succession (Vocontian Basin, southeastern France). Cretac. Res. 45, 25—42.

Chen, X., Wang, C., Wu, H., Kuhnt, W., Jia, J., Holbourn, A., Zhang, L., Ma, C., 2015. Orbitally forced sea-level changes in the upper Turonian—lower Coniacian of the Tethyan Himalaya, southern Tibet. Cretac. Res. 56, 691—701.

Cheng, L., Wang, J., Wan, Y., Fu, X., Zhong, L., 2017. Astrochronology of the middle Jurassic Buqu Formation (Tibet, China) and its implications for the Bathonian time scale. Palaeogeogr. Palaeoclimatol. Palaeoecol. 487, 51—58.

Clemmanson, L.B., Kent, D.V., Jenkins Jr., F.A., 1998. A Late Triassic lake system in East Greenland: facies, depositional cycles and palaeoclimate. Palaeocl. Palaeogeo. Palaeoclimatol. 140, 135—159.

Cozzi, A., Hinnov, L.A., Hardie, L.A., 2005. Orbitally forced Lofer cycles in the Dachstein Limestone of the Julian Alps (northeastern Italy). Geology 33.

Deng, S., Wang, S., Yang, Z., Lu, Y., Li, X., Hu, Q., An, C., Xi, D., Wan, X., 2015. Comprehensive study of the middle—Upper Jurassic Strata in the Junggar Basin, Xinjiang. Acta Geoscientica Sinica 36, 559—574 (In Chinese with English abstract).

Eldrett, J.S., Ma, C., Bergman, S.C., Lutz, B., Gregory, F.J., Dodsworth, P., Phipps, M., Hardas, P., Minisini, D., Ozkan, A., Ramezani, J., Bowring, S.A., Kamo, S.L., Ferguson, K., Macaulay, C., Kelly, A.E., 2015. An astronomically calibrated stratigraphy of the Cenomanian, Turonian and earliest Coniacian from the Cretaceous western Interior Seaway, USA: implications for global chronostratigraphy. Cretac. Res. 56, 316—344.

Fiet, N., 2000. Calibrage temporel de l'Aptien et des sous-étages associés par une approche-cyclostratigraphique appliquée à la série pélagique de Marches-Ombrie, Italie centrale. Bull. Soc. Géol. Fr. 171, 103—113.

Fiet, N., Beaudoin, B., Parize, O., 2001. Lithostratigraphic analysis of Milankovitch cyclicity in pelagic Albian deposits of central Italy: implications for the duration of the stage and substages. Cretac. Res. 22, 265—275.

Fiet, N., Gorin, G., 2000. Lithological expression of Milankovitch cyclicity in carbonate-dominated, pelagic, Barremian deposits in central Italy. Cretac. Res. 21, 457—467.

Fiet, N., Quidelleur, X., Parize, O., Bulot, L., Gillot, P., 2006. Lower Cretaceous stage durations combining radiometric data and orbital chronology: towards a more stable relative time scale? Earth Planet. Sci. Lett. 246, 407—417.

Friedrich, O., Batenburg, S.J., Moriya, K., Voigt, S., Cournède, C., Möbius, I., Blum, P., Bornemann, A., Fiebig, J., Hasegawa, T., Hull, P.M., Norris, R.D., Röhl, U., Westerhold, T., Wilson, P.A., 2016. Maastrichtian carbon isotope stratigraphy and cyclostratigraphy of the Newfoundland Margin (site U1403, IODP Leg 342). Clim. Past Discuss. 1—21.

Fu, W., Jiang, D.Y., Montanez, I.P., Meyers, S.R., Motani, R., Tintori, A., 2016. Eccentricity and obliquity paced carbon cycling in the Early Triassic and implications for postextinction ecosystem recovery. Sci. Rep. 6, 27793.

Furin, S., Preto, N., Rigo, M., Roghi, G., Gianolla, P., Crowley, J.L., Bowring, S.A., 2006. Highprecision U-Pb zircon age from the Triassic of Italy: implications for the Triassic time scale and the Carnian origin of calcareous nannoplankton and dinosaurs. Geology 34, 1009—1012.

Gale, A., 1995. Cyclostratigraphy and correlation of the Cenomanian stage in western Europe. Geol. Soc. Lond. Spec. Publ. 85, 177—197.

Gale, A.S., Bown, P., Caron, M., Crampton, J., Crowhurst, S.J., Kennedy, W.J., Petrizzo, M.R., Wray, D.S., 2011. The uppermost Middle and Upper Albian succession at the Col de Palluel, Hautes-Alpes, France: an integrated study (ammonites, inoceramid bivalves, planktonic foraminifera, nannofossils, geochemistry, stable oxygen and carbon isotopes, cyclostratigraphy). Cretac. Res. 32, 59—130.

Gale, A.S., Hardenbol, J., Hathway, B., Kennedy, W.J., Young, J.R., Phansalkar, V., 2002. Global correlation of Cenomanian (Upper Cretaceous) sequences: Evidence for Milankovitch control on sea level. Geology 30, 291—294.

Giraud, F., Beaufort, L., Cotillon, P., 1995. Periodicities of carbonate cycles in the Valanginian of the Vocontian Trough: a strong obliquity control. Geol. Soc. Lond. Spec. Publ. 85, 143—164.

Goldhammer, R.K., Dunn, P., Hardie, L., 1990. Depositional cycles, composite sea-level changes, cycle stacking patterns, and the hierarchy of stratigraphic forcing: examples from Alpine Triassic platform carbonates. Geol. Soc. Am. Bull. 102, 535—562.

Goldhammer, R.K., Dunn, P.A., Hardie, L.A., 1987. High frequency glacio-eustatic sealevel oscillations with Milankovitch characteristics recorded in Middle Triassic platform carbonates in northern Italy. Am. J. Sci. 287, 853—892.

Gradstein, F.M., Ogg, J.G., Smith, A.G., 2004. A Geologic Time Scale 2004. Cambridge University Press, Cambridge.

Gradstein, F.M., 2012. Introduction. In: Gradstein, F.M., Ogg, J.G., Schmitz, M.D., Ogg, G.M. (Eds.), The Geologic Time Scale 2012. Elsevier B. V., Amsterdam, pp. 1—29.

Grippo, A., Fischer, A.G., Hinnov, L.A., Herbert, T.D., Premoli Silva, I., 2004. Cyclostratigraphy and chronology of the Albian stage (Piobbico core, Italy). In: D'Argenio, B., Fischer, A.G., Premoli Silva, I., Weissert, H., Ferreri, V. (Eds.), Cyclostratigraphy: Approaches and Case Histories: Society for Sedimentary. Geology Special Publication, pp. 57—81.

Guo, G., Tong, J., Zhang, S., Zhang, J., Bai, L., 2008. Cyclostratigraphy of the induan (early Triassic) in West Pingdingshan section, Chaohu, Anhui province. Sci. China Ser. D-Earth Sci. 51, 22—29.

Hansen, H.J., Lojen, S., Toft, P., Dolence, T., Tong, J., Michaelsen, P., 2000. Magnetic susceptibility and organic carbon isotopes of sediments across some marine and terrestrial Permo-Triassic boundaries. In: Yin, H., Dickins, J.M.R.,S.G., Tong, J. (Eds.), Permian-Triassic Evolution of Tethys and Western Circum Pacific. Elsevier, Amsterdam, pp. 271—289.

Hardie, L.A., Hinnov, L.A., Brack, P., Mundil, R., Oberli, F., Meier, M., Rieber, H., 1997. Biostratigraphic and radiometric age data question the Milankovitch characteristics of the Latemar cycle (southern Alps, Italy): Comment and reply. Geology 25, 470—472.

Hennebert, M., Robaszynski, F., Goolaerts, S., 2009. Cyclostratigraphy and chronometric scale in the Campanian — Lower Maastrichtian: the Abiod Formation at Ellès, central Tunisia. Cretac. Res. 30, 325—338.

Hilgen, F.J., 2010. Astronomical dating in the 19th century. Earth-Science Rev. 98, 65—80.

Hinnov, L.A., Goldhammer, R.K., 1991. Spectral analysis of the middle Triassic Latemar limestone. J. Sediment. Res. 61, 1173–1193.

Hinnov, L.A., Hilgen, F.J., 2012. Cyclostratigraphy and astrochronology. In: Gradstein, F., Ogg, J., Ogg, G., Smith, D. (Eds.), The Geologic Time Scale 2012. Elsevier B.V., Amsterdam, pp. 63–83.

Hinnov, L.A., Park, J.J., 1999. Strategies for assessing early-middle (Pliensbachian-Aalenian) Jurassic cyclochronologies. Philosophical Trans. R. Soc. Lond. 357, 1831–1859.

Huang, C., Hesselbo, S.P., 2014. Pacing of the Toarcian oceanic anoxic event (early Jurassic) from astronomical correlation of marine sections. Gondwana Res. 25, 1348–1356.

Huang, C., Hesselbo, S.P., Hinnov, L., 2010a. Astrochronology of the late Jurassic Kimmeridge clay (Dorset, England) and implications for earth system processes. Earth Planet. Sci. Lett. 289, 242–255.

Huang, C., Hinnov, L., Fischer, A.G., Grippo, A., Herbert, T., 2010b. Astronomical tuning of the Aptian Stage from Italian reference sections. Geology 38, 899–902.

Huang, C., Hinnov, L.A., Swientek, O., Smelnor, M., 2010c. Astronomical Tuning of Late Jurassic-early Cretaceous Sediments (Volgian-ryazanian Stages), Greenland-Norwegian Seaway. AAPG Annual Convention, New Orleans, LA.

Huang, C., Tong, J., Hinnov, L., Chen, Z.Q., 2011. Did the great dying of life take 700 k.y.? Evidence from global astronomical correlation of the Permian-Triassic boundary interval. Geology 39, 779–782.

Huang, Z., Ogg, J., Gradstein, F., 1993. A quantitative study of Lower Cretaceous cyclic sequences from the Atlantic ocean and the Vocontian Basin (SE France). Paleoceanography 8, 275–291.

Huret, E., Hinnov, L.A., Galbrun, B., Collin, P.-Y., Gardin, S., Rouget, I., 2008. Astronomical calibration and correlation of the Lower Jurassic, Paris and Lombard basins (Tethys). In: 33nd International Geological Congress, Oslo, Norway.

Hüsing, S.K., Beniest, A., van der Boon, A., Abels, H.A., Deenen, M.H.L., Ruhl, M., Krijgsman, W., 2014. Astronomically-calibrated magnetostratigraphy of the Lower Jurassic marine successions at St. Audrie's Bay and East Quantoxhead (Hettangian–Sinemurian; Somerset, UK). Palaeogeogr. Palaeoclimatol. Palaeoecol. 403, 43–56.

Hüsing, S.K., Deenen, M.H.L., Koopmans, J.G., Krijgsman, W., 2011. Magnetostratigraphic dating of the proposed Rhaetian GSSP at Steinbergkogel (Upper Triassic, Austria): implications for the Late Triassic time scale. Earth Planet. Sci. Lett. 302, 203–216.

Husson, D., Galbrun, B., Laskar, J., Hinnov, L.A., Thibault, N., Gardin, S., Locklair, R.E., 2011. Astronomical calibration of the Maastrichtian (Late Cretaceous). Earth Planet. Sci. Lett. 305, 328–340.

Husson, D., Thibault, N., Galbrun, B., Gardin, S., Minoletti, F., Sageman, B., Huret, E., 2014. Lower Maastrichtian cyclostratigraphy of the Bidart section (Basque Country, SW France): a remarkable record of precessional forcing. Palaeogeogr. Palaeoclimatol. Palaeoecol. 395, 176–197.

Ikeda, M., Bôle, M., Baumgartner, P.O., 2016. Orbital-scale changes in redox condition and biogenic silica/detrital fluxes of the Middle Jurassic Radiolarite in Tethys (Sogno, Lombardy, N-Italy): possible link with glaciation? Palaeogeogr. Palaeoclimatol. Palaeoecol. 457, 247–257.

Ikeda, M., Hori, R.S., 2014. Effects of Karoo–Ferrar volcanism and astronomical cycles on the Toarcian oceanic anoxic events (early Jurassic). Palaeogeogr. Palaeoclimatol. Palaeoecol. 410, 134–142.

Ikeda, M., Tada, R., 2014. A 70 million year astronomical time scale for the deep-sea bedded chert sequence (Inuyama, Japan): implications for Triassic–Jurassic geochronology. Earth Planet. Sci. Lett. 399, 30–43.

Ikeda, M., Tada, R., Ozaki, K., 2017. Astronomical pacing of the global silica cycle recorded in Mesozoic bedded cherts. Nat. Commun. 8, 15532.

Ikeda, M., Tada, R., Sakuma, H., 2010. Astronomical cycle origin of bedded chert: a middle Triassic bedded chert sequence, Inuyama, Japan. Earth Planet. Sci. Lett. 297, 369–378.

Jarvis, I., Mabrouk, A., Moody, R.T.J., Cabrera, S.D., 2002. Late Cretaceous (Campanian) carbon isotope events, sea-level change and correlation of the Tethyan and Boreal realms. Palaeogeogr. Palaeoclimatol. Palaeoecol. 188, 215–248.

Joo, Y.J., Sageman, B.B., 2014. Cenomanian to Campanian carbon isotope chemostratigraphy from the western Interior Basin. U.S.A. J. Sediment. Res. 84, 529–542.

Jung, C., Voigt, S., Friedrich, O., 2012. High-resolution carbon-isotope stratigraphy across the Campanian–Maastrichtian boundary at Shatsky Rise (tropical Pacific). Cretaceous Res. 37, 177–185.

Jung, C., Voigt, S., Friedrich, O., Koch, M.C., Frank, M., 2013. Campanian-Maastrichtian ocean circulation in the tropical Pacific. Paleoceanography 28, 562–573.

Kemp, D.B., Coe, A.L., 2007. A nonmarine record of eccentricity forcing through the Upper Triassic of southwest England and its correlation with the Newark Basin astronomically calibrated geomagnetic polarity time scale from North America. Geology 35, 991–994.

Kemp, D.B., Coe, A.L., Cohen, A.S., Weedon, G.P., 2011. Astronomical forcing and chronology of the early Toarcian (Early Jurassic) oceanic anoxic event in Yorkshire, UK. Paleoceanography 26.

Kent, D.V., Muttoni, G., Brack, P., 2004. Magnetostratigraphic confirmation of a much faster tempo for sea-level change for the Middle Triassic Latemar platform carbonates. Earth Planet. Sci. Lett. 228, 369–377.

Kent, D.V., Olsen, P.E., 2008. Early Jurassic magnetostratigraphy and paleolatitudes from the Hartford continental rift basin (eastern North America): Testing for polarity bias and abrupt polar wander in association with the central Atlantic magmatic province. J. Geophys. Res. Solid Earth (1978–2012) 113.

Kent, D.V., Olsen, P.E., Muttoni, G., 2017. Astrochronostratigraphic polarity time scale (APTS) for the Late Triassic and Early Jurassic from continental sediments and correlation with standard marine stages. Earth-Science Rev. 166, 153–180.

Kietzmann, D.A., Palma, R.M., Iglesia Llanos, M.P., 2015. Cyclostratigraphy of an orbitally-driven Tithonian-Valanginian carbonate ramp succession, southern Mendoza, Argentina: implications for the Jurassic-Cretaceous boundary in the Neuquén Basin. Sediment. Geol. 315, 29–46.

Kietzmann, D.A., Iglesia Llanos, M.P., Kohan Martinez, M., 2018. This volume. Astronomical calibration of the Upper Jurassic — Lower Cretaceous in the Neuquén Basin, Argentina: a contribution from the southern hemisphere to the geologic time scale. In: Stratigraphy and Time Scales, vol. 3. Elsevier Publ. (this volume).

Kuiper, K., Deino, A., Hilgen, F., Krijgsman, W., Renne, P., Wijbrans, J., 2008. Synchronizing rock clocks of Earth history. Science 320, 500–504.

Lanci, L., Muttoni, G., Erba, E., 2010. Astronomical tuning of the Cenomanian Scaglia Bianca Formation at Furlo, Italy. Earth Planet. Sci. Lett. 292, 231–237.

Laskar, J., Fienga, A., Gastineau, M., Manche, H., 2011. La2010: a new orbital solution for the long-term motion of the Earth. Astron. Astrophy. 532.

Laskar, J., Robutel, P., Joutel, F., Gastineau, M., Correia, A., Levrard, B., 2004. A long-term numerical solution for the insolation quantities of the Earth. Astron. Astrophy. 428, 261–285.

Laurin, J., Čech, S., Uličný, D., Štaffen, Z., Svobodová, M., 2014. Astrochronology of the Late Turonian: implications for the behavior of the carbon cycle at the demise of peak greenhouse. Earth Planet. Sci. Lett. 394, 254–269.

Laurin, J., Meyers, S.R., Galeotti, S., Lanci, L., 2016. Frequency modulation reveals the phasing of orbital eccentricity during Cretaceous Oceanic Anoxic Event II and the Eocene hyperthermals. Earth Planet. Sci. Lett. 442, 143–156.

Lefranc, M., Beaudoin, B., Chilès, J.P., Guillemot, D., Ravenne, C., Trouiller, A., 2008. Geostatistical characterization of Callovo–Oxfordian clay variability from high-resolution log data. Phys. Chem. Earth, Parts A/B/C 33, S2–S13.

Lehrmann, D.J., Ramezani, J., Bowring, S.A., Martin, M.W., Montgomery, P., Enos, P., Payne, J.L., Orchard, M.J., Hongmei, W., Jiayong, W., 2006. Timing of recovery from the end-Permian extinction: Geochronologic and biostratigraphic constraints from south China. Geology 34, 1053–1056.

Lehrmann, D.J., Stepchinski, L., Altiner, D., Orchard, M.J., Montgomery, P., Enos, P., Ellwood, B.B., Bowring, S.A., Ramezani, J., Wang, H., Wei, J., Yu, M., Griffiths, J.D., Minzoni, M., Schaal, E.K., Li, X., Meyer, K.M., Payne, J.L., 2015. An integrated biostratigraphy (conodonts and foraminifers) and chronostratigraphy (paleomagnetic reversals, magnetic susceptibility, elemental chemistry, carbon isotopes and geochronology) for the Permian–Upper Triassic strata of Guandao section, Nanpanjiang Basin, south China. J. Asian Earth Sci. 108, 117–135.

Li, M., Huang, C., Hinnov, L., Chen, W., Ogg, J., Tian, W., 2018. Astrochronology of the Anisian stage (middle Triassic) at the Guandao reference section, south China. Earth Planet. Sci. Lett. 482, 591–606.

Li, M., Ogg, J., Zhang, Y., Huang, C., Hinnov, L., Chen, Z.-Q., Zou, Z., 2016. Astronomical tuning of the end-Permian extinction and the early Triassic Epoch of south China and Germany. Earth Planet. Sci. Lett. 441, 10–25.

Li, M., Zhang, Y., Huang, C., Ogg, J., Hinnov, L., Wang, Y., Zou, Z., Li, L., Grasby, S., Zhong, Y.J., Huang, K.K., 2017b. Astronomical tuning and magnetostratigraphy of the Upper Triassic Xujiahe Formation of south China and Newark Supergroup of North America: implications for the Late Triassic time scale. Earth Planet. Sci. Lett. 475, 207–223.

Li, M., Zhang, Y., Huang, C., Ogg, J., Hinnov, L., Wang, Y., Zou, Z., Li, L., 2017c. Astrochronology and magnetostratigraphy of the Xujiahe Formation and Newark Supergroup: implications for the Late Triassic time scale. AGU Abstr.

Li, Y.-X., Montañez, I.P., Liu, Z., Ma, L., 2017. Astronomical constraints on global carbon-cycle perturbation during Oceanic Anoxic Event 2 (OAE2). Earth Planet. Sci. Lett. 462, 35–46.

Liu, Z., Liu, X., Huang, S., 2017. Cyclostratigraphic analysis of magnetic records for orbital chronology of the Lower Cretaceous Xiagou Formation in Linze, northwestern China. Palaeogeogr. Palaeoclimatol. Palaeoecol. 481, 44–56.

Locklair, R.E., Sageman, B.B., 2008. Cyclostratigraphy of the Upper Cretaceous Niobrara Formation, Western Interior, U.S.A.: A Coniacian–Santonian orbital timescale. Earth Planet. Sci. Lett. 269, 540–553.

Ma, C., Meyers, S.R., Sageman, B.B., Singer, B.S., Jicha, B.R., 2014. Testing the astronomical time scale for oceanic anoxic event 2, and its extension into Cenomanian strata of the Western Interior Basin (USA). Geol. Soc. Am. Bull. 126, 974–989.

Maron, M., Rigo, M., Bertinelli, A., Katz, M.E., Godfrey, L., Zaffani, M., Muttoni, G., 2015. Magnetostratigraphy, biostratigraphy, and chemostratigraphy of the Pignola-Abriola section: New constraints for the Norian-Rhaetian boundary. Geol. Soc. Am. Bull. 127, 962–974.

Martinez, M., Deconinck, J.-F., Pellenard, P., Reboulet, S., Riquier, L., 2013. Astrochronology of the Valanginian Stage from reference sections (Vocontian Basin, France) and palaeoenvironmental implications for the Weissert Event. Palaeogeogr. Palaeoclimatol. Palaeoecol. 376, 91–102.

Martinez, M., Deconinck, J.-F., Pellenard, P., Riquier, L., Company, M., Reboulet, S., Moiroud, M., 2015. Astrochronology of the Valanginian—Hauterivian stages (Early Cretaceous): Chronological relationships between the Paraná—Etendeka large igneous province and the Weissert and the Faraoni events. Glob. Planet. Change 131, 158—173.

Martinez, M., Krencker, F.-N., Mattioli, E., Bodin, S., 2017. Orbital chronology of the Pliensbachian — Toarcian transition from the Central High Atlas Basin (Morocco). Newsl. Stratigr. 50, 47—69.

Martinez, M., Pellenard, P., Deconinck, J.F., Monna, F., Riquier, L., Boulila, S., Moiroud, M., Company, M., 2012. An orbital floating time scale of the Hauterivian/Barremian GSSP from a magnetic susceptibility signal (Río Argos, Spain). Cretac. Res. 36, 106—115.

Maurer, F., Hinnov, L., Schlager, W., 2004. Statistical time-series analysis and sedimentological tuning of bedding rhythms in a Triassic basinal succession (Southern Alps, Italy). In: Cyclostratigraphy: Approaches and Case Histories, pp. 83—99.

Mazza, M., Rigo, M., 2012. Taxonomy and biostratigraphic record of the Upper Triassic conodonts of the Pizzo Mondello section (Western Sicily, Italy), GSSP candidate for the base of the Norian. Riv. Ital. Paleontol. Stratigr. 118 (1), 85—130.

Meyers, S.R., Siewert, S.E., Singer, B.S., Sageman, B.B., Condon, D.J., Obradovich, J.D., Jicha, B.R., Sawyer, D.A., 2012. Intercalibration of radioisotopic and astrochronologic time scales for the Cenomanian-Turonian boundary interval. West. Inter. Basin USA. Geol. 40, 7—10.

Midtkandal, I., Svensen, H.H., Planke, S., Corfu, F., Polteau, S., Torsvik, T.H., Faleide, J.I., Gundvag, S.-A., Selnes, H., Kürschner, W., Olaussen, S., 2016. The Aptian (Early Cretaceous) oceanic anoxic event (OAE1a) in Svalbard, Barents Sea, and the absolute age of the Barremian-Aptian boundary. Palaeogeogr. Palaeoclimatol. Palaeoecol. 463, 126—135.

Mietto, P., Manfrin, S., Preto, N., Rigo, M., Roghi, G., Furin, S., Gianolla, P., Posenato, R., Muttoni, G., Nicora, A., Buratti, N., Cirilli, S., Spötl, C., Ramezani, J., Bowring, S.A., 2012. The Global Boundary Stratotype Section and Point (GSSP) of the Carnian Stage (Late Triassic) at Prati di Stuores/Stuores Wiesen Section (Southern Alps, NE Italy). Episodes 35, 414—430.

Milankovitch, M., 1941. Kanon der Erdbestrahlung und seine Anwendung auf das Eiszeitenproblem (1998 reissue in English: Canon of Insolation and the Ice-Age Problem). Royal Serbian Academy, Section of Mathematical and Natural Sciences, Belgrade.

Mitchell, R.N., Bice, D.M., Montanari, A., Cleaveland, L.C., Christianson, K.T., Coccioni, R., Hinnov, L.A., 2008. Oceanic anoxic cycles? Orbital prelude to the Bonarelli Level (OAE 2). Earth Planet. Sci. Lett. 267, 1—16.

Müller, T., Price, G.D., Bajnai, D., Nyerges, A., Kesjár, D., Raucsik, B., Varga, A., Judik, K., Fekete, J., May, Z., Pálfy, J., Hesselbo, S., 2017. New multiproxy record of the Jenkyns Event (also known as the Toarcian Oceanic Anoxic Event) from the Mecsek Mountains (Hungary): Differences, duration and drivers. Sedimentology 64, 66—86.

Mundil, R., Brack, P., Meier, M., Rieber, H., Oberli, F., 1996. High resolution U-Pb dating of Middle Triassic volcaniclastics: Time-scale calibration and verification of tuning parameters for carbonate sedimentation. Earth Planet. Sci. Lett. 141, 137—151.

Mundil, R., Zühlke, R., Bechstädt, T., Peterhänsel, A., Egenhoff, S.O., Oberli, F., Meier, M., Brack, P., Rieber, H., 2003. Cyclicities in Triassic Platform Carbonates: Synchronizing Radio-Isotopic and Orbital Clocks. Terra Nova 15, 81—87.

Muttoni, G., Kent, D.V., Olsen, P.E., Di Stefano, P., Lowrie, W., Bernasconi, S.M., Hernández, F.M., 2004. Tethyan magnetostratigraphy from Pizzo Mondello (Sicily) and correlation to the Late Triassic Newark astrochronological polarity time scale. Geol. Soc. Am. Bull. 116, 1043.

Ogg, J., Ogg, G., Gradstein, F., 2016. A Concise Geologic TimeScale 2016. Elsevier, Amsterdam.

Ogg, J.G., Hinnov, L.A., Huang, C., Jurassic, 2012a. In: Gradstein, F.M., Ogg, J.G., Schmitz, M., Ogg, G. (Eds.), The Geologic Time Scale 2012. Elsevier, pp. 793–853.

Ogg, J.G., Hinnov, L.A., Huang, C., Cretaceous, 2012b. In: Gradstein, F.M.O., Ogg, J.G., Schmitz, M., Ogg, G. (Eds.), The Geologic Time Scale 2012. Elsevier, pp. 731–791.

Olsen, P.E., Kent, D.V., 1996. Milankovitch Climate forcing in the tropics of Pangaea during the late Triassic. Palaeogeogr. Palaeoclimatol. Palaeoecol. 122, 1–26.

Olsen, P.E., Kent, D.V., 1999. Long-period Milankovitch cycles from the Late Triassic and Early Jurassic of eastern North America and their implications for the calibration of the Early Mesozoic time-scale and the long-term behaviour of the planets. Philosophical Trans. R. Soc. A Math. Phys. Eng. Sci. 357, 1761–1786.

Olsen, P.E., Kent, D.V., Whiteside, J.H., 2011. Implications of the Newark Supergroup-based astrochronology and geomagnetic polarity time scale (Newark-APTS) for the tempo and mode of the early diversification of the Dinosauria. Earth Environ. Sci. Trans. R. Soc. Edinburgh 101, 201–229.

Ovtcharova, M., Bucher, H., Schaltegger, U., Galfetti, T., Brayard, A., Guex, J., 2006. New Early to Middle Triassic U–Pb ages from South China: Calibration with ammonoid biochronozones and implications for the timing of the Triassic biotic recovery. Earth Planet. Sci. Lett. 243, 463–475.

Ovtcharova, M., Goudemand, N., Hammer, Ø., Guodun, K., Cordey, F., Galfetti, T., Schaltegger, U., Bucher, H., 2015. Developing a strategy for accurate definition of a geological boundary through radio-isotopic and biochronological dating: The Early–Middle Triassic boundary (South China). Earth-Science Rev. 146, 65–76.

Paillard, D., Labeyrie, L., Yiou, P., 1996. Macintosh Program Performs Time-Series Analysis, Eos, Trans. AGU 77, 379.

Perdiou, A., Thibault, N., Anderskouv, K., van Buchem, F., Arie Buijs, G.J., Bjerrum, C.J., 2016. Orbital calibration of the late Campanian carbon isotope event in the North Sea. J. Geol. Soc. 173, 504–517.

Preto, N., Hinnov, L.A., 2003. Unraveling the origin of carbonate platform cyclothems in the Upper Triassic Durrenstein Formation (Dolomites, Italy). J. Sediment. Res. 73, 774–789.

Preto, N., Hinnov, L.A., Hardie, L.A., De Zanche, V., 2001. Middle Triassic orbital signature recorded in the shallow-marine Latemar carbonate buildup (Dolomites, Italy). Geology 29, 1123–1126.

Preto, N., Hinnov, L.A., Zanche, V.D., Mietto, P., Hardie, L.A., 2004. The Milankovitch interpretation of the Latemar platform cycles (Dolomites, italy): implications for geochronology, biostratigraphy, and Middle Triassic carbonate accumulation. In: Publication, S.S. (Ed.), Cyclostratigraphy: Approaches and Case Histories. Society for Sedimentary Geology, pp. 167–182.

Prokoph, A., Villeneuve, M., Agterberg, F.P., Rachold, V., 2001. Geochronology and calibration of global Milankovitch cyclicity at the Cenomanian-Turonian boundary. Geology 29, 523–526.

Rampino, M.R., Prokoph, A., Adler, A., 2000. Tempo of the end-Permian event: high-resolution cyclostratigraphy at the Permian–Triassic boundary. Geology 28, 643–646.

Rodriguez-Tovar, F.J., Pardo-Iguzquiza, E., Reolid, M., Bartolini, A., 2016. Spectral analysis of Toarcian sediments from the Valdorbia section (Umbria-Marche Apennines): the astronomical input in the foraminiferal record. Res. Paleontol. Stratigraphy 12, 187–197.

Ruebsam, W., Münzberger, P., Schwark, L., 2014. Chronology of the Early Toarcian environmental crisis in the Lorraine Sub-Basin (NE Paris Basin). Earth Planet. Sci. Lett. 404, 273–282.

Ruhl, M., Deenen, M., Abels, H., Bonis, N., Krijgsman, W., Kürschner, W., 2010. Astronomical constraints on the duration of the early Jurassic Hettangian stage and recovery rates following the end-Triassic mass extinction (St Audrie's Bay/East Quantoxhead, UK). Earth Planet. Sci. Lett. 295, 262—276.

Ruhl, M., Hesselbo, S.P., Hinnov, L., Jenkyns, H.C., Xu, W., Riding, J.B., Storm, M., Minisini, D., Ullmann, C.V., Leng, M.J., 2016. Astronomical constraints on the duration of the Early Jurassic Pliensbachian Stage and global climatic fluctuations. Earth Planet. Sci. Lett. 455, 149—165.

Sageman, B.B., Meyers, S.R., Arthur, M.A., 2006. Orbital time scale and new C-isotope record for Cenomanian-Turonian boundary stratotype. Geology 34, 125—128.

Sageman, B.B., Singer, B.S., Meyers, S.R., Siewert, S.E., Walaszczyk, I., Condon, D.J., Jicha, B.R., Obradovich, J.D., Sawyer, D.A., 2014. Integrating 40Ar/39Ar, U-Pb, and astronomical clocks in the Cretaceous Niobrara Formation, Western Interior Basin, USA. Geol. Soc. Am. Bull. 126, 956—973.

Selby, D., Mutterlose, J., Condon, D.J., 2009. U—Pb and Re—Os geochronology of the Aptian/Albian and Cenomanian/Turonian stage boundaries: Implications for timescale calibration, osmium isotope seawater composition and Re—Os systematics in organic-rich sediments. Chemical Geology 265, 394—409.

Schoene, B., Guex, J., Bartolini, A., Schaltegger, U., Blackburn, T.J., 2010. Correlating the end-Triassic mass extinction and flood basalt volcanism at the 100 ka level. Geology 38, 387—390.

Sha, J., Olsen, P.E., Pan, Y., Xu, D., Wang, Y., Zhang, X., Yao, X., Vajda, V., 2015. Triassic-Jurassic climate in continental high-latitude Asia was dominated by obliquity-paced variations (Junggar Basin, Ürümqi, China). Proc. Natl. Acad. Sci. USA 112, 3624—3629.

Schaller, M.F., Wright, J.D., Kent, D.V., 2015. A 30 Myr record of Late Triassic atmospheric pCO_2 variation reflects a fundamental control of the carbon cycle by changes in continental weathering. Geol. Soc. Am. Bull. 127, 661—671.

Shen, C., Schoepfer, S.D., Henderson, C.M., 2017. Astronomical tuning of the Early Triassic Induan Stage in the distal Montney Formation, NE British Columbia, Canada. Geol. Soc. Am. 49 (6) https://doi.org/10.1130/abs/2017AM-300972. gsa.confex.com/gsa/2017AM/webprogram/Paper300972. html.

Shen, S.Z., Bowring, S.A., 2014. The end-Permian mass extinction: a still unexplained catastrophe. Nat. Sci. Rev. 1, 492—495.

Shen, S.-Z., Ramezani, J., Chen, J., Cao, C.-Q., Erwin, D.H., Zhang, H., Xiang, L., Schoepfer, S.D., Henderson, C.M., Zheng, Q.-F., Bowring, S.A., Wang, Y., Li, X.-H., Wang, X.-D., Yuan, D.-X., Zhang, Y.-C., Mu, L., Wang, J., Wu, Y.-S., 2018. A sudden end-Permian mass extinction in South China. Geol. Soc. Am. Bull. in press.

Siewert, S.E., 2011. Integrating 40Ar/39Ar, U-pb and Astronomical Clocks in the Cretaceous Niobrara Formation. University of Wisconsin at Madison, p. 74 (M.S. thesis).

Spahn, Z.P., Kodama, K.P., Preto, N., 2013. High-resolution estimate for the depositional duration of the Triassic Latemar Platform: A new magnetostratigraphy and magnetic susceptibility cyclostratigraphy from basinal sediments at Rio Sacuz, Italy. Geochem. Geophy. Geosys. 14, 1245—1257.

Sprenger, A., Ten Kate, W.G., 1993. Orbital forcing of calcilutite-marl cycles in southeast Spain and an estimate for the duration of the Berriasian stage. Geol. Soc. Am. Bull. 105, 807—818.

Sprovieri, M., Coccioni, R., Lirer, F., Pelosi, N., Lozar, F., 2006. Orbital tuning of a lower Cretaceous composite record (Maiolica Formation, central Italy). Paleoceanography 21.

Sprovieri, M., Sabatino, N., Pelosi, N., Batenburg, S.J., Coccioni, R., Iavarone, M., Mazzola, S., 2013. Late Cretaceous orbitally-paced carbon isotope stratigraphy from the Bottaccione Gorge (Italy). Palaeogeogr. Palaeoclimatol. Palaeoecol. 379-380, 81—94.

Strasser, A., 2007. Astronomical time scale for the Middle Oxfordian to Late Kimmeridgian in the Swiss and French Jura Mountains. Swiss J. Geosci. 100, 407—429.

Suan, G., Mattioli, E., Pittet, B., Lécuyer, C., Suchéras-Marx, B., Duarte, L.V., Philippe, M., Reggiani, L., Martineau, F., 2010. Secular environmental precursors to Early Toarcian (Jurassic) extreme climate changes. Earth Planet. Sci. Lett. 290, 448—458.

Sucheras-Marx, B., Giraud, F., Fernandez, V., Pittet, B., Lecuyer, C., Olivero, D., Mattioli, E., 2013. Duration of the Early Bajocian and the associated $\delta13C$ positive excursion based on cyclostratigraphy. J. Geol. Soc. 170, 107—118.

Swientek, O., 2002. The Greenland Norwegian Seaway: Climatic and Cyclic Evolution of Late Jurassic-early Cretaceous Sediments. PhD Thesis. Mathematisch-Naturwissenschaftliche Fakultät der Universität zu Köln, p. 119.

Swientek, O., 2004. Lightness of Sediment Core IKU-6814/04-U-02. PANGAEA. https://doi.org/10.1594/PANGAEA.141090.

Szurlies, M., 2004. Magnetostratigraphy: the key to a global correlation of the classic Germanic Trias-case study Volpriehausen Formation (Middle Buntsandstein), Central Germany. Earth Planet. Sci. Lett. 227, 395—410.

Szurlies, M., Bachmann, G.H., Menning, M., Nowaczyk, N.R., Kaeding, K.C., 2003. Magnetostratigraphy and high-resolution lithostratigraphy of the Permian—Triassic boundary interval in central Germany. Earth Planet. Sci. Lett. 212, 263—278.

Szurlies, M., Geluk, M.C., Krijgsman, W., Kürschner, W.M., 2012. The continental Permian—Triassic boundary in the Netherlands: Implications for the geomagnetic polarity time scale. Earth Planet. Sci. Lett. 317, 165—176.

Thibault, N., Harlou, R., Schovsbo, N., Schiøler, P., Minoletti, F., Galbrun, B., Lauridsen, B.W., Sheldon, E., Stemmerik, L., Surlyk, F., 2012a. Upper Campanian—Maastrichtian nannofossil biostratigraphy and high-resolution carbon-isotope stratigraphy of the Danish Basin: Towards a standard $\delta13C$ curve for the Boreal Realm. Cretac. Res. 33, 72—90.

Thibault, N., Husson, D., Harlou, R., Gardin, S., Galbrun, B., Huret, E., Minoletti, F., 2012b. Astronomical calibration of upper Campanian—Maastrichtian carbon isotope events and calcareous plankton biostratigraphy in the Indian Ocean (ODP Hole 762C): Implication for the age of the Campanian—Maastrichtian boundary. Palaeogeogr. Palaeoclimatol. Palaeoecol. 337-338, 52—71.

Thibault, N., Galbrun, B., Gardin, S., Minoletti, F., Le Callonnec, L., 2016a. The end-Cretaceous in the southwestern Tethys (Elles, Tunisia): orbital calibration of paleoenvironmental events before the mass extinction. Int. J. Earth Sci. 105, 771—795.

Thibault, N., Jarvis, I., Voigt, S., Gale, A.S., Attree, K., Jenkyns, H.C., 2016b. Astronomical calibration and global correlation of the Santonian (Cretaceous) based on the marine carbon isotope record. Paleoceanography 31, 847—865.

Tian, S., Chen, Z.-Q., Huang, C., 2014. Orbital Forcing and Sea-Level Changes in the Earliest Triassic of the Meishan Section, South China. J. Earth Sci. 25, 64—73.

Vennari, V.V., Lescano, M., Naipauer, M., Aguirre-Urreta, B., Concheyro, A., Schaltegger, U., Armstrong, R., Pimentel, M., Ramos, V.A., 2014. New constraints on the Jurassic—Cretaceous boundary in the High Andes using high-precision U-Pb data. Gondwana Res. 26, 374—385.

Voigt, S., Erbacher, J., Mutterlose, J., Weiss, W., Westerhold, T., Wiese, F., Wilmsen, M., Wonik, T., 2008. The Cenomanian — Turonian of the Wunstorf section — (North Germany): global stratigraphic reference section and new orbital time scale for Oceanic Anoxic Event 2. Newsl. Stratigr. 43, 65—89.

Voigt, S., Gale, A.S., Jung, C., Jenkyns, H.C., 2012. Global correlation of Upper Campanian — Maastrichtian successions using carbon-isotope stratigraphy: development of a new Maastrichtian timescale. Newsl. Stratigr. 45, 25—53.

Voigt, S., Schönfeld, J., 2010. Cyclostratigraphy of the reference section for the Cretaceous white chalk of northern Germany, Lägerdorf—Kronsmoor: A late Campanian—early Maastrichtian orbital time scale. Palaeogeogr. Palaeoclimatol. Palaeoecol. 287, 67—80.

Vollmer, T., Werner, R., Weber, M., Tougiannidis, N., Roehling, H.-G., Hambach, U., 2008. Orbital control on Upper Triassic playa cycles of the Steinmergel-Keuper (Norian): A new concept for ancient playa cycles. Palaeogeogr. Palaeoclimatol. Palaeoecol. 267, 1—16.

Waterhouse, H.K., 1999. Orbital forcing of palynofacies in the Jurassic of France and the United Kingdom. Geology 27, 511—514.

Weedon, G., Jenkyns, H., 1999. Cyclostratigraphy and the Early Jurassic timescale: data from the Belemnite Marls, Dorset, southern England. Geol. Soc. Am. Bull. 111, 1823—1840.

Weedon, G.P., Coe, A.L., Gallois, R.W., 2004. Cyclostratigraphy, orbital tuning and inferred productivity for the type Kimmeridge Clay (Late Jurassic), Southern England. J. Geol. Soc. 161, 655—666.

Weedon, G.P., Jenkyns, H.C., Coe, A.L., Hesselbo, S.P., 1999. Astronomical calibration of the Jurassic time-scale from cyclostratigraphy in British mudrock formations. Philosophical Trans. R. Soc. Lond. Ser. A Mathematical, Physical and Engineering Sciences 357, 1787—1813.

Westerhold, T., Röhl, U., Raffi, I., Fornaciari, E., Monechi, S., Reale, V., Bowles, J., Evans, H.F., 2008. Astronomical calibration of the Paleocene time. Palaeogeogr. Palaeoclimatol. Palaeoecol. 257, 377—403.

Whiteside, J.H., Olsen, P.E., Eglinton, T., Brookfield, M.E., Sambrotto, R.N., 2010. Compound-specific carbon isotopes from Earth's largest flood basalt eruptions directly linked to the end-Triassic mass extinction. Proc. Natl. Acad. Sci. 107, 6721—6725.

Wotzlaw, J.-F., Brack, P., Storck, J.-C., 2018. High-resolution stratigraphy and zircon U—Pb geochronology of the Middle Triassic Buchenstein Formation (Dolomites, northern Italy): precession-forcing of hemipelagic carbonate sedimentation and calibration of the Anisian—Ladinian boundary interval. J. Geol. Soc. 175, 71—85.

Wotzlaw, J.F., Guex, J., Bartolini, A., Gallet, Y., Krystyn, L., McRoberts, C.A., Taylor, D., Schoene, B., Schaltegger, U., 2014. Towards accurate numerical calibration of the Late Triassic: High-precision U-Pb geochronology constraints on the duration of the Rhaetian. Geology 42, 571—574.

Wu, H., Zhang, S., Feng, Q., Jiang, G., Li, H., Yang, T., 2012. Milankovitch and sub-Milankovitch cycles of the early Triassic Daye Formation, South China and their geochronological and paleoclimatic implications. Gondwana Res. 22, 748—759.

Wu, H., Zhang, S., Hinnov, L.A., Jiang, G., Yang, T., Li, H., Wan, X., Wang, C., 2014. Cyclostratigraphy and orbital tuning of the terrestrial upper Santonian—Lower Danian in Songliao Basin, northeastern China. Earth Planet. Sci. Lett. 407, 82—95.

Wu, H., Zhang, S., Jiang, G., Hinnov, L., Yang, T., Li, H., Wan, X., Wang, C., 2013. Astrochronology of the Early Turonian—Early Campanian terrestrial succession in the Songliao Basin, northeastern China and its implication for long-period behavior of the Solar System. Palaeogeogr. Palaeoclimatol. Palaeoecol. 385, 55—70.

Wu, H., Zhang, S., Jiang, G., Huang, Q., 2009. The floating astronomical time scale for the terrestrial Late Cretaceous Qingshankou Formation from the Songliao Basin of Northeast China and its stratigraphic and paleoclimate implications. Earth Planet. Sci. Lett. 278, 308—323.

Yin, H., Zhang, K., Tong, J., Yang, Z., Wu, S., 2001. The global stratotype section and point (GSSP) of the Permian—Triassic boundary. Episodes 24, 102—114.

Zhang, Y., Li, M., Ogg, J.G., Montgomery, P., Huang, C., Chen, Z.-Q., Shi, Z., Enos, P., Lehrmann, D.J., 2015. Cycle-calibrated magnetostratigraphy of middle Carnian from South China: Implications for Late Triassic time scale and termination of the Yangtze Platform. Palaeogeogr. Palaeoclimatol. Palaeoecol. 436, 135—166.

Zühlke, R., 2004. Integrated Cyclostratigraphy of a Model Mesozoic Carbonate Platform— the Latemar (Middle Triassic, Italy), Cyclostratigraphy: Approaches and Case Histories. Society for Sedimentary Geology (SEPM), pp. 183—211.

Zühlke, R., Bechstädt, T., Mundil, R., 2003. Sub-Milankovitch and Milankovitch forcing on a model Mesozoic carbonate platform - the Latemar (Middle Triassic, Italy). Terra. Nova 15, 69—80.

CHAPTER THREE

Cyclostratigraphy of Shallow-Marine Carbonates — Limitations and Opportunities

André Strasser

Department of Geosciences, University of Fribourg, Fribourg, Switzerland
E-mail: andreas.strasser@unifr.ch

Contents

Abstract

The sedimentary record of ancient shallow-marine carbonate platforms commonly displays a stacking of different facies, which reflects repetitive changes of depositional environments through time. These changes can be induced by external factors such as cyclical changes in climate and/or sea level, but also by internal factors such as lateral migration of sediment bodies and/or changes in the ecology of the

Stratigraphy & Timescales, Volume 3
ISSN 2468-5178
https://doi.org/10.1016/bs.sats.2018.07.001

carbonate-producing organisms. If it can be demonstrated that the facies changes formed in tune with the orbital (Milankovitch) cycles of known duration, then a high-resolution time framework can be established. This demonstration is not an easy task because the orbital signal may be too weak to be recorded, or it may be distorted and/or overprinted by local or regional processes. The limitations of the cyclostratigraphical approach are discussed, but a case study from the Oxfordian of the Swiss Jura Mountains also shows its potential. A well-established chrono- and sequence-stratigraphic framework and detailed facies analysis allow identification of elementary, small-scale, and medium-scale depositional sequences that formed in tune with the precession, the short eccentricity, and the long eccentricity cycles, respectively. In the best case, a depositional sequence attributed to the precession cycle with a duration of 20'000 years can be interpreted in terms of sequence stratigraphy. This then allows estimating rates of sea-level change and sedimentation within a relatively narrow time window, thus facilitating comparisons between ancient carbonate platforms and Holocene or Recent shallow-marine environments where such rates are well quantified.

1. INTRODUCTION

The repetitive stacking of beds observed in the sedimentary record has always fascinated geologists. Especially in hemipelagic settings, the regularly arranged limestone–marl alternations suggest that cyclical processes were at work and that these might reflect orbital (Milankovitch) cyclicity. Gilbert (1895) was the first to interpret such alternations in the Late Cretaceous Benton, Niobrara and Pierre formations of Colorado (USA) as being related to the precession cycle and thus could propose a time duration for these stratigraphic units. In shallow–water carbonates, it was Schwarzacher (1947) who first interpreted the hierarchical stacking of beds in the Late Triassic Dachstein (Austria) as having formed in tune with the precession and the short eccentricity cycles. Since these pioneering publications, hundreds of studies have dealt with the analysis and interpretation of repetitive, "cyclical" stacking of beds and facies in carbonates as well as in evaporites and siliciclastics, from shallow to deep, and from lacustrine to fluviatile to marine depositional systems. As the periodicities of the orbital cycles are known for today and for the geological past (e.g., Berger et al., 1989, 1992; Laskar et al., 2011), the cyclostratigraphic analysis allows for establishing astronomical time scales (e.g., Strasser et al., 2006; Hinnov and Ogg, 2007; Hinnov, 2013; Zeeden et al., 2015).

The prerequisite to correctly interpret the sedimentary record in terms of cyclicity and to attribute numerical values for the time represented by this record is to understand the processes that controlled sedimentation. The orbital cycles commonly invoked in cyclostratigraphic studies are caused by the precession of the equinoxes relative to the perihelion and aphelion of the Earth's orbit around the Sun, changes in the obliquity of the Earth's axis, and the shorter- and longer-period changes in the eccentricity of the ellipse on which the Earth travels around the Sun (e.g., Hays et al., 1976; Schwarzacher, 1993, 2000; Strasser et al., 2006; Hinnov, 2013). These cycles are responsible for changes of the solar irradiance on top of the atmosphere (Milankovitch, 1941). From there, this signal is transferred to the sedimentary environment through complex interactions and feedback mechanisms in the atmosphere, the hydrosphere, the cryosphere, and the biosphere (e.g., Pisias and Shackleton, 1984; Strasser, 1991; Sames et al., 2016; Wendler and Wendler, 2016).

In the hemiplegic realm, limestone-marl alternations commonly reflect repetitive changes in the carbonate-clay ratio. Clay input from the hinterland is controlled mainly by the activity of rivers, which in turn is influenced by rainfall and vegetation cover; clay distribution in the ocean then depends on sea level, tides, and oceanic currents. The carbonate ratio depends on the productivity of planktonic organisms (controlled by water temperature and nutrients), and/or on export of carbonate mud from shallow platforms (influenced by sea level, tides, and currents; e.g., Milliman et al., 1993; Pittet et al., 2000). Orbital cycles have a direct control on the climate (via the heat distribution in the atmosphere) and thus on rainfall, vegetation, and glaciation. Latitude, orography, and ocean-land distribution are additional factors. Sea-level changes result from glaciation—deglaciation cycles but also from thermal expansion and contraction of the ocean water, from retention and release of water in aquifers, and from changes in deep-water circulation (Sames et al., 2016). Nutrient distribution in turn depends on terrigenous input and upwelling of deep ocean waters. Once the sediment is deposited, the carbonate-clay ratio may be modified by diagenesis (carbonate being dissolved at the contact with the clay minerals and migrating to the carbonate-rich levels), thus enhancing the limestone-marl contrast (Einsele and Ricken, 1991).

In theory, limestone-marl alternations should reflect the orbital cycles quite faithfully, although some lag may occur before the sedimentary system reacts to the insolation signal. Time-series analysis can be performed on proxies of clay input (gamma ray, magnetic susceptibility, colour), and/or by considering the thickness of the carbonate beds as a proxy of planktonic productivity and/or mud export from the platform. However, there are caveats: If the orbital forcing is not strong enough to pass the signal on to the depositional environment and create a contrast in facies or geochemical signature, orbital cycles will not be recorded (or the record is so faint that it cannot be detected). In condensed intervals, the record of several cycles may be amalgamated and become undecipherable. One climatic or one sea-level cycle in tune with one orbital cycle may have two or more windows within which clays or carbonate mud are shed into the hemipelagic realm, or within which planktonic productivity is stimulated, meaning that one orbital cycle may be represented by two or more limestone-marl couplets (e.g., Pittet and Strasser, 1998a). In addition, diagenesis may distort the limestone-marl contrasts or even create alternations independently of orbital forcing (Westphal et al., 2004).

If there are no tectonic or synsedimentary disturbances (such as slumps), hemipelagic sections are well suited for establishing astronomical time scales (e.g., Cotillon, 1987; Huang et al., 1993, 2010; Martinez et al., 2015). The biostratigraphic and isotopic (mainly $\delta^{13}C$) frameworks generally are well established and allow for basin-wide correlations.

Compared to the relatively simple system of hemipelagic limestone-marl alternations, the deciphering of the shallow-marine record of orbital cycles is much more challenging. In the following chapter, the many difficulties in identifying an orbital signal on shallow carbonate platforms will be discussed. A case study (Chapter 3) will then show how these difficulties may be dealt with and what the potential of the cyclostratigraphic approach can be.

2. THE CYCLICAL RECORD IN SHALLOW-MARINE CARBONATES

2.1 Factors Controlling Carbonate Production and Sedimentation

In shallow-marine, carbonate-dominated sedimentary systems, most of the carbonate is produced by organisms (e.g., Bathurst, 1975; Tucker and Wright, 1990; Flügel, 2004). From the viewpoint of these organisms, important life-sustaining parameters are:

- water temperature (controlled by air temperature and heat distribution through currents);
- water depth (controlled by relative sea level and sediment accumulation);
- water transparency (controlled by terrigenous input and/or planktonic productivity);
- water energy (controlled by waves and wind- or tide-generated currents);

- sunlight for photosynthesis (controlled by latitude, water depth and water transparency);
- chemical composition of the water (pH, Ca^{2+}, Mg^{2+}, CO_3^{2-}, SO_4^{2-}, O_2, and trace-element concentrations controlled by river input, ocean currents, evaporation, or recycling within the environment);
- nutrients (controlled by terrigenous input, upwelling, or recycling within the environment);
- type of substrate (soft, firm or hard depending on sediment type, sedimentation rate, and early diagenesis);
- ecological coexistence or competition (depending on the organisms involved).

Many of the above-mentioned parameters are themselves controlled by climate (hot, cold, humid, arid, seasonal, storm frequency and force), by the hinterland (furnishing siliciclastics and nutrients), by oceanic currents (distributing heat and nutrients), by relative sea level (defining water depth and influencing tidal regime), and by the morphology of the coastline and the platform (modifying currents and sediment distribution). As already mentioned, sea level itself is at least partly connected to climate through water stored in ice caps, alpine glaciers, and aquifers, and through thermally-controlled volume changes in the ocean water. The ecological factors follow the evolution of ecosystems through geological time.

Climate thus has a major, direct or indirect influence on carbonate production and sediment accumulation, and climate itself is controlled by the insolation changes that follow the orbital cycles in the Milankovitch frequency band. It has to be considered, however, that the atmospheric cells (Hadley, Ferrel, and polar) and, consequently, the high-pressure and low-pressure zones shift as a function of insolation (Matthews and Perlmutter, 1994). For example, while climate changes throughout one insolation cycle are small at the equator (very humid to humid), they are more pronounced at around 20 degrees latitude (i.e., from humid at the climatic maximum to arid at the climatic minimum). When interpreting cyclostratigraphic patterns it is therefore important to consider the paleolatitude of the study area. Furthermore, regional climate is influenced by the ocean-land distribution, oceanic circulation, and orography (e.g., Matthews and Perlmutter, 1994; Feng and Poulsen, 2014).

2.2 Repetitive (Cyclic) Sediment Accumulation

Once produced, the carbonate sediment can either stay in place (as in the case of microbial mats, reefs, or low-energy lagoons), may be rolled back

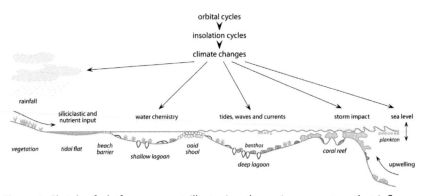

Figure 1 Sketch of platform transect illustrating the main parameters that influence carbonate production by benthic and planktonic organisms, and sediment distribution. Parameters independent of orbital cycles (such as tectonics or volcanic activity) are not considered.

and forth as on ooid and bioclastic shoals, or is removed from the production area by currents. Through time, the sediment will accumulate vertically and/or laterally, will possibly be reworked by waves, currents, and bioturbation, and will finally be stabilized through microbial mats, shallow burial, and/or early diagenesis. The resulting vertical and lateral facies changes are the record of changing environmental conditions as mentioned in Chapter 2.1 (Fig. 1).

A "depositional sequence" is here defined as a facies succession that translates facies changes (deepening- then shallowing–up), or that represents aggradation of sediment without facies change and is delimited by marly seams or distinct surfaces (Strasser et al., 1999). A depositional sequence is independent of scale and can, if facies changes and diagnostic surfaces allow for it, be interpreted in terms of sequence stratigraphy (Mitchum and Van Wagoner, 1991; Posamentier et al., 1992). A priori, a depositional sequence is descriptive and does not imply any interpretation in terms of cyclostratigraphy. The term "cycle" is here used to describe a repetitive process that is more or less cyclical (such as climatic and sea-level changes induced by orbital cycles). A depositional sequence may be created by a sea-level cycle but also by other processes (see Chapter 2.2.1), and a symmetrical sea-level cycle may create an asymmetrical depositional sequence (Chapter 2.2.2).

Unconformities created by absence of sediment accumulation on the sea floor (e.g., current–induced bypass) or by submarine or subaerial erosion will form boundaries that may be recognized in the field and delimit "beds". The same holds for rapid changes in facies where especially clay input creates well-defined bedding planes. Such boundaries, however, may be obliterated by bioturbation as long as the sediment stays soft and the bottom waters are oxygenated. During early diagenesis, diagenetic fronts may also create bedding planes that are recognizable in the field (e.g., Westphal et al., 2004). Stacking of beds means that there were repetitive environmental changes through time, and the analysis of the facies evolution within the beds and of the bedding surfaces allows determining which processes were active. However, a bed as seen in the outcrop does not necessarily correspond to a depositional sequence as defined above.

On a shallow carbonate platform, sea-level changes are an important factor because they control water depth and accommodation potential. A second important parameter is rainfall in the hinterland that delivers clays and other terrigenous material (e.g., quartz, organic matter) to the marine realm. Depending on the energy regime and the morphology of the platform, the clays may be reworked and winnowed by currents and waves, or they are ponded in depressions (Fig. 2). If sea-level changes and terrigenous input are associated with orbitally controlled climate changes and the clays create bedding surfaces, the repetitive stacking of beds may indeed be interpreted in terms of cyclostratigraphy. However, the terrigenous input may be in-phase or out-of-phase with sea-level changes, or it may be independent of sea level (Strasser and Hillgärtner, 1998). For example, during one sea-level cycle, one depositional sequence is created that shows first a deepening-up then a shallowing-up facies evolution. If there are two or more episodes of rainfall in the hinterland during the same time interval, the clays being washed into the system will lead to the formation of two or

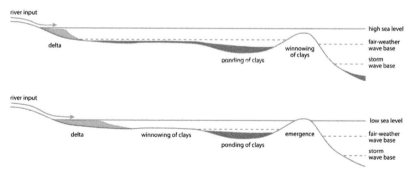

Figure 2 Clay input and distribution as a function of sea level (high versus low), wave base, and platform morphology. The amount of clays is a function of river input (and thus climate in the hinterland), which can vary in tune with sea-level changes, or independently.

more beds within the one depositional sequence. The interpretation of clay seams or marly layers separating carbonate beds therefore has to be carefully evaluated.

2.2.1 Allocyclic versus Autocyclic and Random Processes

Processes that are repetitive (cyclical) and controlled by factors external to the sedimentary basin are called "allocyclic". Examples are eustatic sea-level changes and climate changes that act over the entire carbonate platform and beyond. In contrast, "autocyclic" processes are inherent to the sedimentary system and create facies repetitions independently of external factors. For example, lateral migration of shoals may lead to several, superimposed shallowing-up units at constant sea level (Pratt and James, 1986). Another possibility is the progradation of a tidal flat, which fills in the lagoon and thus stops the carbonate production that feeds the same tidal flat: a shallowing-up sequence has been created. Once subsidence has allowed a new lagoon to form, a new tidal flat will prograde again to create a second sequence (Ginsburg, 1971).

Many shallow carbonate platforms display facies mosaics (e.g., Rankey and Reeder, 2010; Adomat and Gischler, 2015). These may follow substrate morphology or structural elements such as fault lines but may also have a random or self-organized distribution (e.g., Burgess and Wright, 2003; Burgess, 2006). Facies evolution through time thus is difficult to predict as no regular pattern of facies successions can be established (Fig. 3). For example, at rising sea level, an aggrading or even shallowing-up sequence can form contemporaneously to a deepening-up one, on the same platform, if carbonate accumulation keeps up with or outpaces sea-level rise (Fig. 3C). Based on computer simulations, Dexter et al. (2009) stated that stochastic processes could mask an orbitally controlled sea-level signal in the sedimentary record.

If the amplitude of the allocyclic processes is high enough, the autocyclic or random ones will be overprinted. For example, a significant drop in eustatic sea level will subaerially expose the entire platform and create a karst surface; if such drops are repeated, a record of stacked beds with clearly defined boundaries will develop. Repetitive periods of heavy rainfall in the hinterland will flush clays onto the entire platform and create discrete bedding surfaces, independent of the possibly random facies evolution within these beds.

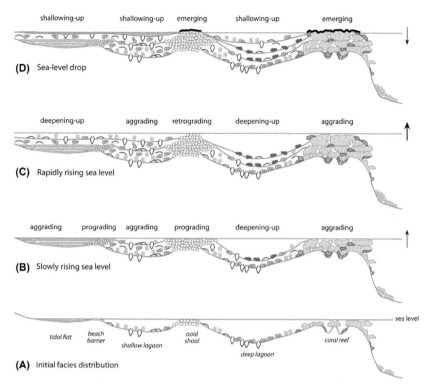

shallowing-up   shallowing-up   emerging   shallowing-up   emerging

(D)   Sea-level drop

deepening-up   aggrading   retrograding   deepening-up   aggrading

(C)   Rapidly rising sea level

aggrading   prograding   aggrading   prograding   deepening-up   aggrading

(B)   Slowly rising sea level

sea level

tidal flat   beach   ooid   coral reef
               barrier   shoal
                       shallow lagoon   deep lagoon

(A)   Initial facies distribution

Figure 3 Sketch of possible facies evolution of different, juxtaposed environments during slow then rapid sea-level rise (A, B, C), and finally during a sea-level drop (D). Note that the resulting vertical depositional sequences may display aggrading, prograding, and deepening-up trends at the same time, depending on their position on the platform and on the potential of carbonate production and accumulation.

2.2.2 Long-Term versus Short-Term Processes

Orbitally-induced environmental changes act on periodicities of a few tens to a few hundreds of thousands of years (Milankovitch frequency band) as a function of the orbital parameters (Berger et al., 1989; Laskar et al., 2011):

- precession cycle with an average of 20 kyr today but somewhat shorter in the past;
- obliquity cycle with a major peak at 41 kyr today but much shorter in the past;
- short and long eccentricity cycles (100 and 405 kyr, respectively), stable throughout Earth history;

- the longer periodicities (1.2 and 2.4 Myr; Hilgen et al., 2003) are not considered here because their impact often is difficult to distinguish from the long-term tectonic and eustatic processes affecting a carbonate platform.

The high-frequency cycles (20 to 405 kyr) are superimposed on million-year scale eustatic changes in sea level induced by mid-ocean-ridge activities, water exchange with the mantle, intraplate deformation, or changes in large-scale sediment input into the ocean basins (e.g., Cloetingh, 1986; Conrad, 2013; Haq, 2014). A long-term sea-level rise will generally create more accommodation on the platform and attenuate short-term sea-level falls, while a long-term sea-level fall will lead to more subaerial exposure (Fig. 4).

The second factor controlling accommodation is subsidence. On passive margins where many carbonate platforms are hosted, subsidence rate generally is low (a few cm/kyr; e.g., Grotzinger, 1986; Husinec and Jelaska, 2006) but may be irregular due to block-faulting (e.g., Wildi et al., 1989). Subsidence mainly acts on the long-term history of a carbonate platform. However, Cisne (1986) and Bosence et al. (2009) have proposed that high-frequency changes in subsidence rate may create stacking patterns of beds similar to those formed under the influence of cyclical sea-level changes.

In Fig. 4, the sedimentary record at one point on the shallow platform resulting from 5 sea-level cycles is simulated. For simplification, the cycles are assumed to be symmetrical as it would be the case in a greenhouse world where insolation changes are more or less directly translated into sea-level changes (e.g., Read et al., 1995), in contrast to the high-amplitude asymmetrical cycles typical for icehouse worlds where there is a slow build-up of polar ice and a fast melting (e.g., Shackleton, 1987). Amplitudes are variable, reflecting the variability of insolation intensity of 5 precession cycles modulated by a short eccentricity cycle (Berger, 1990). The sea-level amplitudes are on the order of one to several meters, as inferred from the typical meter-scale beds in the sedimentary record (see Chapter 3). Subsidence is considered to be constant. The sedimentary record is interpreted in terms of sequence stratigraphy (Fig. 4A), where sequence boundaries form during loss of accommodation, transgressive surfaces at the beginning of creation of new accommodation, and maximum-flooding surfaces or maximum-flooding intervals during the fastest increase in accommodation (Vail et al., 1991; Montañez and Osleger, 1993). Sedimentation rate is variable: slow after subaerial

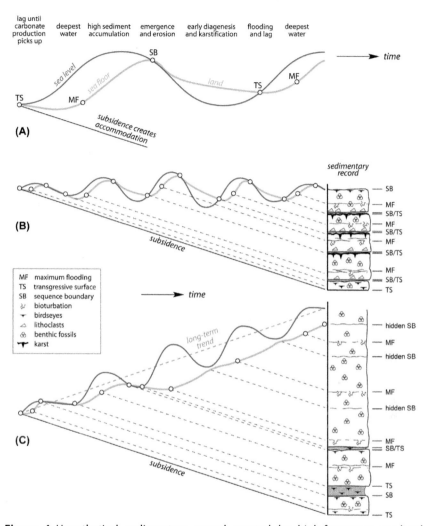

Figure 4 Hypothetical sedimentary record created by high-frequency sea-level fluctuations (amplitudes are assumed to be meter-scale). (A) Relation between sea level and sediment surface, and positions of sequence boundaries, transgressive surfaces, and maximum-flooding intervals. (B) Sedimentary record created by 5 sea-level cycles with varying amplitudes, no long-term trend. (C) Sedimentary record resulting from the same high-frequency fluctuations superimposed on a long-term rising trend of sea level. Sedimentary columns without compaction. For more explanation refer to text.

emergence when the carbonate-producing organisms first have to recolonize the exposure surface (lag time: Tipper, 1997; Kemp and Sadler, 2014), high when the ecological conditions are optimal, and slow again when the water is not deep enough to permit healthy carbonate production. It has to be kept in mind, however, that this simulation concerns only one point on the platform, and that sedimentation rates may vary significantly from one place to the other (Fig. 3). Once sea level drops below the sediment surface, some erosion of loose material may occur before early diagenesis stabilizes the sediment and chemical erosion (karstification) predominates.

In Fig. 4B, long-term accommodation is created only by subsidence. It shows that sequence boundaries related to low-amplitude sea-level cycles are characterized by tidal flats (visualized here by the birdseye symbol), whereas high-amplitude fluctuations lead to development of karst surfaces. If the same high-frequency sea-level curve is superimposed on a long-term rising sea-level trend (Fig. 4C), the resulting accommodation is much greater. Starting from zero, the first two sequences still have a tidal flat at their boundaries, but subsequent, even high-amplitude sea-level drops do not generate subaerial exposure surfaces to create a well-visible sequence boundary. Only a facies change (from deeper to shallower facies and back) allows the identification of the lowest accommodation ("hidden sequence boundary" in Fig. 4C; "subtidal cycles" of Osleger, 1991). If no facies change is discernible, then the sea-level cycle is not recorded at all.

High-frequency fluctuations in accommodation have often been reconstructed by using "Fischer plots" (Fischer, 1964; Read and Goldhammer, 1988; Sadler, 1994; Husinec et al., 2008) whereby the thickness of each unit (with a deepening-shallowing facies trend, and/or delimited by distinctive surfaces) is plotted against time. By this, long-term trends in accommodation change can be detected. However, such reconstructions are only valid as sea-level proxies if each unit shallows upwards into the intertidal zone (i.e., up to zero water depth), if there is no erosion, if subsidence rate is constant, and if each unit represents the same time duration. Unfortunately, these conditions are only rarely fulfilled.

If accommodation is not completely filled by sediment and the beds represent subtidal cycles (Osleger, 1991), estimation of water depth based on facies is very difficult (Immenhauser, 2009; Purkis et al., 2015). The evolution of unit thicknesses through time thus gives information about carbonate production potential and accommodation but cannot be used as an equivalent of sea-level changes.

2.2.3 Fragmentary Sedimentary Record

Orbitally-controlled climate and/or sea-level changes are only fully recorded if the conditions are ideal:

- enough accommodation (as a function of subsidence and eustatic sea level) to avoid prolonged subaerial exposure and erosion;
- high sedimentation rate to avoid condensation;
- absence of high-energy events that destroy previous sedimentary deposits.

On shallow carbonate platforms, these conditions are only rarely fulfilled, and hiatuses and condensation are common phenomena (e.g., Sadler, 1994; Strasser, 2015; Husinec and Read, 2018). For example, if a long-term sea-level drop combines with a slow subsidence rate, even high amplitudes of high-frequency sea-level changes may not be enough to create a sedimentary record on the platform. Cycles that are not recorded on emerged platforms were called "missed beats" by Goldhammer et al. (1990). During an individual sea-level cycle, only part of the history is recorded if sea-level fall outpaces subsidence, thus leading to loss of accommodation (Fig. 4A). Currents and waves may constantly rework the sediment, and the record of high-frequency changes is lost. If sediment accumulation is low, bioturbation may destroy any originally cyclic sedimentation. If the orbital signal (and its translation into climate and/or sea-level changes) is weak, it may be recorded in one depositional environment that is sensitive enough but not in a different one. On tidal flats, even a sea-level change of only a few tens of centimeters will induce important changes in the distribution of tidal channels and microbial mats. In a shallow lagoon, a drop in wave base following a drop in sea level will induce winnowing on the sea floor, while the same drop will not affect a deeper lagoon. It may thus be difficult to correlate depositional sequences from one environment to another.

2.3 The Latemar Controversy

From the above it is clear that the estimation of the time involved in the cyclical stacking of facies and beds on shallow carbonate platforms is not an easy task (e.g., Eberli, 2013). A good example of the difficulties encountered is the controversy about the Middle Triassic Latemar platform in northern Italy (Zühlke, 2004; Meyers, 2008). The individual, meter-scale beds display a shallowing-up facies evolution, implying that they formed driven by sea-level changes. The hierarchical stacking (5 beds commonly form a bundle) and time-series analysis suggest that there was an orbital control (5 precession cycles within one short eccentricity cycle), implying that the cyclic series had a duration of 9 to 12 Myr (e.g., Goldhammer et al., 1990; Satterley, 1996; Preto et al., 2004). In contrast, ammonite biostratigraphy calibrated by U-Pb ages measured in volcanic ash layers suggests a duration of 2 to 4 Myr for the same interval (e.g., Brack et al., 1996; Mundil et al., 1996). This would mean that one bed does not correspond to the precession cycle of 20 kyr but must have formed in tune with a higher-frequency cyclical process with periodicities of 4.2 kyr ("sub-Milankovitch" cycles; Zühlke et al., 2003; Zühlke, 2004). The modelling by Forkner et al. (2010, p. 1) produced "Latemar-like stratigraphy with both pure Milankovitchian and mixed Milankovitchian—sub-Milankovitchian temporal frameworks". Based on magnetostratigraphy, Kent et al. (2004) and Spahn et al. (2013) found that the cyclical stacking was controlled by sea-level changes with periodicities of about 1.7 kyr.

During the Quaternary, cyclicities of a few kyr duration are related to ice-sheet dynamics and oceanic circulation patterns (Dansgaard-Oeschger cycles, Heinrich events, Bond cycles; e.g., Romero et al., 2011; Saha, 2015). Rodrigo-Gamiz et al. (2014, p. 78), analysing sediments of the last 20 kyr in the western Mediterranean, found periodicities of 1.3, 1.5, 2, and 5 kyr and explained these by "climate cycles … coupled with ocean-atmosphere fluctuations". Concerning the Latemar platform, Zühlke (2004, p. 206) wrote: "it is well possible that rapid oceanographic and/or atmospheric changes triggered a strong sub-Milankovitch signal in the Middle Triassic". Sub-Milankovitch periodicities are also seen in ancient deeper-water sections. For example, Rodríguez-Tovar and Pardo-Igúzquiza (2003) detected a 13 to 14-kyr peak in the Kimmeridgian of southern Spain, and De Winter et al. (2014) found a prominent 7-kyr cycle expressed in the Campanian of DSDP Site 516F in the South Atlantic.

3. CASE STUDY: OXFORDIAN OF THE SWISS JURA

In order to show the caveats and limitations when applying the cyclo-stratigraphic approach to shallow–marine carbonates but also to demonstrate its potential to better interpret the sedimentary record, a case study of the Oxfordian (Late Jurassic) in the Swiss Jura Mountains is presented.

3.1 Geological and Stratigraphic Context

During the Oxfordian, the Swiss Jura Mountains were part of a shallow carbonate platform situated at a paleolatitude of 26 to 27°N, north of

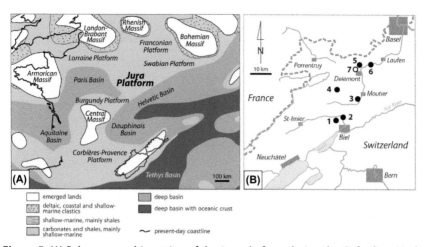

Figure 5 (A) Paleogeographic setting of the Jura platform during the Oxfordian. Modified from Carpentier et al. (2006), based on Enay et al. (1980), Ziegler (1990), and Thierry et al. (2000). (B) Locations of the studied sections in the Jura Mountains of northwestern Switzerland. 1: Forêt de Châtel; 2: Péry-Reuchenette; 3: Gorges de Court; 4: Pichoux; 5: Mettemberg-Soyhières; 6: Liesberg; 7: Vorbourg. In blue (or grey): lakes and rivers.

the Tethys Ocean (Fig. 5A; Dercourt et al., 1993). The platform was structured by synsedimentary faults related to the reactivation of basement faults (Allenbach, 2001). The lithostratigraphy and the ammonite biostratigraphy have been established by Gygi (1995, 2000), and a sequence-stratigraphic interpretation was first proposed by Gygi et al. (1998). For the purpose of this case study, the focus will be on six sections (Fig. 5B) that cover the interval between sequence boundaries Ox 7 and Ox 8 (Fig. 6). These sequence boundaries are labelled according to Hardenbol et al. (1998) and Gygi et al. (1998). The Vorbourg section (number 7 in Fig. 5B) offers an example of detailed analysis of an elementary sequence corresponding to the precession cycle (see Chapter 4).

3.2 Facies Evolution

The section of Gorges de Court, logged in detail by Strasser et al. (2000) and Hug (2003), demonstrates the evolution of depositional environments through time (Fig. 7). The Hauptmumienbank Member is characterised by abundant oncoids but also contains ooids, echinoderms and corals, implying normal-marine shallow-water conditions. The Oolithe rousse Member is rich in quartz sand and plant fragments; its reddish ("rousse") color stems from iron oxides. While the environment still was shallow-marine, a more humid

Chrono-strat.	Biostratigraphy ammonite zones	Sequence boundaries	Lithostratigraphy formations	Lithostratigraphy members
early Kimm.	Hypselocyclum		Reuchenette	
		Kim 2		
early Kimm.	Platynota		Reuchenette	
		Kim 1		
early Kimm. / late Oxfordian	Planula		Balsthal	Verena
		Ox 8		
early Kimm. / late Oxfordian	Bimammatum		Courgenay	La May
			Vellerat	Oolithe rousse
		Ox 7		
late Oxfordian	Hypselum		Vellerat	Hauptmumienbank
		Ox 6		
late Oxfordian	Bifurcatus		Vellerat	Röschenz

Figure 6 Stratigraphic context of the studied sections. Chrono- and biostratigraphy following Wierzbowski et al. (2016). The Oxfordian-Kimmeridgian boundary in the Tethyan realm was first placed at the Planula-Platynota boundary but is now proposed to be situated at the limit between the Hypselum and Bimammatum ammonite zones. Sequence stratigraphy according to Hardenbol et al. (1998) and Gygi et al. (1998); lithostratigraphy according to Gygi (1995).

climate furnished terrigenous material from the emerged lands to the north (Fig. 5A). The overlying La May Member starts with an ooid grainstone, then contains more oncoids and corals. The depositional environment was mainly that of protected, bioturbated lagoons with patch reefs. However, many bed tops are dolomitized (clear-rim-cloudy-center displacive crystals; Hug, 2003), suggesting a more arid climate and the development of sabkhas. The member ends with a dolomitized tidal flat overlain by a layer of coal, indicating a rapid change from a more arid to a more humid climate. The overlying Verena Member is composed of ooid- and peloid-rich grainstones with brachiopods and echinoderms, indicating a fully marine high-energy environment.

The sedimentary record is well structured into individual beds (Fig. 7), the boundaries of which are marked by clay seams.

3.3 Sequence- and Cyclostratigraphic Interpretation

The sequence- and cyclostratigraphic interpretation and the correlation of the six studied sections (Fig. 8) are based on the concepts presented in Strasser et al. (1999); the sequence-stratigraphic nomenclature is that of Vail et al. (1991). Elementary sequences are defined as the smallest units in which facies trends indicate a cycle of environmental change, including sea-level change. When no facies change occurs within a bed, the clay seams at the bed limits indicate that there was terrigenous input from the

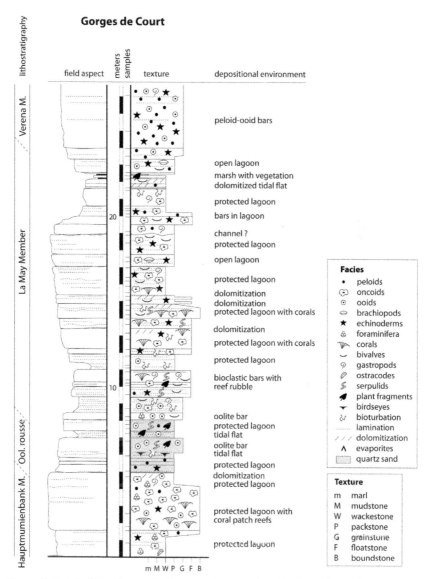

Figure 7 Detail of the Gorges de Court section (number 3 in Fig. 5B). Log, facies analysis and interpretation of depositional environments following Strasser et al. (2000) and Hug (2003).

hinterland (Figs. 2 and 5A; Strasser and Hillgärtner, 1998). Small-scale sequences are composed of 2 to 7 elementary sequences and generally display a deepening then shallowing trend, with the shallowest facies at the boundaries. Four small-scale sequences compose a medium–scale

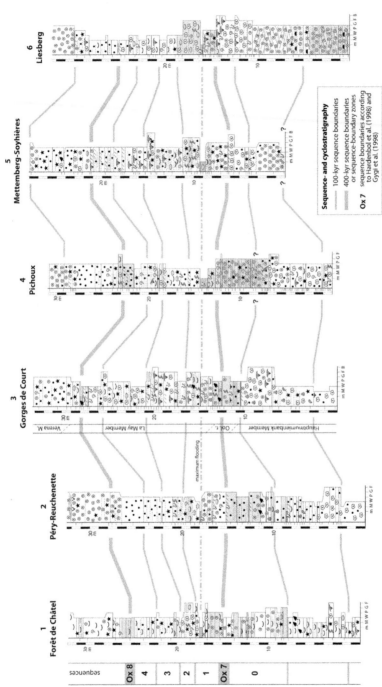

Figure 8 Sequence- and cyclostratigraphic correlation of selected Oxfordian sections, concentrating on the interval between sequence boundaries Ox 7 and Ox 8 (see Fig. 6). The geographic position of the sections is shown in Fig. 5B. The log of Forêt de Châtel after Gsponer (1999), Péry-Reuchenette, Mettemberg-Soyhières and Liesberg after Hug (2003), Gorges de Court after Hug (2003), Gorges de Court after Strasser et al. (2000) and Hug (2003), Pichoux after Pittet (1996). Symbols are explained in Fig. 7. For discussion refer to text.

sequence, which again displays a general deeping-shallowing trend of facies evolution and the relatively shallowest facies at its boundaries. In the case presented here, the sequence defined by the boundaries Ox 7 and Ox 8 represents a medium-scale sequence (Strasser et al., 2000). The relatively deepest and/or most open-marine facies of this sequence formed at the time of maximum flooding (taken as datum in Fig. 8). If several elementary sequences compose an interval of shallowest or deepest facies, a sequence-boundary zone respectively a maximum-flooding interval is defined (Montañez and Osleger, 1993). The concept of parasequences bounded by marine flooding surfaces (Van Wagoner et al., 1990) is not applied here because many of the elementary, small-scale, and medium-scale sequences are clearly bounded by surfaces expressing the shallowest depositional environment, and because these sequences can be further subdivided into deepening and shallowing parts. Each section is first interpreted independently, then a best-fit solution is sought for that respects the chronostratigraphic framework and shows the least contradiction in sequence- and cyclostratigraphic interpretation (Fig. 8).

According to Gygi et al. (1998), sequence boundary Ox 7 is situated at the base of the Bimammatum ammonite zone, Ox 8 at the limit between the Bimammatum and Planula zones (Fig. 6). Hardenbol et al. (1998), who established a sequence-stratigraphic framework for European basins, attributed an age of 155.15 Ma to Ox 7 and of 154.63 Ma to Ox 8. These numerical values are interpolations based on the duration of the Oxfordian of about 5.3 Myr, from 159.4 ± 3.6 to 154.1 ± 3.2 Ma (Berggren et al., 1995). Gradstein et al. (2012) and the International Chronostratigraphic Chart (2017) indicate a duration of about 6.2 Myr, from 163.5 ± 1.0 to 157.3 ± 1.0 Ma. In the Geological Time Scale 2012 (Gradstein et al., 2012, p. 756), the sequence boundaries are indicated only for the Boreal realm: Ox 7 is dated at 158.8, Ox 8 at 158.2 Ma. Despite these uncertainties related to the Boreal-Tethyan correlation (e.g., Colombié and Rameil, 2007; Wierzbowski et al., 2016), the time interval between sequence boundaries Ox 7 and Ox 8 can be estimated to be in the order of 500 to 600 kyr. However, considering the hierarchical stacking (bundling) of 2 to 7 elementary sequences into a small-scale sequence, and of 4 small-scale sequences into a medium-scale one, it can be assumed as a working hypothesis that the medium-scale sequence discussed here corresponds to the 405-kyr long eccentricity cycle. In fact, it has been shown that in several cases the "third-order" sequences of classical sequence stratigraphy (e.g., Vail et al., 1991) actually formed in tune with the long eccentricity cycle or multiples thereof (e.g., Strasser et al., 2000; Gale et al., 2002; Boulila et al., 2008, 2010, 2011; Ogg et al., 2010). Consequently, the small-scale sequences would represent the 100-kyr short eccentricity cycle. The elementary sequences often are difficult or impossible to correlate. They may correspond to the 20-kyr precession cycle, or else they formed through autocyclic processes (see Chapter 2.2.1). The obliquity cycle did not leave any clear signal on the Jura platform that could be interpreted in terms of sea-level variations (Pittet, 1996).

The correlation presented in Fig. 8 shows significant facies variability, vertically as well as horizontally along the time lines given by the cyclostratigraphic interpretation. From this, Strasser et al. (2015) reconstructed facies mosaics for different time slices and monitored the lateral migration of sediment bodies.

3.4 Reconstruction of Orbitally Controlled Sea-Level Changes

In order to reconstruct high- and low-frequency sea-level changes that controlled carbonate production and deposition on an ancient shallow-marine platform, several conditions have to be fulfilled:

- the sedimentary record has to be decompacted according to facies, in order to estimate the sediment thickness accumulated during a certain time interval prior to burial;
- the water depth has to be estimated for each facies;
- erosion of previously deposited sediment has to be minimal;
- the time steps have to be as short as possible in order to obtain a curve of highest-frequency sea-level changes.

Two sections are chosen to attempt such a reconstruction: Gorges de Court and Mettemberg-Soyhières, where the well-structured sedimentary record allows for a relatively straightforward cyclostratigraphic interpretation. The smallest stratigraphic units are the elementary sequences that for these two sections are interpreted to have formed in tune with the orbital precession cycle of 20 kyr. The methodology follows that developed by Pittet (1994) for the Oxfordian and by Hillgärtner and Strasser (2003) for the Berriasian and Valanginian of the Jura Mountains.

3.4.1 Compaction and Decompaction

According to Moore (1989), mechanical reorganization of grains and dewatering in carbonate mud leads to a porosity loss of 10% to 30% after the first 100 m of burial. Shinn and Robbin (1983) showed in experiments that mechanical and dewatering compaction leads to a 20% to 70% loss of volume. However, these values are less if the sediment is stabilized by early cementation (Halley and Harris, 1979). Goldhammer (1997) estimated a compaction of slightly over 50% for carbonate mud buried at 1000 m, and of about 15% for carbonate sand at the same burial depth. According to Enos (1991), muddy terrigenous and muddy carbonate sediments do not have significantly different compaction rates, but pressure solution along clay seams may enhance chemical compaction in carbonates (Bathurst, 1987). With deeper burial, chemical compaction becomes important. In the Jura Mountains, however, burial was at the most 2000 m (Trümpy, 1980); nevertheless, pressure solution at grain contacts shows that some compaction must have occurred.

Based on these published values, the Gorges de Court and the Mettemberg-Soyhières sections are decompacted with the following factors (Figs. 9 and 10): 1.2 for grainstones and 2.5 for mudstones. For packstones and wackestones, the intermediate values of 1.5 and 2 are used, for marls the factor 3. Boundstones are thought to have resisted major compaction and a factor of 1.2 is applied. Floatstones with their micritic matrix are treated like packstones.

3.4.2 Estimation of Water Depth

On Great Bahama Bank, which is often taken as analogue for ancient carbonate platforms, Harris et al. (2015) found that there is no clear relationship between water depth and facies, and that facies distribution (i.e., muddy versus grainy facies) is rather controlled by currents. They also stated that 60% of the Bank surface lie in water depths of 5 m or less. In the studied sections, only tidal flats and bedding surfaces that exhibit penecontemporaneous dolomitization or evaporites can be used as tie points, and a water depth of zero is attributed to them (tidal range is not considered here). For the other facies, the estimation of water depth includes wide error margins. Nevertheless, for the purpose of this case study, the following values are assumed:

Packstones and wackestones with restricted fauna are thought to have formed in protected, shallow bays and are considered to represent about 1 m water depth. Grainstones often result from winnowing by tidal currents, and also for such facies a water depth of 1 m is assumed. Packstones and floatstones with abundant fauna are set at 2 m. Coral patch reefs possibly grew up to sea level but are assumed to have needed 2 m water depth to develop. When an open-marine fauna (e.g., brachiopods) is present, then a water depth of 3 m is assumed.

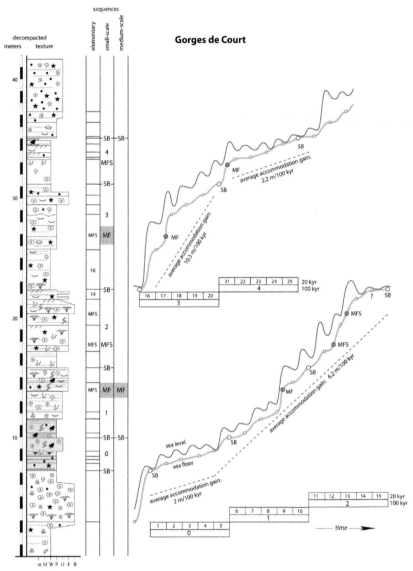

Figure 9 Decompacted section of Gorges de Court (compare with Fig. 7; for decompaction factors see text). Elementary, small-scale, and medium-scale sequences are interpreted following Strasser et al. (2000). SB: sequence boundary; MFS: maximum-flooding surface; MF: maximum-flooding interval. Aggradation of the sea floor interpreted for equal time steps of the elementary (20-kyr) sequences. Sea-level curve interpreted according to the water depth in which the different facies formed (see text for explanation). Average accommodation gain (i.e., subsidence + long-term sea-level rise) calculated between tie points where water depth was zero. The sequence boundary at meter 31.2 of the section is not at zero but estimated to have formed at ca. 1.5 m water depth.

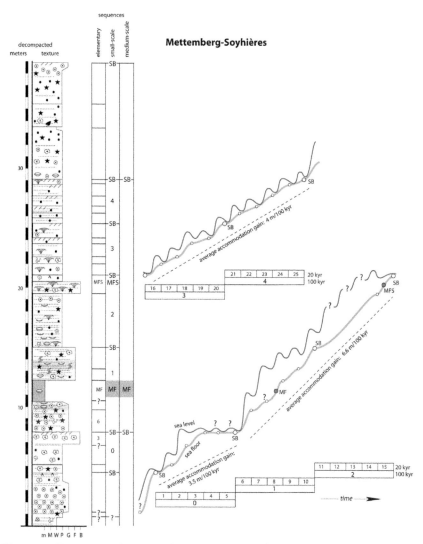

Figure 10 Decompacted section of Mettemberg-Soyhières. For explanation see legend to Fig. 9 and text.

No signs of deep erosion or karstification are visible in the studied sections, with the exception of the irregular surface at meter 19 in Gorges de Court (Fig. 7) that possibly represents a channel floor. This implies that there was a continuous gain in accommodation, which allowed for the recording of most of the sediment produced on the platform. Consequently, the model of Fig. 4C is applied, and the water depth values mentioned above correspond to the highest position of sea level. The input of clays to form the bedding surfaces in subtidal facies is assumed to be related to a sea-level drop that exposed the hinterland to increased erosion. This is indicated by the fact that — with the exception of marly maximum-flooding intervals with marine fauna — the marl seams separating the beds are commonly related to the shallowest facies (Pittet and Strasser, 1998b).

3.4.3 Reconstruction of Accommodation and Sea-Level Changes

Based on the decompacted sections and the estimated water depths, and within the time frame given by the cyclostratigraphic interpretation, the evolution through time of the sea floor as well as of the sea level can be plotted (Figs. 9 and 10). The time steps are given by the 20 kyr attributed to the elementary sequences. For each of these sequences, the decompacted thickness is added to represent the sea floor, and the estimated water depth for the dominant facies in each sequence is added to approximate the sea level. Furthermore, it is assumed that the sea-level fluctuations were symmetrical (greenhouse conditions; Fig. 4). In the cases where one 100-kyr sequence contains less than 5 elementary sequences, it is assumed that the missing ones were not recorded due to low accommodation at a small-scale sequence boundary (elementary sequence 15 at Gorges de Court and elementary sequences 4 and 5 at Mettemberg-Soyhières). Small-scale sequence 2 at Mettemberg-Soyhières does not exhibit well-defined beds but - in comparison with the same sequences at Gorges de Court - it is assumed that it nevertheless spans 100 kyr.

An average accommodation gain is visualized by a line that connects two tie points with zero water depth. For the boundary between the 100-kyr sequences 3 and 4 at Gorges de Court, a water depth of 1.5 m is assumed (Fig. 9).

3.4.4 Sea-Level Amplitudes and the Orbital Signal

In a first step, the average accommodation gain is subtracted from the sea-level curves reconstructed for both sections (Fig. 11). This visualizes the high-frequency sea-level changes and reflects in some cases the relation to the 100-kyr cyclicity (cycles 0, 1, and 2 for Mettemberg-Soyhières, cycle 3 for Gorges de Court). However, there are wide discrepancies between the two curves, the most striking one in the first part of 100-kyr cycle 1 (rising trend at Mettemberg-Soyhières versus falling trend at Gorges de Court).

In a second step, only the high-frequency (i.e., 20-kyr) cycles and their amplitudes are considered by subtracting the 100-kyr trends. The

Sea-level deviation from average accommodation gain

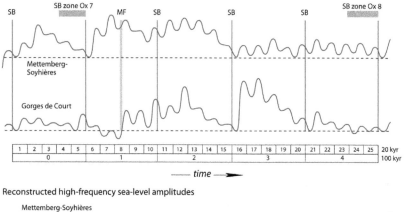

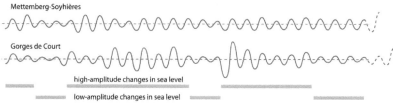

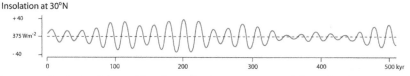

Figure 11 Analysis of the sea-level amplitudes reconstructed from the decompacted sections of Gorges de Court and Mettemberg-Soyhières. In a first step (top of figure) the deviation of sea level from the average accommodation gain is plotted, representing an approximation of high-frequency sea-level change disregarding subsidence and long-term sea-level rise. Sequence boundaries related to the 100-kyr cyclicity are indicated, as are the maximum-flooding interval and the sequence-boundary zones of the 405-kyr cycle. In the lower part of the figure, the amplitudes of the 20-kyr cyclicity are reconstructed by subtracting the 100-kyr trends. Intervals of relatively high and relatively low amplitudes can be identified, comparable to the amplitude changes of insolation at 30°N for the last 500 kyr (Berger, 1990). For both Oxfordian sections the sea-level amplitudes increase above SB zone Ox 8 (to the right in the figure), but they cannot be reconstructed because information about the accommodation gain is missing. Note that the amplitudes of insolation also increase at around 500 kyr. For more explanation refer to text.

correspondence between the two sections is better, and intervals of rather high-amplitude changes versus intervals of rather low-amplitude changes can be distinguished. However, there are still discrepancies between the two curves, namely for the 20-kyr cycle 16 (Fig. 11).

The third step concerns the comparison of the reconstructed high-frequency amplitude changes with the curve of insolation changes calculated by Berger (1990) for 30°N latitude at the top of the atmosphere, which shows that the amplitudes created by the 20-kyr precession cycle vary significantly through time. The 405-kyr long eccentricity cycle is well defined by the intervals of low amplitude, while the expression of the 100-kyr short eccentricity cycle is more subtle.

4. DISCUSSION

Despite the caveats mentioned in Chapter 2, and despite the large error margins inherent in the sea-level reconstructions proposed in Chapter 3, some interesting trends can be distilled from Figs. 9–11.

The interval between sequence boundaries Ox 7 and Ox 8 is part of a long-term transgressive trend (Hardenbol et al., 1998), which explains the absence of major hiatuses and allows for a relatively faithful recording of high-frequency sea-level fluctuations. The long-term rising trend of eustatic sea level certainly was synchronous across the Jura platform, which means that the differences in accommodation gain (Figs. 9 and 10) must be due to differences in subsidence rate. The distance between the Gorges de Court and the Mettemberg-Soyhières sections is 16 km today (Fig. 5B) but was in the order of 30 km before the folding of the Jura Mountains (Strasser et al., 2015). Allenbach (2001) has shown that synsedimentary tectonic movements were common on the Jura platform during the Oxfordian, and it is reasonable to assume that subsidence rates differed between sections. The general trend of accommodation gain is first similar for both sections, and this can be attributed to a rising trend in eustatic sea level: slow during the deposition of small-scale sequence 0, then rapidly rising for sequences 1 and 2. Sequences 3 and 4 differ: a continuous gain of 4 m/100 kyr at Mettemberg-Soyhières (Fig. 10) versus 10.3 m/100 kyr then 2.2 m/100 kyr at Gorges de Court (Fig. 9). This rapid accommodation gain at Gorges de Court may be due to a pulse of increased subsidence at this location.

It is interesting to note that the maximum-flooding interval of the medium-scale (405-kyr) sequence formed at the beginning of the rapidly rising trend of eustatic sea level, within small-scale sequence 1 (Figs. 8—10). In the sequence-stratigraphic model this would mean that this particular 405-kyr sequence has a thin transgressive and a thick highstand systems tract. Wildi et al. (1989) gave an estimate of Late Jurassic subsidence rates in the southern Jura Mountains of 20 to 40 m/Myr, i.e., 2 to 4 m per 100 kyr. Subtracting these values from the accommodation gain of 6.2 to 6.6 m during 100-kyr cycles 1 and 2 (Figs. 9 and 10), there results a eustatic sea-level rise of 2.2 to 4.6 m per 100 kyr over this time interval.

The uncertainties increase when dealing with the high-frequency sea-level changes. As it can be assumed that high-frequency eustatic sea-level changes were the same over the Jura platform, the discrepancies between the two reconstructed sea-level curves (Fig. 11) must be due to misidentification of the elementary sequences, incorrect decompaction factors, and/or incorrect water depth estimates. During the Oxfordian greenhouse climate, insolation changes probably translated more or less directly into sea-level changes, mainly through thermal expansion and contraction of the volume in the oceanic surface waters (Sames et al., 2016). Sames et al. (2016) suggest that thermal expansion and retraction of ocean water operates on time scales of 1 to 10'000 years and causes sea-level changes of 5 to 10 m amplitude. This is well within the time frame of the 20-kyr precession cycle of the Earth's orbit and also corresponds to the meter-scale amplitudes needed to accommodate meter-scale depositional sequences. The insolation curve shown in Fig. 11 is calculated for 30°N (Berger, 1990), which is close to the paleolatitude of the Jura platform (26—27°N; Dercourt et al., 1993). There is no significant change in the frequencies and amplitudes between this curve and the one calculated by Forkner et al. (2010) for their Latemar study at an equatorial position. The curve displays intervals of high amplitudes and intervals of low amplitudes, as it is seen in the reconstructed sea-level curves. The low-amplitude intervals in insolation correspond to the boundaries of the 405-kyr long eccentricity cycle, and also the sea-level amplitudes around medium-scale sequence boundaries Ox 7 and Ox 8 have relatively low amplitudes.

This comparison now allows for a better interpretation of the sedimentary record. It supports the assumption that the elementary sequences are related to the 20-kyr precession cycle and the small-scale sequences to the 100-kyr short eccentricity cycle, and that the interpretation of missing elementary sequences at small-scale sequences boundaries probably is correct. However, the large amplitude change in elementary sequence 16 at Gorges de Court (Fig. 11), that is not seen in Mettemberg-Soyhières, may at least partly stem from a misinterpretation of water depth. Following the dolomitization of the sequence-boundary (top of elementary sequence 14) and sequence 15 missing, a lag time probably occurred and the corals at the base of sequence 16 grew in very shallow water. However, also a localized tectonic pulse could have contributed to this discrepancy.

One step further is the analysis of selected elementary sequences, assuming that they formed during 20'000 years and that facies evolution was controlled by a meter-scale sea-level change. Of course, not every elementary sequence can be interpreted in detail, especially if there are no discernible deepening or shallowing facies trends. A good example where an interpretation is possible occurs in the Röschenz Member of the Vorbourg section (Figs. 5 and 6), which has been analysed by Védrine and Strasser (2009), Stienne (2010), and Strasser et al. (2012). Fig. 12 shows the decompacted elementary sequence (from originally 1.1 m to 2.2 m) where both sequence boundaries and the maximum-flooding surface are well developed. The facies analysis is based on continuously sampled rock slabs and 14 thin-sections (Stienne, 2010). The transgressive deposits are carbonate-dominated and contain several discontinuities attributed to reactivation surfaces and incipient hardgrounds. The maximum-flooding surface

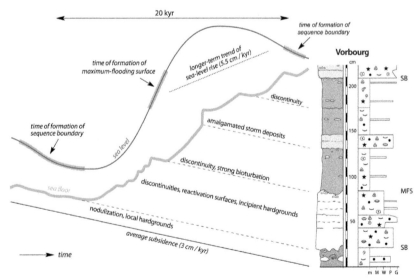

Figure 12 Analysis of one decompacted elementary sequence of the Vorbourg section and interpretation of the sea-level cycle responsible for its deposition (modified from Strasser et al., 2012). Facies symbols as in Fig. 7. MFS, maximum-flooding surface; SB, sequence boundary. For discussion refer to text.

is strongly bioturbated, suggesting a reduced sedimentation rate. The high-stand deposits are marl-dominated and contain an interval of amalgamated storm deposits. The evolution of the sea floor has been reconstructed by assuming varying sedimentation rates depending on the facies interpretation (e.g., slow where hardgrounds formed and at the bioturbated maximum-flooding surface, fast for the storm deposits).

The sea-level cycle is assumed to have been symmetrical. Water depth has to be estimated as no intertidal or supratidal facies are present. Subsidence is assumed to have been around 3 cm/kyr (Wildi et al., 1989), and a longer-term sea-level rise of 5.5 cm/kyr is needed to accommodate the decompacted sequence. The graphical interpretation in Fig. 12 suggests that the fastest rate of short-term sea-level rise was in the order of 30 cm/kyr (time of formation of the maximum-flooding surface). This corresponds to 3 cm/100 yr, which is still slow when compared to the 20 cm/100 yr rise in modern times (globally averaged sea-level rise over the last 100 years; IPCC, 2014).

Based on the interpretation of sea-floor evolution (Fig. 12), sedimentation rates can be estimated. The 65-cm thick limestone interval (decompacted) at the base of the sequence took about 6000 years to accumulate; the average rate of sediment accumulation thus was in the order of 0.1 mm/yr. However, sedimentation was interrupted by reactivation events and hardground formation, meaning that sediment production rate certainly was higher because time was lost in the discontinuities (Strasser, 2015). The 110-cm thick marly part of the sequence (without the 15-cm thick amalgamated storm deposits) accumulated within some 10,000 years, giving again an average sedimentation rate of about 0.1 mm/yr. Sediment production rate, however, must have been higher because time was lost at the discontinuity. Despite the many uncertainties in interpretation, the estimation of sedimentation rates within the short time window offered by cyclostratigraphy gives relatively realistic values that can be compared with the Holocene rates for lagoons of 0.1 to 1 mm/yr (Enos, 1991; Strasser and Samankassou, 2003).

5. CONCLUSIONS

Modern shallow-water carbonate platforms are complex systems where facies distribution is controlled by multiple hydrological and ecological factors. There is no reason to believe that such platforms were less complex in the geological past, and the challenge is to decipher their evolution from the sedimentary record. To do this, the time resolution has to be as high as possible in order to allow for realistic reconstructions and comparisons between ancient and modern environments. Cyclostratigraphy is the tool that has the potential to offer a resolution of 20,000 years (the duration of the orbital precession cycle), which is still low compared to the resolution obtained by, e.g., ^{14}C dating in the Holocene but high for more ancient rock records.

The cyclostratigraphic interpretation of ancient shallow-marine carbonates is not straightforward. Limitations are:

- The insolation signal of the orbital cycles does not necessarily translate directly into climate and/or sea-level changes that are then recorded in the sediment.
- The depositional sequences observed in the sedimentary record may be created by processes other than orbitally controlled climate and/or sea-level changes.
- Hiatuses may lead to "missed beats", and condensations may cause amalgamation of orbitally controlled sequences.
- Depending on the sensitivity of the depositional system, orbital cycles may be recorded faithfully in one place of the platform but not in another, making platform-wide correlations difficult.
- Time-series analyses are difficult to perform because of the common combination of autocyclic and allocyclic processes and the resulting complex facies patterns.
- Chronostratigraphic time control may not be good enough to demonstrate that the observed sequences and their stacking patterns correspond to orbital cyclicity.

Nevertheless, if it can be shown that the observed depositional sequences really formed in tune with orbital cycles, a time window is opened that offers several opportunities:

- The high time-resolution allows estimating the rates of climate and sea-level changes, and comparing with modern rates.
- Sedimentation rates can be estimated.
- The durations of hiatuses and condensations can be evaluated.
- Because eustatic sea-level changes are synchronous across the platform, accommodation changes due to differential subsidence can be identified.
- The evolution of ecosystems and individual organisms can be monitored at a high time resolution.

- Well-developed depositional sequences can be interpreted in terms of sequence stratigraphy, which gives a dynamic picture of the sedimentary processes.

Based on a good sedimentary record and a detailed facies analysis, and with the many caveats in mind, the cyclostratigraphic analysis of shallow-marine carbonates can thus greatly improve the understanding of their history.

ACKNOWLEDGMENTS

This study would not have been possible without the yearlong financial support of the Swiss National Science Foundation and the results produced by numerous Msc and PhD students at the University of Fribourg. I thank Michael Montenari for having invited me to write this book chapter. The constructive comments of Antun Husinec and two anonymous reviewers greatly helped improving the manuscript.

REFERENCES

Adomat, F., Gischler, E., 2015. Sedimentary patterns and evolution of coastal environments during the Holocene in Central Belize, Central America. J. Coast. Res. 31, 802–826.

Allenbach, R.P., 2001. Synsedimentary tectonics in an epicontinental sea: a new interpretation of the Oxfordian basins of Northern Switzerland. Eclogae Geol. Helv. 94, 265–287.

Bathurst, R.G.C., 1975. Carbonate sediments and their diagenesis. Dev. Sedim. 12, 658.

Bathurst, R.G.C., 1987. Diagenetically enhanced bedding in argillaceous platform limestones: stratified cementation and selective compaction. Sedimentology 34, 749–778.

Berger, A., 1990. Paleo-insolation at the Plio-Pleistocene boundary. Paléobiol. Cont. 17, 1–24.

Berger, A., Loutre, M.F., Dehant, V., 1989. Astronomical frequencies for pre-Quaternary palaeoclimate studies. Terra nova. 1, 474–479.

Berger, A., Loutre, M.F., Laskar, J., 1992. Stability of the astronomical frequencies over the Earth's history for paleoclimate studies. Science 255, 560–566.

Berggren, W.A., Kent, D.V., Aubry, M.P., Hardenbol, J. (Eds.), 1995. Geochronology, Time Scales and Global Stratigraphic Correlation. SEPM Spec. Publ., vol. 54, 386 pp.

Bosence, D., Procter, E., Aurell, M., Bel Kahla, A., Boudagher-Fadel, M., Casaglia, F., Cirilli, S., Mehdie, M., Nieto, L., Rey, J., Scherreiks, R., Soussi, M., Waltham, D., 2009. A dominant tectonic signal in high-frequency, peritidal carbonate cycles? A regional analysis of Liassic platforms from western Tethys. J. Sed. Res 79, 389–415.

Boulila, S., Hinnov, L.A., Huret, E., Collin, P.-Y., Galbrun, B., Fortwengler, D., Marchand, D., Thierry, J., 2008. Astronomical calibration of the early Oxfordian (Vocontian and Paris basins, France): consequences of revising the Late Jurassic time scale. Earth Planet. Sci. Lett. 276, 40–51.

Boulila, S., Galbrun, B., Hinnov, L.A., Collin, P.Y., Ogg, J.G., Fortwengler, D., Marchand, D., 2010. Milankovitch and sub-Milankovitch forcing of the Oxfordian (Late Jurassic) Terres Noires formation (SE France) and global implications. Basin Res. 22, 717–732.

Boulila, S., Galbrun, B., Miller, K.G., Pekar, S.F., Browning, J.V., Laskar, J., Wright, J.D., 2011. On the origin of Cenozoic and Mesozoic "third-order" eustatic sequences. Earth-Sci. Rev. 109, 94–112.

Brack, P., Mundil, R., Oberli, F., Meier, M., Rieber, H., 1996. Biostratigraphic and radiometric age data question the Milankovitch characteristics of the Latemar cycles (Southern
Alps, Italy). Geology 24, 371—375.

Burgess, P.M., 2006. The signal and the noise: forward modeling of allocyclic and
autocyclic processes influencing peritidal carbonate stacking patterns. J. Sed. Res. 76,
962—977.

Burgess, P.M., Wright, V.P., 2003. Numerical forward modeling of carbonate platform
dynamics: an evaluation of complexity and completeness in carbonate strata. J. Sed.
Res. 73, 637—652.

Carpentier, C., Martin-Garin, B., Lathuilière, B., Ferry, S., 2006. Correlation of reefal
Oxfordian episodes and climatic implications in the eastern Paris Basin (France). Terra
Nova. 18, 191—201.

Cisne, J.L., 1986. Earthquakes recorded stratigraphically on carbonate platforms. Nature 323,
320—322.

Cloetingh, S., 1986. Intraplate stresses: a new tectonic mechanism for fluctuations of relative
sea level. Geology 14, 617—620.

Colombié, C., Rameil, N., 2007. Tethyan-to-boreal correlation in the Kimmeridgian using
high-resolution sequence stratigraphy (Vocontian basin, Swiss Jura, Boulonnais, Dorset).
Int. J. Earth Sci. 96, 567—591.

Conrad, C.P., 2013. The solid Earth's influence on sea level. Bull. Geol. Soc. Am. 125,
1027—1052.

Cotillon, P., 1987. Bed-scale cyclicity of pelagic Cretaceous successions as a result of
world-wide control. Mar. Geol. 78, 109—123.

De Winter, N.J., Zeeden, C., Hilgen, F.J., 2014. Low-latitude climate variability in the
Heinrich frequency band of the Late Cretaceous greenhouse world. Clim. Past. 10,
1001—1015.

Dercourt, J., Ricou, L.E., Vrielynck, B. (Eds.), 1993. Atlas: Tethys Palaeoenvironmental
Maps. Gauthier-Villars, Paris.

Dexter, T.A., Kowalewski, M., Read, J.F., 2009. Distinguishing Milankovitch-driven processes in the rock record from stochasticity using computer-simulated stratigraphy. J.
Geol. 117, 349—361.

Eberli, G.P., 2013. The uncertainties involved in extracting amplitude and frequency of orbitally driven sea-level fluctuations from shallow-water carbonate cycles. Sedimentology
60, 64—84.

Einsele, G., Ricken, W., 1991. Limestone-marl alternation — an overview. In: Einsele, G.,
Ricken, W., Seilacher, A. (Eds.), Cycles and Events in Stratigraphy. Springer, Heidelberg, pp. 23—47.

Enay, R., Cariou, E., Debrand Passard, S., Menot, J.-C., Rioult, M., 1980. Middle-
Oxfordian. In: Enay, R., Mangold, C. (Eds.), Synthèse paléogéographique du Jurassique
français, Doc. Lab. Geol. Lyon h.s., vol. 5, pp. 181—184.

Enos, P., 1991. Sedimentary parameters for computer modeling. In: Franseen, E.K.,
Watney, W.L., Kendall, C.G.StC., Ross, W. (Eds.), Sedimentary Modeling: Computer
Simulations and Methods for Improved Parameter Definition, Kansas Geol. Survey
Mem., vol. 233, pp. 63—99.

Feng, R., Poulsen, C.J., 2014. Andean elevation control on tropical Pacific climate and
ENSO. Paleoceanography 29, 795—809.

Fischer, A.G., 1964. Lofer cyclothems of the alpine Trias. Kans. Geol. Surv. Bull. 169, 107—148.

Flügel, E., 2004. Microfacies of Carbonate Rocks. Springer, 976 pp.

Forkner, R.M., Hinnov, L.A., Smart, P., 2010. Use of insolation as a proxy for high-
frequency eustasy in forward modeling of platform carbonate cyclostratigraphy - a promising approach. Sed. Geol. 231, 1—13.

Gale, A.S., Hardenbol, J., Hathway, B., Kennedy, W.J., Young, J.R., Phansalkar, V., 2002. Global correlation of Cenomanian (Upper Cretaceous) sequences: evidence for Milankovitch control on sea level. Geology 30, 291—294.

Gilbert, G.K., 1895. Sedimentary measurement of Cretaceous time. J. Geol. 3, 121—127.

Ginsburg, R.N., 1971. Landward movement of carbonate mud: new model for regressive cycles (abstr.). Bull. Amer. Assoc. Pet. Geol. 55, 340.

Goldhammer, R.K., 1997. Compaction and decompaction algorithms for sedimentary carbonates. J. Sed. Res. 67, 26—35.

Goldhammer, R.K., Dunn, P.A., Hardie, L.A., 1990. Depositional cycles, composite sea-level changes, cycle stacking patterns, and the hierarchy of stratigraphic forcing: examples from Alpine Triassic platform carbonates. Bull. Geol. Soc. Am. 102, 535—562.

Gradstein, F., Ogg, J., Schmitz, M., Ogg, G., 2012. The Geologic Time Scale 2012. Elsevier, 1144 pp.

Grotzinger, J.P., 1986. Cyclicity and paleoenvironmental dynamics, Rocknest platform, northwest Canada. Bull. Geol. Soc. Am. 97, 1208—1231.

Gsponer, P., 1999. Etude géologique et sédimentologique de l'anticlinal du Chasseral dans la région de La Heutte (Unpubl. diploma thesis). Univ. Fribourg, 107 pp.

Gygi, R.A., 1995. Datierung von Seichtwassersedimenten des Späten Jura in der Nordwestschweiz mit Ammoniten. Eclogae Geol. Helv. 88, 1—58.

Gygi, R.A., 2000. Integrated stratigraphy of the Oxfordian and Kimmeridgian (Late Jurassic) in northern Switzerland and adjacent southern Germany. Mem. Swiss Acad. Sci. 104, 152.

Gygi, R.A., Coe, A.L., Vail, P.R., 1998. Sequence stratigraphy of the Oxfordian and Kimmeridgian stages (Late Jurassic) in northern Switzerland. SEPM Spec. Publ. 60, 527—544.

Halley, R.B., Harris, P.M., 1979. Fresh-water cementation of a 1,000-year-old oolite. J. Sed. Pet. 49, 969—988.

Haq, B.U., 2014. Cretaceous eustasy revisited. Glob. Planet. Change 113, 44—58.

Hardenbol, J., Thierry, J., Farley, M.B., Jacquin, T., de Graciansky, P.-C., Vail, P.R., 1998. Jurassic sequence chronostratigraphy. In: de Graciansky, P.-C., Hardenbol, J., Jacquin, T., Vail, P.R. (Eds.), Mesozoic and Cenozoic Sequence Stratigraphy of European Basins, SEPM Spec. Publ., 60 chart.

Harris, P.M., Purkis, S.J., Ellis, J., Swart, P.K., Reijmer, J.J.G., 2015. Mapping bathymetry and depositional facies on Great Bahama Bank. Sedimentology 62, 566—589.

Hays, J.D., Imbrie, J., Shackleton, N.J., 1976. Variations in the Earth's orbit: pacemakers of the ice ages. Science 194, 1121—1132.

Hilgen, F.J., Abdul Aziz, H., Krijgsman, W., Raffi, I., Turco, E., 2003. Integrated stratigraphy and astronomical tuning of the Serravallian and lower Tortonian at Monte dei Corvi (Middle-Upper Miocene, northern Italy). Palaeogeo. Palaeoclim. Palaeoeco. 199, 229—264.

Hillgärtner, H., Strasser, A., 2003. Quantification of high-frequency sea-level fluctuations in shallow-water carbonates: an example from the Berriasian-Valanginian (French Jura). Palaeogeo. Palaeoclim. Palaeoeco. 200, 43—63.

Hinnov, L.A., 2013. Cyclostratigraphy and its revolutionizing applications in the Earth and planetary sciences. Geol. Soc. Amer. Bull. 125, 1703—1734.

Hinnov, L.A., Ogg, J.G., 2007. Cyclostratigraphy and the astronomical time scale. Stratigraphy 4, 239—251.

Huang, Z., Ogg, J.G., Gradstein, F.M., 1993. A quantitative study of lower Cretaceous cyclic sequences from the Atlantic Ocean and the Vocontian basin (SE France). Paleoceanography 8, 275—291.

Huang, C., Hinnov, L., Fischer, A.G., Grippo, A., Herbert, T., 2010. Astronomical tuning of the Aptian Stage from Italian reference sections. Geology 38, 899—902.

Hug, W., 2003. Sequenzielle Faziesentwicklung der Karbonatplattform des Schweizer Jura im Späten Oxford und frühesten Kimmeridge. GeoFocus 7. Fribourg, 156 pp.

Husinec, A., Basch, D., Rose, B., Read, J.F., 2008. FISCHERPLOTS: an Excel spreadsheet for computing Fischer plots of accommodation change in cyclic carbonate successions in both the time and depth domains. Comput. Geosci. 34, 269–277.

Husinec, A., Jelaska, V., 2006. Relative sea-level changes recorded on an isolated carbonate platform: Tithonian to Cenomanian succession, southern Croatia. J. Sed. Res. 76, 1120–1136.

Husinec, A., Read, J.F., 2018. Cyclostratigraphic and δ^{13}C record of the lower Cretaceous Adriatic platform, Croatia: assessment of Milankovitch-forcing. Sed. Geol. 373, 11–31.

Immenhauser, A., 2009. Estimating palaeo-water depth from the physical rock record. Earth-Sci. Rev. 96, 107–139.

International Chronostratigraphic Chart, 2017. www.stratigraphy.org.

IPCC, 2014. Climate Change 2014. Intergovernmental Panel on Climate Change. Synthesis Report. Summary for Policy Makers. http://www.ipcc.ch/report/ar5/syr/.

Kemp, D.B., Sadler, P.M., 2014. Climatic and eustatic signals in a global compilation of shallow marine carbonate accumulation rates. Sedimentology 61, 1286–1297.

Kent, D.V., Muttonic, G., Brack, P., 2004. Magnetostratigraphic confirmation of a much faster tempo for sea-level change for the Middle Triassic Latemar platform carbonates. Earth Planet. Sci. Lett. 228, 369–377.

Laskar, J., Fienga, A., Gastineau, M., Manche, H., 2011. La2010: a new orbital solution for the long-term motion of the Earth. Astronomy Astrophysics 532 (A89), 1–15.

Martinez, M., Deconinck, J.-F., Pellenard, P., Riquier, L., Company, M., Reboulet, S., Moiroud, M., 2015. Astrochronology of the Valanginian-Hauterivian stages (early Cretaceous): Chronological relationships between the Parana-Etendeka large igneous province and the Weissert and Faraoni events. Glob. Planet. Change 131, 158–173.

Matthews, M.D., Perlmutter, M.A., 1994. Global cyclostratigraphy: an application to the Eocene Green River basin. In: de Boer, P., Smith, D.G. (Eds.), Orbital Forcing and Cyclic Sequences, IAS Spec. Publ., vol. 19, pp. 459–481.

Meyers, S.R., 2008. Resolving Milankovitchian controversies: the Triassic Latemar limestone and the Eocene Green River formation. Geology 36, 319–322.

Milankovitch, M., 1941. Kanon der Erdbestrahlung und seine Anwendung auf das Eiszeitenproblem. Acad. Roy. Serbe 133, 633.

Milliman, J.D., Freile, D., Steinen, R.P., Wilber, R.J., 1993. Great Bahama Bank aragonitic muds: mostly inorganically precipitated, mostly exported. J. Sed. Pet. 63, 589–595.

Mitchum Jr., R.M., Van Wagoner, J.C., 1991. High-frequency sequences and their stacking patterns: sequence-stratigraphic evidence of high-frequency eustatic cycles. Sed. Geol. 70, 131–160.

Montañez, I.A., Osleger, D.A., 1993. Parasequence stacking patterns, third-order accommodation events, and sequence stratigraphy of Middle to Upper Cambrian platform carbonates, Bonanza King Formation, southern Great Basin. AAPG Mem. 57, 305–326.

Moore, C.H., 1989. Carbonate diagenesis and porosity. Dev. Sedimentol. 46, 338.

Mundil, R., Brack, P., Meier, M., Rieber, H., Oberli, F., 1996. High resolution U-Pb dating of middle Triassic volcaniclastics: time-scale calibration and verification of tuning parameters for carbonate sedimentation. Earth Planet. Sci. Lett. 141, 137–151.

Ogg, J.G., Coe, A.L., Przybylski, P.A., Wright, J.K., 2010. Oxfordian magnetostratigraphy of Britain and its correlation to Tethyan regions and Pacific marine magnetic anomalies. Earth Planet. Sci. Lett. 289, 433–448.

Osleger, D., 1991. Subtidal carbonate cycles: implications for allocyclic vs. autocyclic controls. Geology 19, 917–920.

Pisias, N.G., Shackleton, N.J., 1984. Modelling the global climate response to orbital forcing and atmospheric carbon dioxide changes. Nature 310, 757–759.

Pittet, B., 1994. Modèle d'estimation de la subsidence et des variations du niveau marin: Un exemple de l'Oxfordien du Jura suisse. Eclogae Geol. Helv. 87, 513—543.

Pittet, B., 1996. Contrôles climatiques, eustatiques et tectoniques sur des systèmes mixtes carbonates-siliciclastiques de plate-forme: exemples de l'Oxfordien (Jura suisse, Normandie, Espagne) (Ph.D. thesis). Univ. Fribourg, 258 pp.

Pittet, B., Strasser, A., 1998a. Depositional sequences in deep-shelf environments formed through carbonate-mud export from the shallow platform (Late Oxfordian, German Swabian Alb and eastern Swiss Jura. Eclogae Geol. Helv. 91, 149—169.

Pittet, B., Strasser, A., 1998b. Long-distance correlations by sequence stratigraphy and cyclostratigraphy: examples and implications (Oxfordian from the Swiss Jura, Spain, and Normandy). Geol. Rundsch. 86, 852—874.

Pittet, B., Strasser, A., Mattioli, E., 2000. Depositional sequences in deep-shelf environments: a response to sea-level changes and shallow-platform-carbonate productivity (Oxfordian, Germany and Spain). J. Sed. Res. 70, 392—407.

Posamentier, H.W., Allen, G.P., James, D.P., 1992. High-resolution sequence stratigraphy — the East Coulee delta, Alberta. J. Sed. Pet. 62, 310—317.

Pratt, B.R., James, N.P., 1986. The St George Group (Lower Ordovician) of western Newfoundland: tidal flat island model for carbonate sedimentation in shallow epeiric seas. Sedimentology 33, 313—343.

Preto, N., Hinnov, L.A., De Zanche, V., Mietto, P., Hardie, L.A., 2004. The Milankovitch interpretation of the Latemar platform cycles (Dolomites, Italy): implications for geochronology, biostratigraphy, and Middle Triassic carbonate accumulation. SEPM Spec. Publ. 81, 167—182.

Purkis, S.J., Rowlands, G.P., Kerr, J.M., 2015. Unravelling the influence of water depth and wave energy on the facies diversity of shelf carbonates. Sedimentology 62, 541—565.

Rankey, E.C., Reeder, S.L., 2010. Controls on platformscale patterns of surface sediments, shallow Holocene platforms. Bahamas. Sedimentol. 57, 1545—1565.

Read, J.F., Kerans, C., Weber, L.J., Sarg, J.F., Wright, F.M., 1995. Milankovitch sea-level changes, cycles, and reservoirs on carbonate platforms in greenhouse and icehouse worlds. Soc. Sed. Geol. Short. Course 35.

Read, J.F., Goldhammer, R.K., 1988. Use of Fischer plots to define third-order sea-level curves in Ordovician peritidal cyclic carbonates. Appalachians. Geol. 16, 895—899.

Rodrigo-Gamiz, M., Martinez-Ruiz, F., Rodriguez-Tovar, F.J., Jimenez-Espejo, F.J., Pardo-Iguzquiza, E., 2014. Millenial- to centennial-scale climate periodicities and forcing mechanisms in the westernmost Mediterranean for the past 20,000 years. Quat. Res. 81, 78—93.

Rodríguez-Tovar, F.J., Pardo-Igúzquiza, E., 2003. Strong evidence of high-frequency (sub-Milankovitch) orbital forcing by amplitude modulation of Milankovitch signals. Earth Planet. Sci. Lett. 210, 179—189.

Romero, O.E., Leduc, G., Vidal, L., Fischer, G., 2011 Millennial variability and long-term changes of the diatom production in the eastern equatorial Pacific during the last glacial cycle. Paleoceanography 26. PA2212.

Sadler, P.M., 1994. The expected duration of upward-shallowing peritidal carbonate cycles and their terminal hiatuses. Bull. Geol. Soc. Am. 106, 791—802.

Saha, R., 2015. Millennial-scale oscillations between sea ice and convective deep water formation. Paleoceanography 30, 1540—1555.

Sames, B., Wagreich, M., Wendler, J.E., Haq, B.U., Conrad, C.P., Melinte-Dobrinescu, M.C., Hug, X., Wendler, I., Wolfgring, E., Yilmaz, I.Ö., Zorina, S.O., 2016. Review: short-term sea-level changes in a greenhouse world - a view from the Cretaceous. Palaeogeo. Palaeoclim. Palaeoeco. 441, 393—411.

Satterley, A.K., 1996. The interpretation of cyclic successions of the middle and Upper Triassic of the northern and southern Alps. Earth-Sci. Rev. 40, 181—207.

Schwarzacher, W., 1947. Über die sedimentäre Rhytmik der Dachsteinkalkes von Lofer. Verh. Geol. Bundesanstalt H10—12, pp. 175—188.

Schwarzacher, W., 1993. Cyclostratigraphy and the Milankovitch theory. Dev. Sedimentol. 52, 225.

Schwarzacher, W., 2000. Repetitions and cycles in stratigraphy. Earth-Sci. Rev. 50, 51—75.

Shackleton, N.J., 1987. Oxygen isotopes, ice volume and sea level. Quat. Sci. Rev. 6, 183—190.

Shinn, E.A., Robbin, D.M., 1983. Mechanical and chemical compaction in fine-grained shallow-water limestones. J. Sed. Pet. 53, 595—618.

Spahn, Z.P., Kodama, K.P., Preto, N., 2013. High-resolution estimate for the depositional duration of the Triassic Latemar Platform: a new magnetostratigraphy and magnetic susceptibility cyclostratigraphy from basinal sediments at Rio Sacuz. Italy. Geochem. Geophys. Geosyst. 14, 1—14.

Stienne, N., 2010. Paléoécologie et taphonomie comparative en milieux carbonatés peu profonds (Oxfordien du Jura Suisse et Holocène du Belize). GeoFocus 22. Fribourg, 248 pp.

Strasser, A., 1991. Lagoonal-peritidal sequences in carbonate environments: autocyclic and allocyclic processes. In: Einsele, G., Ricken, W., Seilacher, A. (Eds.), Cycles and Events in Stratigraphy. Springer, Heidelberg, pp. 709—721.

Strasser, A., 2015. Hiatuses and condensation: an estimation of time lost on a shallow carbonate platform. Dep. Rec. 1, 91—117.

Strasser, A., Hilgen, F.J., Heckel, P.H., 2006. Cyclostratigraphy — concepts, definitions, and applications. Newsl. Stratigr. 42, 75—114.

Strasser, A., Hillgärtner, H., 1998. High-frequency sea-level fluctuations recorded on a shallow carbonate platform (Berriasian and lower Valanginian of Mount Salève, French Jura). Eclogae Geol. Helv. 91, 375—390.

Strasser, A., Hillgärtner, H., Hug, W., Pittet, B., 2000. Third-order depositional sequences reflecting Milankovitch cyclicity. Terra Nova. 12, 303—311.

Strasser, A., Pittet, B., Hillgärtner, H., Pasquier, J.-B., 1999. Depositional sequences in shallow carbonate-dominated sedimentary systems: concepts for a high-resolution analysis. Sed. Geol. 128, 201—221.

Strasser, A., Pittet, B., Hug, W., 2015. Palaeogeography of a shallow carbonate platform: the case of the middle to late Oxfordian in the Swiss Jura Mountains. J. Palaeogeogr. 4, 251—268.

Strasser, A., Samankassou, E., 2003. Carbonate sedimentation rates today and in the past: Holocene of Florida bay, Bahamas, and Bermuda vs. Upper Jurassic and lower Cretaceous of the Jura Mountains (Switzerland and France). Geol. Croat. 56, 1—18.

Strasser, A., Védrine, S., Stienne, N., 2012. Rate and synchronicity of environmental changes on a shallow carbonate platform (Late Oxfordian, Swiss Jura Mountains). Sedimentology 59, 185—211.

Thierry, J., 41 Co-authors, 2000. Early Kimmeridgian (146 — 144 Ma). In: Dercourt, J., et al. (Eds.), Atlas Peri-Tethys. CCGM/CGMW, pp. 85—97.

Tipper, J.C., 1997. Modeling carbonate platform sedimentation — lag comes naturally. Geology 25, 495—498.

Trümpy, R., 1980. Geology of Switzerland, a Guide-book. Part A: An Outline of the Geology of Switzerland. Wepf & Co., Basel, 104 pp.

Tucker, M.E., Wright, V.P., 1990. Carbonate Sedimentology. Blackwell, 482 pp.

Vail, P.R., Audemard, F., Bowman, S.A., Eisner, P.N., Perez-Cruz, C., 1991. The stratigraphic signatures of tectonics, eustasy and sedimentology - an overview. In: Einsele, G., Ricken, W., Seilacher, A. (Eds.), Cycles and Events in Stratigraphy. Springer, pp. 617—659.

Van Wagoner, J.C., Mitchum, R.M., Campion, K.M., Rahmanian, V.D., 1990. Siliciclastic sequence stratigraphy in Well logs, Cores, and outcrops. Am. Assoc. Pet. Geol. Methods Explor. 7, 55.

Védrine, S., Strasser, A., 2009. High-frequency palaeoenvironmental changes on a shallow carbonate platform during a marine transgression (Late Oxfordian, Swiss Jura Mountains). Swiss J. Geosci. 102, 247–270.

Wendler, J.E., Wendler, I., 2016. What drove sea-level fluctuations during the mid-Cretaceous greenhouse climate? Palaeogeo. Palaeoclim. Palaeoeco. 441, 412–419.

Westphal, H., Böhm, F., Bornholdt, S., 2004. Orbital frequencies in the carbonate sedimentary record: distorted by diagenesis? Facies 50, 3–11.

Wierzbowski, A., Atrops, F., Grabowski, J., Hounslow, M.W., Matyja, B.A., Olóriz, F., Page, K.N., Parent, H., Mikhail, A., Rogov, M.A., Schweigert, G., Villaseñor, A.B., Wierzbowski, H., WrightJ, K., 2016. Towards a consistent Oxfordian/Kimmeridgian global boundary: current state of knowledge. Vol. Jurassica 14, 15–50.

Wildi, W., Funk, H., Loup, B., Amato, E., Huggenberger, P., 1989. Mesozoic subsidence history of the European marginal shelves of the Alpine Tethys (Helvetic realm, Swiss Plateau and Jura). Eclogae Geol. Helv. 82, 817–840.

Zeeden, C., Meyers, S.R., Lourens, L.J., Hilgen, F.J., 2015. Testing astronomically tuned age models. Paleoceanography 30, 369–383.

Ziegler, P.A., 1990. Geological Atlas of Western and Central Europe. Shell Intern. Petroleum Maatshappij, The Hague, 233 pp.

Zühlke, R., 2004. Integrated cyclostratigraphy of a model Mesozoic carbonate platform — the Latemar (middle Triassic, Italy). SEPM Spec. Publ. 81, 183–211.

Zühlke, R., Bechstädt, T., Mundil, R., 2003. Sub-milankovitch and Milankovitch forcing on a model Mesozoic carbonate platform — the Latemar (middle Triassic, Italy). Terra Nova. 15, 69–80.

CHAPTER FOUR

Mechanisms of Preservation of the Eccentricity and Longer-term Milankovitch Cycles in Detrital Supply and Carbonate Production in Hemipelagic Marl-Limestone Alternations

Mathieu Martinez
Univ Rennes, CNRS, Géosciences Rennes, Rennes, France
E-mail: mathieu.martinez@univ-rennes1.fr

Contents

Abstract

Eccentricity cycles often exert a strong influence on sedimentary series, while they only have weak powers in the insolation series. Three case studies previously published in Early Cretaceous hemipelagic marl-limestone alternations of the Tethyan area are reviewed here to evaluate possible mechanisms of transfer of power from the precession to the eccentricity band. In all cases, the sedimentation rates vary cyclically, following the 405-kyr and the 2.4-myr eccentricity cycles. Maximums of sedimentation rates occur in more clayey intervals, suggesting a strong eccentricity forcing on detrital supply and consequently on sedimentation rates. In all sections, proxies related to purely detrital input show an overwhelming influence of the 405-kyr and the 2.4-myr eccentricity cycle compared to the other Milankovitch components. Proxies related

Stratigraphy & Timescales, Volume 3
ISSN 2468-5178
https://doi.org/10.1016/bs.sats.2018.08.002

189

to the carbonate production show lower amplitudes of the 405-kyr cycle and enhanced 100-kyr eccentricity cycle, which can dominate the power in the Milankovitch band.

Memory effects of erosional and pedogenetic processes are proposed as powerful mechanisms of transfer of amplitude from the precession to the eccentricity band. Highly evolved pedogenesis under more humid climates have longer memory effects, which favor the 405-kyr and the 2.4-myr cycles in sedimentary series. Changes in phasing between carbonate production and insolation can suppress the low frequencies and favor the dominance of the 100-kyr eccentricity cycle. Consequently, in Tethyan Early Cretaceous deposits, clay-dominated marl-limestone alternations are associated to higher sedimentation rates and higher amplitudes of the long eccentricity cycles. Conversely, carbonate-dominated marl-limestone alternations are associated to lower sedimentation rates and higher amplitudes of the 100-kyr eccentricity cycles.

1. INTRODUCTION

Hemipelagic marl–limestone alternations of the peri–Tethyan area constitute spectacular testimonies of the orbital imprint on sedimentation during the Early Cretaceous (Fig. 1). The analysis of the clay mineralogy (Cotillon, 1987; Cotillon et al., 1980; Deconinck and Chamley, 1983), of the palynofacies (Kujau et al., 2013), the radiolarian assemblages (Darmedru et al., 1982; De Wever, 1987), the calcareous nannofossil and foraminiferal assemblages (Giraud et al., 2013; Mutterlose and Ruffell, 1999) all show a cyclic behavior following the marl–limestone alternations and allowing the discard of an early diagenetic origin of these alternations. The analysis of clay mineral assemblages between marl and limestone beds has been particularly crucial to validate the hypothesis of an orbital origin (Westphal et al., 2010). In sections having experienced low levels of burial (Deconinck and Debrabant, 1985), marl beds are enriched in kaolinite and illite, while limestone beds are enriched in smectite (Fig. 2). Enrichment in kaolinite and illite indicate high levels of continental weathering and runoff, at the origin of a high-level hydrolysis, erosion of soils and detrital input to the basin (Deconinck and Chamley, 1983; Moiroud et al., 2012; Mutterlose and Ruffell, 1999). Conversely, enrichment of smectite in limestone beds indicate low levels of continental weathering and runoff leading to lower detrital input to the basin under semi-arid conditions. Cycles in palynofacies assemblages also indicate higher humidity levels in marly intervals than in calcareous intervals, in agreement with the clay mineral assemblages (Charbonnier et al., 2016; Kujau et al., 2013). Neritic and pelagic floro-faunal communities respond to these changes in climate and detrital supply.

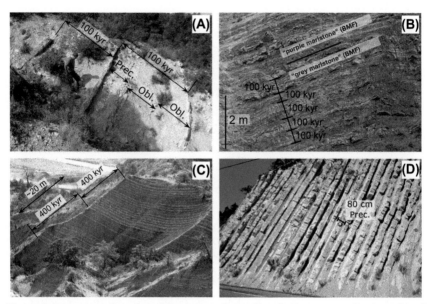

Figure 1 Four examples of marl-limestone alternations in the case studies introduced in this paper. (A) Bundles of 100-kyr eccentricity in the Río Argos section (SE Spain, Hauterivian-Barremian boundary). (B) Bundles of 100-kyr eccentricity in the Bersek Quarry (Hungary, uppermost Valanginian). (C) Bundles of 405-kyr eccentricity in the La Charce section (SE France, Upper Valanginian). (D) Marl-limestone alternations in the La Charce section (Lower Hauterivian). Abbreviations: *Obl.*, Obliquity; *Prec.*, Precession.

In pelagic environments, mesotrophic communities are found in marl intervals while oligotrophic communities are observed in limestone beds suggesting a strong positive coupling of the detrital and the nutrient input to the basin (Giraud et al., 2013; Mutterlose and Ruffell, 1999). Interestingly, the flux of nannofossils is more important in marl beds than in limestone beds, suggesting that nannoplankton productivity was higher during deposition of the marls (Gréselle et al., 2011), while exports from carbonate platforms to the basin led to the deposit of limestone beds in hemipelagic settings (Colombié and Strasser, 2003; Pittet, 2006; Pittet and Mattioli, 2002; Schlager et al., 1994).

Analyses of the stacking and of the frequency content of the marl-limestone alternations commonly evidence the influence of orbital forcing, which controlled the sea-level and climatic changes at the origin of the lithologic alternation (Gilbert, 1895; Giraud et al., 1995; Hinnov and Park, 1999; Moiroud et al., 2012; Pittet and Mattioli, 2002). Spectral analyses of sedimentary series evidence a strong direct control of the eccentricity

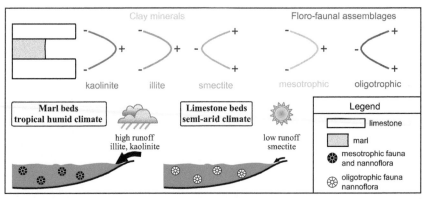

Figure 2 Model of orbital forcing on humid—arid cycles, marl-limestone alternations, clay minerals and floro-faunal assemblages. *Modified from Mutterlose, J., Ruffell, A., 1999. Milankovitch-scale palaeoclimate changes in pale—dark bedding rhythms from the Early Cretaceous (Hauterivian and Barremian) of eastern England and northern Germany. Palaeogr. Palaeocl. 154, 133—160.*

cycles to the deposit of the marl-limestone alternations (Fig. 1), while the power of the eccentricity cycles is nearly null in the insolation cycles. This has arisen the question of the mechanisms of transfer of power from the precession to the eccentricity cycles in the absence of major episode of glaciation during the Cretaceous (Herbert, 1994; Laurin et al., 2015). In these studies, distortion of power, by varying the carbonate production in pelagic environments, and threshold effects at high latitudes have been suggested as mechanisms of transfer from short to long Milankovitch cycles. Surprisingly, while hemipelagic alternating deposits are commonly used to produce orbital time scales (Giraud et al., 1995; Martinez et al., 2017; Meyers and Sageman, 2004), the potential mechanisms of transfer of power from short (precession) to long Milankovitch cycles (eccentricity) in sedimentary basins have not been studied.

In order to highlight such potential mechanisms of power transfer in Early Cretaceous hemipelagic environments, this paper aims at deciphering the impact of the detrital supply and carbonate production on the genesis of the marl-limestone alternations and on the expression of the eccentricity cycles. To address this problem, three case studies are considered: (i) the control of purely detrital proxies vs. mixed carbonate-detrital proxies on the expression of eccentricity cycles at the Hauterivian-Barremian transition of the Subbetic Domain (SE Spain); (ii) the link between detrital supply and bulk carbonate carbon isotopes in the Geresce Mountains (Hungary); (iii) the long-term trend in the detrital supply throughout the

Valanginian-Hauterivian in the Vocontian Basin (South-Eastern France). Through these case-studies, I show that the cycles in detrital supply tend to act as natural low-pass filters, diminishing the amplitude of the precession and obliquity cycles. The phasing between carbonate $\delta^{13}C$ and the lithology can change, likely in function of the balance between the neritic export of carbonate to the basin, the pelagic production and the type of carbonate producers in neritic environments, enhancing the expression of the 100-kyr cycle.

2. GEOLOGICAL SETTINGS

The three cases studies considered here are on Early Cretaceous hemipelagic deposits that were all situated on the Northern margin of the Tethys Ocean (Fig. 3). Their paleolatitude ranged from 20° to 30°N (Dercourt et al., 1993).

The Subbetic Domain is located in the external part of the Betic Range in Southeastern Spain. This hemipelagic domain was separated from the neritic Prebetic Domain during a phase of extensional tectonics during the Early Jurassic, linked to the rifting of the Alpine Tethys (Martín-Algarra

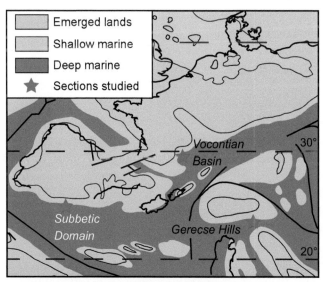

Figure 3 Paleogeographic map of the Western Tethys in the Early Cretaceous with locations of the sections studied. *From Dercourt, J., Ricou, L.E., Vrielynck, B., 1993. Atlas Tethys Paleoenvironmental Maps. Gauthier-Villars, Paris, France.*

et al., 1992; Vera, 2001). During the Early Cretaceous, the Iberian bloc experienced a transtensional motion related to the opening of the North Atlantic and the Alpine oceans (Vera, 2001). The section shown here corresponds to the Hauterivian-Barremian transition of the Río Argos section. The section is composed of marl-limestone alternations deposited in a hemipelagic environment as testified by the numerous ammonites, belemnites, calpionellids and calcareous nannofossils encountered there (Hoedemaeker and Leereveld, 1995). The section has been accurately dated by ammonites and is the current candidate for the Global Stratotype Section and Point (GSSP) of the Barremian Stage (Ogg et al., 2016).

The Vocontian Basin is part of the Dauphinois Zone in the external western Alps. The basin is surrounded by the Provencal carbonate platform to the South, the Ardeche Margin to the West, and the Jura-Dauphiné carbonate platform to the North (Cotillon, 1971). The Vocontian Basin experienced high subsidence rates during the middle Jurassic linked to the rifting phase of the Alpine Tethys (Lemoine et al., 1986; Stampfli and Borel, 2002). The shift from rift to drift tectonics occurred during the Late Jurassic (Roux et al., 1988). Consequently, during the Early Cretaceous, basins in the Northern Tethyan margin experienced decreased subsidence rates (Lemoine et al., 1986; Wilpshaar et al., 1997). The marl–limestone alternations of the Vocontian Basin were deposited in a hemipelagic environment, as evidenced by the ammonites, the belemnites, the calpionellids and the calcareous nannofossils found in this area (Blanc, 1996; Bulot et al., 1993; Thierstein, 1973). Here, I focus on two sections of the basin, the Vergol-Morénas section and the La Charce-Pommerol section (Martinez et al., 2015). Numerous ammonites provided a reference stratigraphic framework in these sections (Bulot et al., 1993; Reboulet and Atrops, 1999) and the sections can be correlated using a series of marker beds as references (Cotillon, 1971; Reboulet and Atrops, 1999). These marker beds form a series of bundles which correspond to the record of the 405-kyr eccentricity cycle (Martinez et al., 2015, Fig. 1). I therefore grouped the two sections as a unique composite series. Detailed bulk carbonate $\delta^{13}C$ curves have been provided in both sections (Gréselle et al., 2011; Hennig et al., 1999; Kujau et al., 2012; van de Schootbrugge et al., 2000), providing a reference curve for stratigraphic correlations (Föllmi, 2012).

The Gerecse Hills are in the NE part of the Transdanubian Mounts, between the Vértes Hills to the South and the Pilis Hills to the East. They are part of the ALCAPA Terrane in the Apulian bloc (Csontos and Vörös, 2004). During the Early Cretaceous, the Gerecse Hills were in the southern side of a flexural basin delimited to the East and the North by prism wedges related to the subduction of the Vardar Ocean (Fodor and Főzy, 2013). The Gerecse Hills remained nonetheless far from the deformation zones during the Valanginian-Hauterivian times (Fodor et al., 2013). The sedimentation in the eastern Gerecse Hills is composed of mixed marl-limestone alternations, while the western part of the Gerecse is marked by deposits of pelagic limestone until the end of the Valanginian, probably at a higher relief position (Császár and Árgyelán, 1994). The section studied here is an outcrop in the Bersek Quarry in the eastern part of the Gerecse Hills. This outcrop is the type-section for the Bersek marl formation and is composed of 31.2-m thick alternations between marl and mudstone beds, in which cm-thick glauconitic sandstone beds intercalate. The sediments are rich in ammonite and calcareous nannofossils, indicating a hemipelagic environment (Fogarasi, 2001; Főzy and Janssen, 2009). The fossil content and bulk carbonate $\delta^{13}C$ allowed the section to be dated as Late Valanginian, the base of the section starting in the Plateau phase of the Weissert Event (Bajnai et al., 2017).

3. RÍO ARGOS

Clay mineral assemblages, Magnetic Susceptibility (MS) and %CaCO$_3$ were acquired in the Río Argos section (Martinez et al., 2012; Moiroud et al., 2012, Fig. 4). Clay mineral assemblages are composed of illite, R0-type smectite-illite mixed layers (thereafter assimilated to smectite), kaolinite and chlorite. Illite and smectite compose more than 90% of the clay mineral (Moiroud et al., 2012). Within a marl-limestone alternation, marls are enriched in kaolinite and illite, while limestones are enriched in smectite, as commonly observed in the marl-limestone alternations of the peri-Tethyan area (Fig. 2). Kaolinite is a mineral forming in a context of strong hydrolysis of silicate from the continental crust under tropical, annually humid conditions, while smectite are formed in context of moderate hydrolysis of silicates from the continental crust, under temperate or semi-arid conditions (Chamley, 1989). Illite and chlorite are observed in endogenous rocks, and their export to the sedimentary basins are thus linked to the mechanical erosion of soils. Enrichment of illite and kaolinite together in marls signs a tropical humid climate marked both by strong levels of chemical hydrolysis and continental weathering, while such processes are only moderate during the deposition of the limestone beds. The MS signal shows

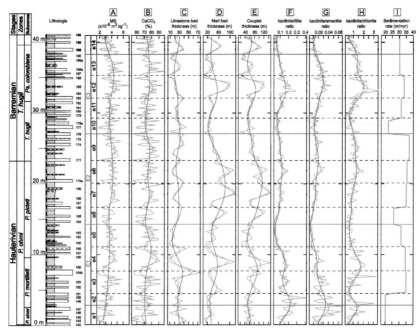

Figure 4 Magnetic susceptibility (MS), CaCO₃ content, clay mineral content ratios and bed thicknesses in the Río Argos sections. The original clay mineral assemblages can be found in Moiroud et al. (2012). The *red curves* represent the Taner low-pass filters of the 405-kyr eccentricity band in each proxy while the *orange curves* represent the Taner low-pass filters of the whole eccentricity band (405 + 100-kyr eccentricity) in each proxy.

a strong inverse correlation with %CaCO₃ (r = −0.94) and has an intercept MS value of 9.7 x 10⁻⁹ m³/kg when %CaCO₃ is 0 (Fig. 5). This is close to the specific MS value of illite or smectite (Hunt et al., 1995), which are the dominant minerals in the clay assemblages. In other words, the MS signal is essentially controlled by paramagnetic clay minerals, and its fluctuations are controlled by dilution effects of the carbonate content. Clay minerals are thus regarded here as proxies for the fluctuations in the detrital supply while MS and %CaCO₃ are proxies for the change in lithology.

Spectra were generated using the multi-taper method (MTM; Thomson, 1982, 1990) applying three 2π-tapers with confidence levels calculated with the Mann and Lees (1996) method modified in Meyers (2014). The filters were performed using Taner low-pass filters (Taner, 2003). The settings of all filters are indicated in Fig. 5. The interpretations in terms of orbital forcing are from Martinez et al. (2012, 2015) and Moiroud et al. (2012). The 405-kyr eccentricity band has periods ranging

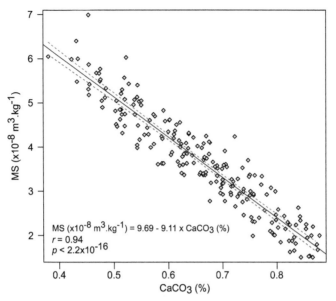

Figure 5 Cross-plot between the magnetic susceptibility (MS) and the CaCO$_3$ content in the Río Argos section. The *blue line* represents the best-fit linear regression while the *red curves* represent the 95% confidence intervals.

from 20 m to 11.8 m (Fig. 6). This large range is due to the intensification of the continental weathering during the Faraoni Oceanic Anoxic Event (Baudin and Riquier, 2014; Moiroud et al., 2012; Sauvage et al., 2013), leading to an increase in kaolinite content in the clay minerals and deviating their period of the 405-kyr to lower frequencies (Martinez et al., 2015). The average period of the 405-kyr eccentricity band without the spectra of the clay minerals is 14.0 m. The 100-kyr eccentricity shows periods ranging from 2.7 to 3.4 m with an average period of 3.2 m. The obliquity has periods ranging from 1.3 to 1.0 m, with an average period of 1.1 m. The precession band show periods ranging from 0.74 to 0.54 m, with an average period of 0.63 m. The spectra of the MS series and %CaCO$_3$ show higher power in the 100-kyr eccentricity band than in the 405-kyr eccentricity band, while the clay mineral ratios show much higher powers in the 405-kyr eccentricity than in other bands (Fig. 6). This is clearly shown in the filter outputs of the whole eccentricity band in Fig. 4. The filter outputs of the eccentricity band in MS and %CaCO$_3$ show a higher amplitude at the 100-kyr band while the filter outputs of the eccentricity band in the clay mineral ratios have higher amplitudes in the 405-kyr band.

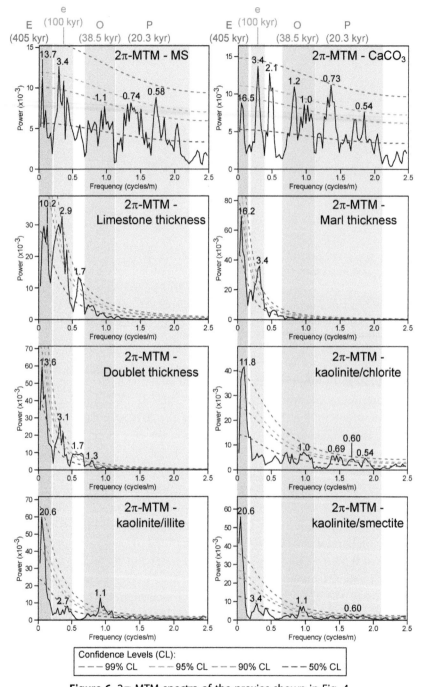

Figure 6 2π-MTM spectra of the proxies shown in Fig. 4.

The orbital tuning of the MS series to the 100-kyr eccentricity shows a cyclicity of 405 kyr, with lower sedimentation rates in calcareous intervals also marked by lower kaolinite/illite, smectite or chlorite ratio (Fig. 4). Higher levels of continental weathering are thus associated to higher detrital supply and higher sedimentation rate. I report here the thicknesses of marl beds, limestone beds, and marl–limestone couplets (Fig. 4) and calculate their 2π-MTM spectra (Fig. 6). The spectra of the marl and marl–limestone couplet thicknesses show a clear dominance of the 405-kyr eccentricity cycle compared to the 100-kyr cycle, while the spectrum of the limestone thickness shows higher powers in the band on the 100-kyr eccentricity. The filter outputs of the whole eccentricity band of the marl thickness and marl–limestone couplet thickness are clearly dominated by the 405-kyr band, in phase with the variations in sedimentation rate calculated from the orbital tuning of the MS series and in phase with the kaolinite/illite, smectite and chlorite ratios (Fig. 4). In contrast, the filter output of the whole eccentricity in the limestone bed thicknesses shows a dominance of the 100-kyr eccentricity, but the phase relationship between this filter output and the other filter outputs changes throughout the whole time series, showing sometimes in phase (e13 and e14, Fig. 4), sometimes in anti-phase (e4, e7, e8, Fig. 4), or dephased. Therefore, the carbonate flux is not necessarily in anti-phase with the detrital flux. Such a change in phasing was also observed in the Valanginian of the Vocontian Basin (Pittet, 2006). At timescales of the eccentricity cycles, the sedimentation rates are controlled mainly by the detrital fluxes, themselves controlled by cyclic changes of levels of continental weathering and runoff. The carbonate flux controls the expression of the marl–limestone alternations at higher frequencies (100-kyr and higher).

4. BERSEK QUARRY

Gamma Ray Spectrometry (GRS), MS and bulk carbonate $\delta^{13}C$ ($\delta^{13}C_{carb}$) have been measured in the Bersek Marl Formation in the type locality of Bersek (Gerecse Hills) (Bajnai et al., 2017, Fig. 7). The lithology of the 31.2-m studied section evolves from a carbonate-dominated to marl-dominated alternations in the Purple Marl Formation. $\delta^{13}C_{carb}$ have been measured throughout the entire interval studied every 10 cm from 0 to 16.2 m to detect the Milankovitch cycles in this series, and then every 20 cm in order to observe the trend of the series and to correlate to other Tethyan sections. The $\delta^{13}C_{carb}$ series has a trend to stable values around

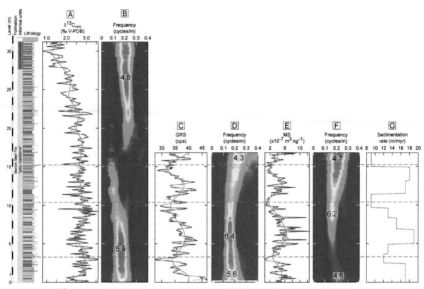

Figure 7 $\delta^{13}C_{carb}$, gamma-ray spectrometry (GRS) and magnetic susceptibility (MS) signals from the Bersek Quarry section together with evolutive spectral analyses and sedimentation rates. The *red* and the *orange curves* respectively represent the Taner low-pass filters of the 405-kyr and the whole eccentricity cycles. (A) $\delta^{13}C$ measured on bulk carbonate. (B) Spectrogram performed on 15-m width windows on the $\delta^{13}C$ signal. Periods are labeled in meters. (C) GRS signal. (D) Spectrogram performed on 15-m width windows on the MS signal. Periods are labeled in meters. (E) MS signal. (F) Spectrogram performed on 15-m width windows on the MS signal. Periods are labeled in meters. (G) Sedimentation rates calculated from the orbital calibration performed on the 100-kyr cycle on the MS signal (see Bajnai et al., 2017). The cut-off frequencies of the Taner low-pass filters are as follows: $\delta^{13}C_{carb}$: red curve: 0.3185 - cycles/m, orange curve: 1.1465 cycles/m; GRS: red curve: 0.3614 cycles/m, orange curve: 1.3855 cycles/m; MS: red curve: 0.3550 cycles/m, orange curve: 1.3018 cycles/m. The roll-off rate of all filters is set at 10^{36}.

2.7‰ (V–PDB) from 0 to 19.2 m, and then progressively decreases to mean values around 1.4‰ at the top of the studied section, in the Purple Marl Formation. The MS and GRS have been measured every 10 cm and 20 cm respectively with the aim to detect the orbital forcing in these marl–limestone alternations on the lowermost 16 m of the series. Both series show an obvious cyclic trend of 5 to 7 m in phase. The GRS signal depends on the gamma emission from the decay of Potassium 40 (^{40}K), Uranium 238 (^{238}U) and Thorium 232 (^{232}Th). These three radioactive elements are abundant in the continental crust (Serra, 1979). Erosion and chemical alteration during the pedogenesis phase concentrate K and Th in the clay minerals while U is concentrated in clay minerals and in organic matter (Quirein et al., 1982; Schmoker, 1981). In organic-poor sediments, high GRS values indicate higher content in clay in the sediment. MS in phase with GRS also indicates high clay content in the sediment.

The spectra of these three series are calculated with the MTM using three 2π-tapers with confidence levels calculated with the Mann and Lees (1996) method modified in Meyers (2014). The spectra were generated after detrending the series as described in Bajnai et al. (2017). Evolutive spectral analyses were performed using the Time-Frequency Weighted Fast Fourier Transform (T-F WFFT, Martinez et al., 2015) on window width of 15 m to detect the low frequencies. The result of this spectral analysis is a 3-dimension spectrum, namely spectrogram, in which the high powers are shown in red colors while the spectral background is in blue colors (Fig. 7).

The 2π-MTM spectra show that the 405-kyr eccentricity cycle is expressed by the spectral peak at 5 m in all proxies (Fig. 8), the 100-kyr eccentricity by a band of periods ranging from 2.0 m to 0.8 m, the obliquity by the band of frequencies ranging from 0.8 to 0.44 m and the precession by the band of frequencies ranging from 0.32 to 0.21 m (Bajnai et al., 2017). In the low frequencies, the spectra of the MS and the GRS signals show a dominant peak in the 405-kyr band, which overwhelms the other bands (Fig. 8), while in the spectrum of the $\delta^{13}C_{carb}$, the 405-kyr band has a much lower amplitude and remains in the same level of the other bands. The high redness level of the spectrum of the GRS signal is due to the fact the GR probe receives the gamma signal over a radius of 20–30 cm, which smooths the high frequencies like a low-pass filter. The MS signal was measured in laboratory from discrete samples collected in the field and is thus not subject to this type of bias. The strong amplitude of the 405-kyr band thus reflects the strong response of lithological changes to this cycle. This strong impact of the 405-kyr cycle on the lithology is also observed in the sedimentation rates. Once the series is calibrated to the 100-kyr cycles observed in the MS signal (Fig. 7), the sedimentation rate shows an obvious cyclicity which follows the filters of the 405-kyr band in the MS and the GRS signals (Fig. 7). The higher the sedimentation rate is, the higher are the MS and the GRS values. As both signals carry a detrital signal, it implies that higher sedimentation rate occur when the detrital supply increases during more humid periods (Bajnai et al., 2017).

In the basal 10 m of the series, the $\delta^{13}C_{carb}$ shows an inverse correlation with MS and GRS at the scale of the 405-kyr cycle (Fig. 7). This inverse correlation stops around 10 m as the amplitude of the 405-kyr in the

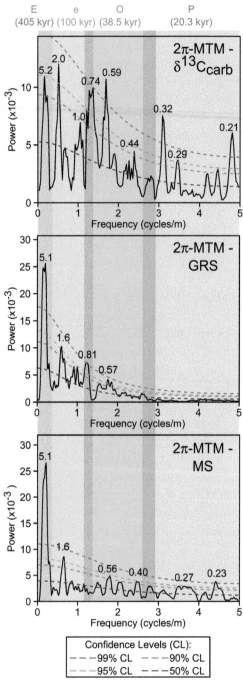

Figure 8 2π-MTM spectra of the $\delta^{13}C_{carb}$, GRS and the MS.

$\delta^{13}C_{carb}$ decreases. At 15 m, the filter output of 405-kyr cycle in the $\delta^{13}C_{carb}$ even shows a minimum in phase with the filter ouputs of the 405-kyr cycle of the the MS and GRS signals (Fig. 7). Thus, from 10 to 15 m, the filter output of the 405-kyr cycle in the $\delta^{13}C_{carb}$ evolves from an anti-phase to an in-phase relationship with the filter outputs of the 405-kyr cycle in the MS and GRS series. This interval from 10 to 15 m, where changes the phasing between the $\delta^{13}C_{carb}$ and the MS, also corresponds to a loss of amplitude and a node in the power spectrogram of the $\delta^{13}C_{carb}$ signal. Ultimately, the 405-kyr band loses its dominant power on the other cycles of the Milankovitch band in the spectrum of the $\delta^{13}C_{carb}$ series (Fig. 8). In hemipelagic areas, the $\delta^{13}C$ measured on bulk carbonate is a versatile proxy, because of the various sources of carbonate, such as export of photozoan or heterozoan carbonate platforms, in-situ pelagic carbonate productivity, in addition to dissolved bicarbonate being potentially influenced by the export of organic matter from the continent (Föllmi et al., 2006). All these sources have different $\delta^{13}C$ values and have different response times to the orbital forcing as they behave differently to humid or arid conditions. In the Western Transdanubian Range, important pelagic carbonate productivity took place from the Tithonian to the late Valanginian, while from the Barremian to the Albian, an important Urgonian-type carbonate platform developed northward of the Gerecse Hills, thus influencing carbonate deposition in the basin (Csaszar and Argyelan, 1994). The Bersek Marl and the Purple Marl Formations may represent a time of transition between various modes of carbonate input to the basin, which explains this change of phasing of the 405-kyr eccentricity in the $\delta^{13}C$ and its much lower amplitude than in the MS and the GRS signals. Such a phase shift triggered by change in carbonate input mode was notably observed in the Early Toarcian in Morocco (Ait-Itto et al., 2018) and resulted in a distortion of the expression of the orbital forcing on the $\delta^{13}C_{carb}$. Thus, much attention must be paid when calibrating time series with the $\delta^{13}C_{carb}$ in mixed carbonate-siliciclastic environments.

5. COMPOSITE VERGOL-MORÉNAS AND LA CHARCE-POMMEROL

GRS measurements have been carried out in the Vergol-Morénas and La Charce-Pommerol sections every 20 cm from the Early Valanginian to the Late Hauterivian (Fig. 9A; Martinez et al., 2013, 2015). The GRS signal increases form the Early to the Late Valanginian and from the early to the

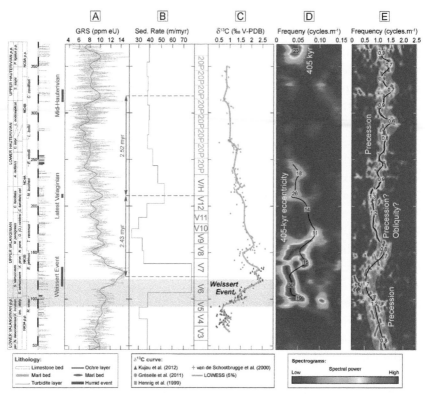

Figure 9 Gamma-ray spectrometry, spectral analyses and sedimentation rates of the composite Vocontian Basin series. (A) Gamma ray spectrometry curve (in gray), with the Taner low-pass filter of the (2.4-myr + 405-kyr) eccentricity band (in red). The Taner low-pass filter is performed on the GRS signal with a cut-off frequency of 0.1168 cycles/m. (B) Sedimentation rate calculated from the orbital calibration. (C) compilation of $\delta^{13}C_{carb}$ measured in the Vergol-Morénas, La Charce and Angles sections (data from Hennig et al., 1999; van de Schootbrugge et al., 2000; Gréselle et al., 2011; Kujau et al., 2012). The *orange curve* represents the Locally Weighted Scatterplot Smoothing (LOWESS; Cleveland, 1979) performed over 5% of the $\delta^{13}C$ data. (D) Spectrograms performed on 40-m width windows on the GRS signal. (E) Spectrograms performed on 15-m width windows on the GRS signal. Periods are labeled in meters.

Late Hauterivian, in two intervals where the marl-limestone alternations enrich in clay. Conversely, the GRS signal decreases from the Late Valanginian to the Early Hauterivian while the alternations become more carbonate-rich. The GRS signal thus follows the lithology and increases with the clay content. The spectral analyses were performed on the GRS signal (Martinez et al., 2013, 2015) using the T–F WFFT evolutive spectrogram (Martinez et al., 2015) on window width of 40 m to detect the low frequencies (Fig. 9D) and a window width of 10 m to follow the evolution of the high frequencies (Fig. 9E).

The spectrograms show cycles from 10 to 30 m corresponding to the 405-kyr eccentricity and cycles from 0.7 to 1.7 m corresponding to the precession, except from 160 to 220 m, where it may reflect the obliquity. The series is calibrated in the Valanginian with the 405-kyr cycle, well expressed in this interval (Martinez et al., 2013; see also Fig. 1A). From the Radiatus Zone in the Hauterivian to the top of the series, the 405-kyr cycle is much less well expressed and I chose instead to calibrate the series on the precession, whose amplitude is much higher as shown on the evolutive spectrogram (Fig. 9E; see also Fig. 1). The sedimentation rate was calculated over two minima of the 405-kyr cycles until the top of cycle V/H (Fig. 9), and over 20 precession cycles, which represents a duration of 406 kyr, assuming a mean duration of the precession of 20.3 kyr during the Hauterivian (Waltham, 2015). This choice of calibration, in anti-phase with the "V" cycles of Martinez et al. (2013), makes it easier to compare the maxima of sedimentation rate to the maxima of the GRS signal. The sedimentation rate shows two maxima during the second half of the Weissert Event (cycles V6-V7) and at the Valanginian-Hauterivian boundary (cycles V12-V/H). Lower frequencies of the 405-kyr and the precession cycles are also observed in these intervals, confirming the increase in the sedimentation rate (Fig. 9). These two periods of high sedimentation rate coincide with higher GRS values, so that, once again, higher sedimentation rates correspond to higher detrital input. This relationship is also observed at the scale of the long-term trend: the maximum of sedimentation is observed during the second half of the Weissert Event, where the series reaches the maximum content in clay (Giraud et al., 1995). Then, the sedimentation rate decreases while the section becomes more carbonated. In the Late Hauterivian, the sedimentation rate remains stable while the GRS values increase. This interval is marked by a drowning event which decreases the neritic carbonate productivity (Föllmi et al., 2006). The thickness of the limestone beds decreases while the thickness of the marlstone beds increases, stabilizing the sedimentation rate. Such a stabilization of the sedimentation is not observed during the drowning phase of the mid-Valanginian Weissert Event (Fig. 9). During the mid-Valanginien Event, the kaolinite content of the clay fraction reaches 40% (Charbonnier et al., 2016; Fesnau, 2008), while it only reaches 10% in the early Late Hauterivian (Godet et al., 2008). During the Weissert Event, the acceleration of the hydrological cycle, likely a consequence of widespread volcanism from the Paraná-Etendeka large igneous province

(Charbonnier et al., 2017; Martinez et al., 2015), triggered large increase in the levels of continental weathering and detrital export to the basin, which explains why the sedimentation rate increase during the mid-Valanginian despite the loss in carbonate export.

The GRS values are maximal during the mid-Valanginian Event (cycles V6-V7), at the end of the Valanginian (cycles V12-V/H) and in the early Late Hauterivian (Sayni Zone) (Fig. 9). These maximums in GRS are separated by 2.4 to 2.5 myr, suggesting that the 2.4-myr eccentricity cycle controlled the recurrence of these peaks in GR. The 2.4-myr cycle also controlled the variation in the sedimentation rate during the Valanginian, as two of the maxima in the GRS signal also coincide with higher sedimentation rates (Fig. 9). Pollen contents and clay mineralogy also show higher humidity levels during the mid-Valanginian and the latest Valanginian paced by the 2.4-myr cycle (Charbonnier et al., 2016; Kujau et al., 2013). The 2.4-myr eccentricity cycle thus paced higher humidity levels, marked by higher content in spore, kaolinite and clay in the sediment. The pacing of the 2.4-myr cycle continues to the Hauterivian, as it paces the occurrence of the mid-Hauterivian humid peak. Thus the 2.4-myr cycle appears to be the pacemaker of the occurrence of transient humid events, at the Berraisian-Valanginian boundary, in the mid-Valanginian, in the latest Valanginian and in the mid-Hauterivian.

6. ORBITAL IMPACT ON DETRITAL INPUT: SIMULATIONS

6.1 Effect of Memory in Erosion Processes

Several lessons can be obtained from our three case studies. In the Río Argos section, lithological proxies have shown a dominance of the 100-kyr cycle while the detrital proxies have shown a dominance of the 405-kyr cycle (Figs. 4 and 6). In the Bersek Quarry, the MS and the GR have shown a dominance of the 405-kyr cycle while the $\delta^{13}C_{carb}$ has shown a suppression of the amplitude of the cycle due to a phase shift with the MS and the GRS (Figs. 7 and 8). In the Vocontian Basin, the detrital input has shown a cyclicity linked to the 405-kyr and to the 2.4-myr cycle (Fig. 9). Therefore, these three examples show that the detrital input is controlled by the eccentricity cycles with a much higher amplitude than the carbonate export to the basin.

It can be surprising to see such a strong influence of the eccentricity cycles in the detrital input despite its very weak power and the predominance of the precession cycles in the power spectrum of the insolation series at low latitudes (Berger and Loutre, 1994). The dominance of the eccentricity cycles in the detrital supply to the basin implies non-linear mechanisms. Pedogenesis, erosion and sediment transport are strong potential sources of transfer of power from the high to the low frequencies (Armitage et al., 2013). Notably, pedogenesis processes are the consequence of the state of the environment at a time t, but also the state of the soil at time t-n (Targulian and Krasilnikov, 2007). Changes in soil, erosion and sediment transport are thus progressive, time averaging, and take into account the state of the environment at previous times. As soil development may take up to 10^6 yr for the most developed soils (Lin, 2011), continental weathering and erosion may act as a moving average over 10^4 to 10^6 yr.

Fig. 10 shows the insolation solution at 25°N from 0.5 to 3.5 Ma (Fig. 10A), with the spectrum of this insolation series calculated over the last 50 myr (Fig. 10B). The power spectrum is dominated by the record of the precession cycles, the other cycles being hardly observable. I applied to the insolation signal a Gaussian-weighted Moving Average (GMA) oriented to the past to simulate the memory effect of the erosion processes to the insolation (Figs. 10C, F, and I). Applying a GMA over 75 kyr disrupts the amplitude modulation pattern on the precession cycles, leading to a power transfer to the eccentricity cycles (Figs. 10C–E). With a GMA over 75 kyr, the 405-kyr cycle has an amplitude as high as the precession cycles, while the amplitude of the 100-kyr cycle is clearly visible in the spectrum (Fig. 10D–E). With a GMA over 500 kyr, the 405-kyr cycle clearly dominates the power spectrum (Fig. 10G and H), while the residual insolation forms clear sequences of ~400 kyr (Fig. 10F). In addition, a sequence of ~2 myr can be observed in the insolation series while a peak at 2.4 myr is visible in the spectrum. With a GMA over 1 myr (maximum of memory effect if I only take into account pedogenetic effects), a clear sequence of ~2 myr dominates the insolation series and the spectrum, while the amplitude of the precession cycles is very weak in comparison to the eccentricity (Figs. 10I–K).

Longer memory effects tend to favor the expression of the 405-kyr and the 2.4-myr eccentricity cycles. As more humid climates led to more detrital inputs to the basin and to more profound pedogenetic processes, they also led to long memory effects in clayey formations which favored a much stronger expression of the 405-kyr and the 2.4-myr eccentricity cycles

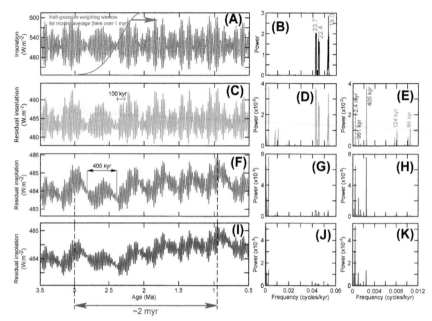

Figure 10 Tests of memory effects applying Gaussian weigthed Moving Averages (GMA) on the insolation series on 21st June at 25°N. (A) Insolation at 25°N on June 21st. The *red curve* represents the shape of the gaussian-distributed weigthed coefficients used to apply the moving averages. (C), (F) and (I) Residual insolation after applying GMAs, respectively at 75 kyr (panel C.), 500 kyr (panel F.) and 1 myr (panel I.). (B), (D), (G) and (J) Spectra of the insolation of residual insolation series shown over the full Milankovitch band. (E), (H), (K) Details of the spectra of the residual insolation series on the eccentricity band.

compared to carbonate-dominated formations. This interpretation can explain why striking bundles of 405 kyr are observed in the Upper Valanginian of the Vocontian Basin but not in the Early Hauterivian (see Figs. 1C–D). Other such examples can be found in the Aptian 'Marnes Bleues' Formation (Vocontian basin; Ghirardi et al., 2014) or in the Late Jurassic 'Terres Noires' Formation (Vocontian basin; Boulila et al., 2010).

6.2 Effect of Phase Shifting on the Insolation Forcing

It is noteworthy that in the previous simulations, the 405-kyr eccentricity cycle has always a stronger amplitude than the 100-kyr eccentricity cycle, while in sedimentary series, stronger amplitudes can be observed in the 100-kyr band (Figs. 6 and 8). Higher amplitudes in the 100-kyr band are notably observed at Río Argos in the lithological proxies, while the proxies of detrital input preserve stronger amplitudes in the 405-kyr band (Fig. 6). In the Bersek Quarry, the $\delta^{13}C_{carb}$ shows higher amplitudes in the 100-kyr band than the MS and the GRS, as the $\delta^{13}C_{carb}$ changes its phasing to the MS and the GRS (Figs. 7 and 8). The response of the carbonate producers to eustatism and climate is still a matter of debate and the variations of the carbonate exports or $\delta^{13}C_{carb}$ to insolation changes can have opposite trends depending on the carbonate producers or the morphology of the platforms surrounding the basin (Föllmi et al., 2006; Pittet and Strasser, 1998; Pittet, 2006; Schlager et al., 1994). The flooding of a carbonate platform during a transgression increases the neritic carbonate production and the export of this production to the basin (Schlager et al., 1994). On a ramp morphology, changes in the sea level moves the neritic carbonate production farther from or closer to the basin but does not change the amount of neritic carbonate produced (Pittet, 2006), leading for instance to less export of neritic carbonate to the basin at times of sea-level highs. This change in the response of the carbonate productivity to an external forcing triggers a change of phasing between carbonate and detrital exports, which notably explains the change in phasing between the thickness of the limestone beds and the marl beds in the Río Argos section (Fig. 4). The phasing between $\delta^{13}C$ and the orbital forcing can also change depending on the source of the carbonate (Föllmi et al., 2006). Photozoan carbonate-platforms, usually aragonitic producers, have higher $\delta^{13}C$ values than heterozoan calcitic platforms. The $\delta^{13}C$ values of hemipelagic carbonates thus reflect the isotopic composition of the carbonates produced in the neritic environments. As photozoan and heterozoan environments develop in different climatic settings, the change in neritic carbonate producer can change the phasing between the $\delta^{13}C_{carb}$ and the orbital forcing. This possibly explains the change of phasing between the $\delta^{13}C_{carb}$ and the MS and GRS in the Bersek Quarry (Fig. 7). It may also explain the change of phasing between the thicknesses of the limestone and the marl beds in the Río Argos section (Fig. 4).

I tested the impact of the change in phasing in the insolation series (Fig. 11). For this, I selected the interval of the residual insolation after applying a GMA over 75 kyr (Fig. 11A) from 2.4 to 0.9 Ma. The interval covers 1.5 myr as in the Río Argos of Bersek Quarry sections and avoids specific intervals of minimums of the 2.4-myr cycle, where the 100-kyr cycle reaches a minimum in amplitude. I selected an interval covering 0.4 myr from 2 to 1.6 Ma in which I inversed the residual insolation to simulate the change of phasing of the proxy to the insolation (labeled as

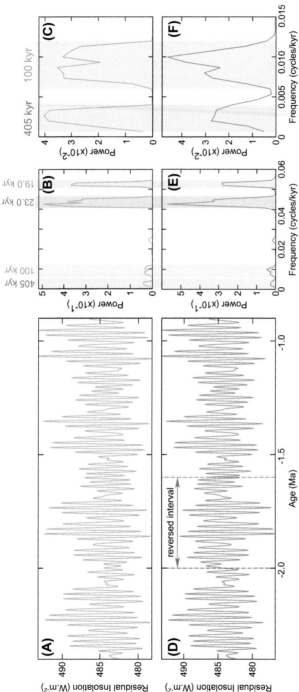

Figure 11 Impact of changing the phasing of the insolation on the spectral analyses in the Milankovitch band. (A) Residual insolation series after applying a Gaussian weighted Moving Average over 75 kyr (same as in Fig. 10C). (B) Spectrum of the residual insolation shown in panel (A). (C) Focus of the spectrum in panel (B) on the eccentricity band. (D) Residual insolation series after changing the phasing of the curve in panel (A) at 2 Ma and 1.6 Ma. The *dash red lines* indicate where the change of phasing have been made (called "reversed interval"). (E) Spectrum of the residual insolation shown in panel (D). (F) Focus of the spectrum in panel (E) on the eccentricity band.

"reversed interval" in Fig. 11D). The interval from 2 to 1.6 Ma is thus in anti-phase with the residual insolation while the rest of the series remains in phase. Compared to the spectrum after applying a GMA over 75 kyr, the spectrum of the series after applying phase changes displays higher amplitudes in the 100-kyr band than in the 405-kyr band (Figs. 11E and F). However, the precession cycle remains dominant in the power spectrum (Fig. 11E). Our experiment thus shows that the amplitude of the 100-kyr cycle can be significantly enhanced with respect to the 405-kyr cycle by changing in-phase relationship between a proxy and the insolation.

Other phenomena must apply to decrease the amplitude of the precession cycles and make the 100-kyr eccentricity the dominant cycle in the power spectrum. Orbital cycles impact the detrital input to the basin, which impact in turn the sedimentation rates (Figs. 4, 7 and 9). I imposed cyclical variations of the sedimentation rates on the residual insolation series after applying the phase shifts (Fig. 12A). The experiment thus simulates the same cyclical variations of the sedimentation rates as observed in Río Argos. The average sedimentation rate is 2.85 cm/kyr. It varies with a mean amplitude of 0.85 cm/kyr with a period of 405 kyr. The resulting spectrum (Fig. 12A) still shows a dominant power in the precession band, so the variations in the sedimentation rates cannot explain alone the dominance of the 100-kyr cycle.

The bioturbation can decrease the amplitude of the high frequencies in the power spectrum below a sedimentation rate of 3 cm/kyr (Pisias, 1983). I simulated the impact of bioturbation in a sedimentary series by applying a simple moving average over 19 kyr on the residual insolation series after applying changing in phasing (Fig. 12B). As a result, the power of the precession decreases and the band of the 100-kyr cycle becomes dominant in the power spectrum. Bioturbation processes alone are thus enough to explain the decrease in power of the precession band, allowing the 100-kyr eccentricity to become dominant in case of a weak effect memory from erosional processes and phase shifts in the low frequencies. Applying bioturbation processes over one precession period after having applied variations in the sedimentation rates on the residual insolation series does not completely erase the power of the precession cycles, though. However, it is noteworthy that only the precession cycle at 23 kyr (namely P1) is observed (Fig. 12C), the power of the 19-kyr precession cycle (P2) being completely flattened out.

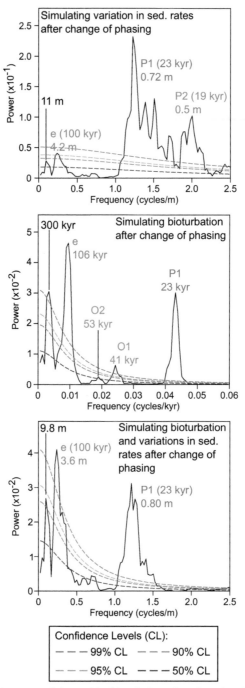

Figure 12 2π-MTM spectra of the residual insolation after changing the phasing (curve in Fig. 11D) and applying variations in sedimentation rates and bioturbation. The confidence levels (CL) are calculated using the Mann and Lees (1996) method modified in Meyers (2014).

7. CONCLUSIONS

In this paper, three case-studies were reviewed to explore the potential mechanisms of transfer of power from the precession to the eccentricity band in Early Cretaceous hemipelagic marl-limestone alternations of the Tethyan area. Higher sedimentation rates are observed in the most clayey intervals of the series showing the highest contents in kaolinite. More carbonated intervals show various phasing with the detrital input and thus do not preserve the lowest frequencies. More humid climate leads to higher detrital input to the basin and higher sedimentation rates in hemipelagic marl-limestone alternations. Memory effects of the erosional and pedogenetic processes are proposed as sedimentary mechanisms responsible for the transfer of power from the precession to the eccentricity band. Higher memory effects due to evolved pedogenetic processes during more humid climate conditions reinforce the 405-kyr and the 2.4-myr eccentricity cycles and can explain their high amplitudes in the sedimentary series in contrast to their weak amplitudes in the insolation series. Conversely, weak memory effects during more arid phases associated to changes in the phasing between orbital cycles and the carbonate production can decrease the amplitude of the long eccentricity cycles and explain the dominance of the 100-kyr cycle in the eccentricity band. Lower sedimentation rates during these intervals can favor the impact of the bioturbation on the sedimentation, which decreases the amplitude of the precession cycles and reinforces the dominance of the 100-kyr eccentricity cycles in the power spectrum of the sedimentary series.

ACKNOWLEDGMENTS

This is a contribution from Team PALEO2D from Geosciences Rennes. I acknowledge anonymous reviewers for their comments. I warmly acknowledge Nicolas Thibault for his positive feedback and his suggestions to improve the quality of the manuscript.

REFERENCES

Ait-Itto, F.-Z., Martinez, M., Price, G.D., Ait Addi, A., 2018. Synchronization of the astronomical time scales in the Early Toarcian: a link between anoxia, carbon-cycle perturbation, mass extinction and volcanism. Earth Planet Sci. Lett. 493, 1—11.

Armitage, J.J., Jones, T.D., Duller, R.A., Whittaker, A.C., Allen, P.A., 2013. Temporal buffering of climate-driven sediment flux cycles by transient catchment response. Earth Planet Sci. Lett. 369-370, 200—210.

Bajnai, D., Pálfy, J., Martinez, M., Price, G.D., Nyerges, A., Főzy, I., 2017. Multi-proxy record of orbital-scale changes in climate and sedimentation during the Weissert event in the Valanginian Bersek Marl formation (Gerecse Mts., Hungary). Cretac. Res. 75, 45—60.

Baudin, F., Riquier, L., 2014. The late Hauterivian Faraoni 'oceanic anoxic event': an update. B. Soc. Geol. Fr. 185, 359—377.

Berger, A., Loutre, M.-F., 1994. Precession, eccentricity, obliquity, insolation and paleoclimates. In: Duplessy, J.-C., Spyridakis, M.-T. (Eds.), Long-term Climatic Variations, NATO ASI Series, vol. I 22. Springer-Verlag, Berlin, pp. 107—151.

Blanc, E., 1996. Transect plate-forme - bassin dans les séries carbonatées du Berriasien supérieur et du Valanginien inférieur (domaines jurassien et nord-vocontien). Geol. Alp., Mémoire H.S. 25 (Grenoble, France).

Boulila, S., Galbrun, B., Hinnov, L.A., Collin, P.-Y., Ogg, J.G., Fortwengler, D., Marchand, D., 2010. Milankovitch and sub-Milankovitch forcing of the Oxfordian (Late Jurassic) Terres Noires formation (SE France) and global implications. Basin Res. 22, 717—732.

Bulot, L.G., Thieuloy, J.-P., Blanc, E., Klein, J., 1993. Le cadre stratigraphique du Valanginien supérieur et de l'Hauterivien du Sud-Est de la France: définition des biochronozones et caractérisation de nouveaux biohorizons. Geol. Alp. 1992, 13—56.

Cleveland, W.S., 1979. Robust locally weighted regression and smoothing scatterplots. J. Am. Stat. Assoc. 74, 829—836.

Cotillon, P., 1971. Le Crétacé inférieur de l'Arc subalpin de Castellane entre l'Asse et le Var : stratigraphie et sédimentologie, Mémoires du B.R.G.M. 68. Editions B.R.G.M., Paris, France.

Cotillon, P., 1987. Bed-scale cyclicity of pelagic Cretaceous successions as a result of world-wide control. Mar. Geol. 78, 109—123.

Cotillon, P., Ferry, S., Gaillard, C., Jautée, E., Latreille, G., Rio, M., 1980. Fluctuation des paramètres du milieu marin dans le domaine vocontien (France Sud-Est) au Crétacé inférieur : mise en évidence par l'étude des formations marno-calcaires alternantes. B. Soc. Geol. Fr., t. XXII (5), 735—744.

Chamley, H., 1989. Clay Sedimentology. Springer-Verlag, Berlin, Germany.

Charbonnier, G., Duchamp-Alphonse, S., Adatte, T., Föllmi, K.B., Spangenberg, J.E., Gardin, S., Galbrun, B., Colin, C., 2016. Eccentricity pacedmonsoon-like systemalong the northwestern Tethyan margin during the Valanginian (Early Cretaceous): new insights from detrital and nutrient fluxes into the Vocontian Basin (SE France). Palaeogr. Palaeocl. 443, 145—155.

Charbonnier, G., Morales, C., Duchamp-Alphonse, S., Westermann, S., Adatte, T., Föllmi, K.B., 2017. Mercury enrichment indicates volcanic triggering of Valanginian environmental change. Sci. Rep. 7, 40808.

Colombié, C., Strasser, A., 2003. Depositional sequences in the Kimmeridgian of the Vocontian Basin (France) controlled by carbonate export from shallow-water platforms. Geobios 675—683.

Császár, G., Árgyelán, T., 1994. Stratigraphic and micromineralogic investigations on Cretaceous formations of the Gerecse Mountains, hungary and their palaeogeographic implications. Cretac. Res. 15, 417—434.

Csontos, L., Vörös, A., 2004. Mesozoic plate reconstruction of the Carpathian region. Palaeogr. Palaeocl. 210, 1—56.

Darmedru, C., Cotillon, P., Rio, M., 1982. Rythmes climatiques et biologiques en milieu marin pélagique : leur relation dans les dépôts crétacés alternants du Bassin Vocontien (S.E. France). B. Soc. Geol. Fr., Série XXIV 7, 627—640.

Deconinck, J.-F., Chamley, H., 1983. Héritage et diagenèse des minéraux argileux dans les alternances marno-calcaires du Crétacé inférieur du domaine subalpin. C. R. Acad. Sci. II 297, 589—594.

Deconinck, J.-F., Debrabant, P., 1985. Diagenèse des argiles dans le domaine subalpin : rôles respectifs de la lithologie, de l'enfouissement et de la surcharge tectonique. Rev. Geol. Dyn. Geogr. 26, 321—330.

Dercourt, J., Ricou, L.E., Vrielynck, B., 1993. Atlas Tethys Paleoenvironmental Maps. Gauthier-Villars, Paris, France.

De Wever, P., 1987. Radiolarites rubanées et variations de l'orbite terrestre. B. Soc. Geol. Fr., Série VIII 3, 957—960.

Fesnau, C., 2008. Enregistrement des changements climatiques dans le domaine téthysien au Valanginien (PhD Thesis). Université de Bourgogne, UMR Biogéosciences, Dijon, France.

Fodor, L., Főzy, I., 2013. Late middle Jurassic to earliest Cretaceous evolution of basin geometry of the Gerecse Mountains. In: Főzy, I. (Ed.), Late Jurassic—early Cretaceous Fauna, Biostratigraphy, Facies and Deformation History of the Carbonate Formations in the Gerecse and Pilis Mountains (Transdanubian Range, Hungary). Institute of Geosciences, University of Szeged, GeoLitera Publishing House, Szeged, pp. 117—136.

Fodor, L., Sztanó, O., Köver, S., 2013. Mesozoic deformation of the northern Transdanubian range (Gerecse and Vértes Hills). Acta Mineral. Petrogr., Field Guide Series 31, 1—34.

Fogarasi, A., 2001. Lower Cretaceous Calcareous Nannoplankton Stratigraphy of the Transdanubian Range. PhD Thesis. Eötvös University, Department of General and Historical Geology, Budapest, Hungary.

Föllmi, K.B., Godet, A., Bodin, S., Linder, P., 2006. Interactions between environmental change and shallow water carbonate buildup along the northern Tethyan margin and their impact on the Early Cretaceous carbon isotope record. Paleoceanography 21, PA4211.

Föllmi, K.B., 2012. Early Cretaceous life, climate and anoxia. Cretac. Res. 35, 230—257.

Főzy, I., Janssen, N.M.M., 2009. Integrated lower Cretaceous biostratigraphy of the Bersek Quarry, Gerecse Mountains, Transdanubian range, Hungary. Cretac. Res. 30, 78—92.

Ghirardi, J., Deconinck, J.-F., Pellenard, P., Martinez, M., Bruneau, L., Amiotte-Suchet, P., Pucéat, E., 2014. Multi-proxy orbital chronology in the aftermath of the Aptian ocean Anoxic event 1a: Palaeoceanographic implications (Serre Chaitieu section, Vocontian basin, SE France). Newsl. Stratigr. 47, 247—262.

Gilbert, G.K., 1895. Sedimentary measurement of Cretaceous time. J. Geol. 3, 121—127.

Giraud, F., Beaufort, L., Cotillon, P., 1995. Periodicities of carbonate cycles in the Valanginian of the Vocontian Trough: a strong obliquity control. In: House, M.R., Gale, A.S. (Eds.), Orbital Forcing Timescales and Cyclostratigraphy, Geol. Soc. Sp. Publ., vol. 85, pp. 143—164. London.

Giraud, F., Reboulet, S., Deconinck, J.-F., Martinez, M., Carpentier, A., Bréziat, C., 2013. The Mid-Cenomanian event in Southeastern France: evidence from palaeontological and clay mineralogical data. Cretac. Res. 46, 43—58.

Godet, A., Bodin, S., Adatte, T., Föllmi, K.B., 2008. Platform-induced clay-mineral fractionation along a northern Tethyan basin-platform transect: implications for the interpretation of Early Cretaceous climate change (Late Hauterivian-Early Aptian). Cretac. Res. 29, 830—847.

Gréselle, B., Pittet, B., Mattioli, E., Joachimski, M., Barbarin, N., Riquier, L., Reboulet, S., Pucéat, E., 2011. The Valanginian isotope event: a complex suite of palaeoenvironmental perturbations. Palaeogr. Palaeocl. 306, 41—57.

Hennig, S., Weissert, H., Bulot, L., 1999. C-isotope stratigraphy, a calibration tool between ammonite and magnetostratigraphy: the Valanginian-Hauterivian transition. Geol. Carpath. 50, 91—96.

Herbert, T.D., 1994. Reading orbital signals distorted by sedimentation: models and examples. In: de Boer, P.L., Smith, D.G. (Eds.), Orbital Forcing and Cyclic Sequences, Sp. Publ. Int. Ass. Sediment, vol. 19, pp. 483—507. Oxford.

Hinnov, L.A., Park, J.J., 1999. Strategies for assessing early—middle (Pliensbachian—Aalenian) Jurassic cyclochronologies. Phil. Trans. R. Soc. A 357, 1831—1859.

Hoedemaeker, P.J., Leereveld, H., 1995. Biostratigraphy and sequence stratigraphy of the Berriasian-lowest Aptian (lower Cretaceous) of the Río Argos succession, Caravaca, SE Spain. Cretac. Res. 16, 195−230.

Hunt, C.P., Moskowitz, B.M., Banerjee, S.K., 1995. Magnetic properties of rocks and minerals. In: Ahrens, T.J. (Ed.), Rock Physics & Phase Relations, AGU Reference Shelf Series, vol. 3, pp. 189−204. Washington D.C.

Kujau, A., Heimhofer, U., Ostertag-Henning, C., Gréselle, B., Mutterlose, J., 2012. No evidence for anoxia during the Valanginian carbon isotope event—an organic-geochemical study from the Vocontian Basin, SE France. Global Planet. Change 92-93, 92−104.

Kujau, A., Heimhofer, U., Hochuli, P.A., Pauly, S., Morales, C., Adatte, T., Föllmi, K., Ploch, I., Mutterlose, J., 2013. Reconstructing Valanginian (Early Cretaceous) mid-latitude vegetation and climate dynamics based on spore-pollen assemblages. Rev. Palaeobot. Palynol. 197, 50−69.

Laurin, J., Meyers, S.R., Uličný, D., Jarvis, I., Sageman, B.B., 2015. Axial obliquity control on the greenhouse carbon budget through middle- to high-latitude reservoirs. Paleoceanography 30. https://doi.org/10.1002/2014PA002736.

Lemoine, M., Bas, T., Arnaud-Vanneau, A., Arnaud, H., Dumont, T., Gidon, M., Bourbon, M., de Graciansky, P.-C., Rudkiewicz, J.-L., Megard-Galli, J., Tricart, P., 1986. The continental margin of the Mesozoic Tethys in the western Alps. Mar. Petrol. Geol. 3, 179−199.

Lin, H., 2011. Three principles of soil change and pedogenesis in time and space. Soil Sci. Soc. Am. J. 75, 2049−2070.

Mann, M.E., Lees, J.M., 1996. Robust estimation of background noise and signal detection in climatic time series. Climatic Change 33, 409−445.

Martín-Algarra, A., Ruiz-Ortiz, P.A., Vera, J.A., 1992. Factors controlling Cretaceous turbidite deposition in the Betic Cordillera. Rev. Soc. Geol. Esp. 5, 53−80.

Martinez, M., Deconinck, J.-F., Pellenard, P., Reboulet, S., Riquier, L., 2013. Astrochronology of the Valanginian stage from reference sections (Vocontian basin, France) and palaeoenvironmental implications for the Weissert event. Palaeogr. Palaeocl. 376, 91−102.

Martinez, M., Deconinck, J.-F., Pellenard, P., Riquier, L., Company, M., Reboulet, S., Moiroud, M., 2015. Astrochronology of the Valanginian-Hauterivian stages (Early Cretaceous): chronological relationships between the Paraná-Etendeka large igneous province and the Weissert and the Faraoni events. Global Planet. Change 131, 158−173.

Martinez, M., Krencker, F.-N., Mattioli, E., Bodin, S., 2017. Orbital chronology of the Pliensbachian−Toarcian transition from the central high Atlas basin (Morocco). Newsl. Stratigr. 50, 47−69.

Martinez, M., Pellenard, P., Deconinck, J.-F., Monna, F., Riquier, L., Boulila, S., Moiroud, M., Company, M., 2012. An orbital floating time scale of the Hauterivian/Barremian GSSP from a magnetic susceptibility signal. Cretac. Res. 36, 106−115.

Meyers, S.R., 2014. Astrochron: An R Package for Astrochronology. Available at. https://cran.r-project.org/web/packages/astrochron/index.html.

Meyers, S.R., Sageman, B.B., 2004. Detection, quantification, and significance of hiatuses in pelagic and hemipelagic strata. Earth Planet Sci. Lett. 224, 55−72.

Moiroud, M., Martinez, M., Deconinck, J.-F., Monna, F., Pellenard, P., Riquier, L., Company, M., 2012. High-resolution clay mineralogy as a proxy for orbital tuning: example of the Hauterivian−Barremian transition in the Betic Cordillera (SE Spain). Sediment. Geol. 282, 336−346.

Mutterlose, J., Ruffell, A., 1999. Milankovitch-scale palaeoclimate changes in pale−dark bedding rhythms from the Early Cretaceous (Hauterivian and Barremian) of eastern England and northern Germany. Palaeogr. Palaeocl. 154, 133−160.

Ogg, J.G., Ogg, G.M., Gradstein, F.M., 2016. A Concise Geologic Time Scale 2016. Elsevier B.V., Amsterdam, The Netherlands.

Pisias, N.G., 1983. Geologic time series from deep-sea sediments: time scales and distortion by bioturbation. Mar. Geol. 51, 99–113.

Pittet, B., 2006. Les alternances marno-calcaires ou l'enregistrement de la dynamique de production et d'export des plates-formes carbonatées (Habilitation Thesis). Université Claude Bernard Lyon 1, UMR PaléoEnvironnements et PaléobioSphère, 79 pp.

Pittet, B., Mattioli, E., 2002. The carbonate signal and calcareous nannofossil distribution in an Upper Jurassic section (Balingen-Tieringen, Late Oxfordian, Southern Germany). Palaeogr. Palaeocl. 179, 71–96.

Pittet, B., Strasser, A., 1998. Depositional sequences in deep-shelf environments formed through carbonate-mud import from the shallow platform (Late Oxfordian, German Swabian Alb and eastern Swiss Jura). Eclogae geol. Helv. 91, 149–169.

Quirein, J.A., Gardner, J.S., Watson, J.T., 1982. Combined natural gamma-ray spectral lithodensity measurements applied to clay mineral identification. AAPG Bull. 66, 1446.

Reboulet, S., Atrops, F., 1999. Comments and proposals about the Valanginian-Lower Hauterivian ammonite zonation of south-eastern France. Eclogae Geol. Helv. 92, 183–197.

Roux, M., Bourseau, J.-P., Bas, T., Dumont, T., de Graciansky, P.-C., Lemoine, M., Rudkiewicz, J.-L., 1988. Bathymetric evolution of the Tethyan margin in the western Alps (data from stalked crinoids): a reappraisal of eustatism problems during the Jurassic. B. Soc. Geol. Fr., tome 4 (4), 633–641.

Sauvage, L., Riquier, L., Thomazo, C., Baudin, F., Martinez, M., 2013. The late Hauterivian Faraoni "oceanic anoxic event" at Río Argos (southern Spain): an assessment on the level of oxygen depletion. Chem. Geol. 340, 77–90.

Schmoker, J.W., 1981. Determination of organic-matter content of Appalachian Devonian shales from gamma-ray logs. AAPG Bull. 65, 1285–1298.

Serra, O., 1979. Diagraphies différées. Bases de l'interprétation. Tome 1 : Acquisition des données diagraphiques. Bulletin du Centre de Recherche Exploration-Production Elf-Aquitaine, Pau, France.

Schlager, W., Reijmar, J.J.G., Droxler, A., 1994. Highstand shedding of carbonate platforms. J. Sediment. Res. B64, 270–281.

Stampfli, G.M., Borel, G.D., 2002. A plate tectonic model for the Paleozoic and Mesozoic constrained by dynamic plate boundaries and restored synthetic oceanic isochrons. Earth Planet Sci. Lett. 196, 17–33.

Taner, M.T., 2003. Attributes Revisited. Technical Publication, Rock Solid Images, Inc., Houston, TX. rocksolidimages.com/pdf/attrib_revisited.htm.

Targulian, V.O., Krasilnikov, P.V., 2007. Soil system and pedogenic processes: self-organization, time scales, and environmental significance. Catena 71, 373–381.

Thierstein, H.R., 1973. Lower Cretaceous Calcareous Nannofossil Biostratigraphy. In: Abhandlungen der Geologischen Bundensanstalt, vol. 29. Geologische Bundensanstalt, Vienna, Austria.

Thomson, D.J., 1982. Spectrum estimation and harmonic analysis. Proc. IEEE 70, 1055–1096.

Thomson, D.J., 1990. Quadratic-inverse spectrum estimates: applications to palaeoclimatology. Phil. Trans. R. Soc. A 332, 539–597.

van de Schootbrugge, B., Föllmi, K.B., Bulot, L.G., Burns, S.J., 2000. Paleoceanographic changes during the Early Cretaceous (Valanginian-Hauterivian): evidence from oxygen and carbon stable isotopes. Earth Planet Sci. Lett. 181, 15–31.

Vera, J.A., 2001. Evolution of the South Iberian continental margin. In: Ziegler, P.A., Cavazza, W., Robertson, A.H.F., Crasquin-Soleau, S. (Eds.), Peri-tethys Memoir 6: Peri-tethyan Rift/Wrench Basins and Passive Margins. Mémoires du Museum National d'Histoire Naturelle, Paris, pp. 109–143.

Waltham, D., 2015. Milankovitch period uncertainties and their impact on cyclostratigraphy. J. Sediment. Res. 85, 990—998.

Westphal, H., Hilgen, F., Munnecke, A., 2010. An assessment of the suitability of individual rhythmic carbonate successions for astrochronological application. Earth-Sci. Rev. 99, 19—30.

Wilpshaar, M., Leereveld, H., Visscher, H., 1997. Early Cretaceous sedimentary and tectonic development of the Dauphinois basin (SE France). Cretac. Res. 18, 457—468.

CHAPTER FIVE

Arabian Orbital Sequences

Moujahed Al-Husseini

Gulf PetroLink-GeoArabia, Manama, Bahrain
E-mail: moujaheda@gmail.com

Contents

Abstract

The orbital scale of glacio-eustasy adopts *stratons* as the fundamental time-rock units of sequence stratigraphy, numbers them as consecutive integers and approximates their duration as 405 kyr (Matthews and Al-Husseini, 2010). It predicts 12 stratons form periodic *dozons* (4.86 myr) and 3 dozons an *orbiton* (14.58 myr), and introduces a concise alphanumeric code for naming them. Importantly the scale predicts the ages of major and minor sequence boundaries (SB), lowstands and flooding intervals (MFI) within these long-period sequences using arithmetical formulas. The scale is based on a simplified model of orbital-forcing of glacio-eustasy (Matthews and Frohlich, 2002),

Stratigraphy & Timescales, Volume 3
ISSN 2468-5178
https://doi.org/10.1016/bs.sats.2018.08.001

which was calibrated with Arabian transgressive-regressive (T-R) sequences (Al-Husseini, 2015, and references therein). The calibration indicated Orbiton 1 was deposited between SB 1 at 16.166 Ma and SB 0 at 1.586 Ma, and SB 37 predicted at 541.046 Ma (1.586 + 37 × 14.58), correlates to the Precambrian/Cambrian Boundary. In many Phanerozoic intervals dozon- and orbiton-scale SBs correlate closely to age estimates of the maximum values in major global positive $\delta^{13}C$ excursions, and negative $\delta^{13}C$ cycles resemble predicted T-R sequences. In this chapter examples of Early Paleozoic, Early Triassic, Late Cretaceous and Miocene Arabian sequences are compared to global $\delta^{13}C$ patterns and both data sets are interpreted as glacio-eustatic proxies and dated with the orbital scale.

1. INTRODUCTION

During the 20th Century Arabia's Phanerozoic lithostratigraphic units were named, defined, ranked, dated and mapped on a national basis by geological surveys, oil companies and academics and most of these units were published in lexicons by the Centre National de la Recherche Scientifique (CNRS) as well as in other books and journals. At the turn of the Century Sharland et al. (2001) tied many of these units with 63 time lines represented by 3rd-order maximum flooding surfaces (MFS) approximately dated using the international biostages of the geological time scale GTS 1996 (Gradstein and Ogg, 1996). The regional Arabian Plate MFS framework has been widely adopted, augmented and recalibrated in more accurate GTS vintages by numerous authors (e.g., Forbes et al., 2010; Haq and Al-Qahtani, 2005; van Buchem et al., 2010a,b).

Also at the turn of the Century Matthews and Frohlich (2002) applied their model of orbital-forcing of glacio-eustasy to tie the Cretaceous and Jurassic MFSs in Arabia (Sharland et al., 2001). The model predicted sequences with periods of 0.4, 2.0 and 2.8 myr that resembled those in Arabia but data/model correlations in absolute time were inconclusive (e.g., Immenhauser and Matthews, 2004). Subsequent comparisons between model and observed T-R sequences dated with more accurate GTS vintages led R.K. Matthews to an important insight. By slightly tuning the orbital formulation of Laskar et al. (2004) the model predicted much longer periodic sequences that closely correlated to Arabia's megasequences (Al-Husseini, 2015; Matthews and Al-Husseini, 2010, and references therein).

To avoid confusion with sequence-stratigraphic nomenclature Matthews and Al-Husseini (2010) named the fundamental unit of orbital stratigraphy the *straton* and cited its age to the nearest thousand years so as to remain synchronized with its correlative long-eccentricity 405-kyr orbital cycle (Fig. 1). Groups of 12 stratons forming long-period sequences were named *dozons* (4.86 myr) and three dozons an *orbiton* (14.58 myr) and the predicted ages of their sequence boundaries (SB), lowstands and flooding intervals (MFI) are shown in the periodic orbital scale (Fig. 1).

Since 2015 further testing of the scale took an interesting turn for the better when it became evident that globally correlative Carbon-13 isotope ($\delta^{13}C$) patterns reflect key aspects of the predicted glacio-eustatic signal. Specifically, the maximum values in major positive $\delta^{13}C$ excursions are empirically dated close to where the scale predicts the lowest sea levels (SB, unconformity). Moreover globally correlative decreasing and increasing $\delta^{13}C$ trends (falling and rising limbs) highlight predicted sea-level rises and falls, respectively, and reflect glacio-eustatic cycles.

These relationships are significant because $\delta^{13}C$ curves are increasingly being used to establish global correlations but their patterns are not recognized as proxies for glacio-eustasy and sequence stratigraphy. Additionally the Astronomical Time Scale (ATS, Hinnov and Hilgen, 2012) is typically constructed by applying spectral analysis to $\delta^{13}C$ records to determine the number of 405-kyr cycles in a stratigraphic interval but its age depends on generally imprecise radiometric dates. Therefore stratons calibrated in the orbital scale can be used to identify and date their corresponding 405-kyr cycles in ATS.

The present study describes several examples of how Arabian sequences and global $\delta^{13}C$ curves can be independently interpreted as glacio-eustatic proxies and deterministically dated using the predictions of the orbital scale. Throughout the study the predicted and empirical age estimates of isochronous surfaces are compared based on GTS 2017 of the International Commission on Stratigraphy (ICS, www.stratigraphy.org) and Time Scale Creator (TSCreator, Ogg et al., 2017) as calibrated by Ogg et al. (2016). These few examples are intended to illustrate the application of this novel technique, which is believed to be applicable in other stratigraphic intervals in Arabia and elsewhere.

2. ARABIAN ORBITONS, DOZONS AND STRATONS

2.1 Seven Lower Paleozoic Orbitons in Oman

Identifying orbitons in Arabia's Lower Paleozoic succession only became possible after Oman's megasequences dated with GTS 2008 (Figs. 2 and 3; Forbes et al., 2010, and references therein) were re-calibrated

Orbiton 1

Age @ Start (Ma)	Cycle Number	Straton	Predicted Stratigraphic Observation	Dozon	Orbiton
0.371	1	0A-4			
0.776	2	0A-3		Orbiton Zero	
1.181	3	0A-2	Major Flooding Interval		
1.586	4	0A-1	Orbiton sequence boundary	SB Zero	
1.991	5	1C-12	(Major unconformity)		
2.396	6	1C-11			
2.801	7	1C-10	Minor Flooding Interval		
3.206	8	1C-9	Minor Lowstand		
3.611	9	1C-8	Minor Flooding Interval		
4.016	10	1C-7		Dozon 1C	
4.421	11	1C-6	Lesser unconformities can occur		
4.826	12	1C-5			
5.231	13	1C-4			
5.636	14	1C-3			
6.041	15	1C-2	Major Flooding Interval		
6.446	16	1C-1	Dozon unconformities	SB 1C	
6.851	17	1B-12	can be Orbiton-scale SBs		
7.256	18	1B-11			
7.661	19	1B-10	Minor Flooding Interval		
8.066	20	1B-9	Minor Lowstand		
8.471	21	1B-8	Minor Flooding Interval		
8.876	22	1B-7		Dozon 1B	Orbiton 1
9.281	23	1B-6	Lesser unconformities can occur		
9.686	24	1B-5			
10.091	25	1B-4			
10.496	26	1B-3			
10.901	27	1B-2	Major Flooding Interval		
11.306	28	1B-1	Dozon unconformities	SB 1B	
11.711	29	1A-12	can be Orbiton-scale SBs		
12.116	30	1A-11			
12.521	31	1A-10	Minor Flooding Interval		
12.926	32	1A-9	Minor Lowstand		
13.331	33	1A-8	Minor Flooding Interval		
13.736	34	1A-7		Dozon 1A	
14.141	35	1A-6	Lesser unconformities can occur		
14.546	36	1A-5			
14.951	37	1A-4			
15.356	38	1A-3			
15.761	39	1A-2	Major Flooding Interval		
16.166	40	1A-1	Orbiton sequence boundary	SB1 Orbiton 2	
16.571	41	2C-12	(Major unconformity)		
16.976	42	2C-11			

Figure 1 Orbital scale of glacio-eustasy (Matthews and Al-Husseini, 2010) consists of periodic dozons (4.86 myr) and orbitons (14.58 myr) and third-order sequence (2.0 and 2.8 myr). The scale is calibrated by stratigraphic units that tracked the long-eccentricity 405-kyr signal named stratons, and it predicts the ages of lowstands containing major or lesser unconformities, and flooding intervals (MFI stratons). Sequence boundaries (SB) correlate to the starts of post-glacial transgressions.

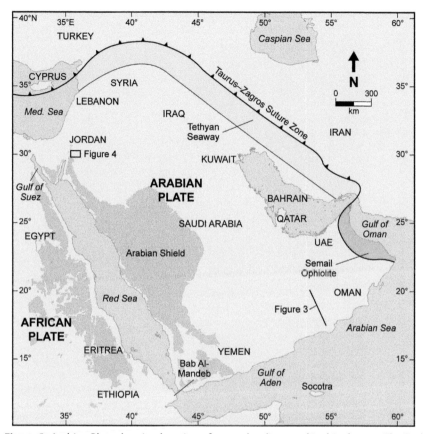

Figure 2 Arabian Plate showing location of examples discussed in this chapter. The Red Sea and Gulf of Aden rift system separated the Arabian Plate from Africa starting in Late Oligocene. During the Mesozoic and most of the Cenozoic the Tethys Ocean was situated along the northeastern margin of the Arabian Plate. During the late Miocene Eurasia collided with the Arabian Plate marked by the present-day Taurus-Zagros Suture Zone. The Tethyan Seaway represents the Miocene foredeep that connected the Arabian and Mediterranean Seas. In the Late Cretaceous the Semail and other ophiolites were obducted onto Arabia's Tethyan margins. For a review of Arabia's geology, see Sharland et al. (2001).

using GTS 2015 (Al-Husseini, 2015). Six orbiton-scale SBs when positioned according to international biostages dated in GTS 2015/2017 fall in hiatuses or correlate to regional unconformities as depicted in Forbes et al. (2010, Fig. 3). One exception is the poorly dated Arabia-wide Angudan Unconformity estimated at 525 ± 5 Ma (Al-Husseini, 2015 and references therein, Fig. 3), which cannot be more precisely correlated to SB 36 (526.446 Ma).

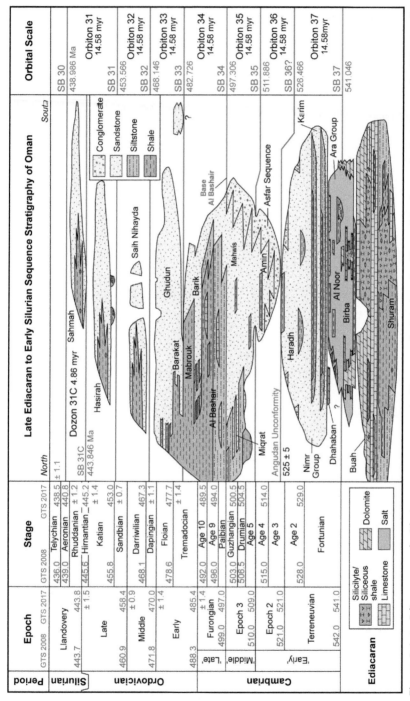

Figure 3 Late Ediacaran–early Silurian megasequences in Oman as depicted and calibrated in the geological time scale GTS 2008 by Forbes et al. (2010, ages shown in red in left-side columns). The boundaries between orbitons and dozons are aligned according to GTS 2017 (blue; www.stratigraphy.org). The Angudan Unconformity is dated between 530 and 520 Ma, and is believed to correlate to SB 36.

At present only dozon-scale SB 31C (443.846 Ma) is recognized in Oman's Early Paleozoic. It correlates to the start of the latest Ordovician post-glacial transgression that accompanied the retreating Hirnantian ice sheets on Gondwana near the Ordovician/Silurian Boundary (Fig. 3; 443.8 ± 1.5 Ma, GTS, 2017). The correlations in Fig. 3 however are not a critical test of the scale's predictions because significant hiatuses occur in Oman's Early Paleozoic. For example the Darriwilian Saih Nihyada Formation falls in middle of Orbiton 32 but the Dapingian and Sandbian stages are hiatuses and the expression of dozons and stratons cannot be demonstrated at this resolution. In contrast, 24 stratons forming two dozons in Orbiton 32 can be identified in Jordan.

2.2 Two-Dozen Stratons in Jordan's Ordovician

Turner et al. (2012) presented a detailed sedimentological and sequence-stratigraphic study of the Middle Darriwilian—Sandbian Hiswah and Dubaydib Formations in Jordan (Fig. 2). They identified 24 4th-order sequences forming 3rd-order Sequences I to VI, bounded by laterally extensive erosional discontinuities SB I to SB VII (Fig. 4). Their hypothesis that the 24 sequences are linked to 405-kyr, 15—50 m, glacio-eustatic cycles can be tested by comparing their field observations to the scale's predictions.

Starting at the top, SB VII is overlain by the Tubeiliyat Member of the Mudawwara Formation (Powell, 1989), which passes laterally to the Raan Member of the Qasim Formation in Saudi Arabia, both containing Arabian Plate MFS O40 (Figs. 4 and 5; Sharland et al., 2001). The Raan Member occurs in the *D. clingani* Graptolite Biozone of the British Caradoc Stage (Vaslet, 1990), which is correlated to North America's *P. tenuis* Conodont Biozone in lowermost Katian Stage (Fig. 4; Goldman et al., 2007). Therefore SB VII occurs near base Katian at 453.0 ± 0.7 Ma (GTS, 2017) and close to SB 31 predicted at 453.566 Ma (1.586 + 31 × 14.58). This preliminary calibration suggests sequences 1—24 are probably Stratons 32B-1 to 32C-12, and SB I probably SB 32B (Fig. 4).

The first test involves comparing the description of lowermost SB I to a post-glacial transgression predicted at dozon-scale SB 32B. SB I is expressed as a sharp change from Hiswah graptolitic siltstones and sandstones to the underlying braided fluvial sandstones of the Umm Sahm Formation (Figs. 4 and 5; Turner et al., 2012, and reference therein). Basal Hiswah contains an intra-formational conglomerate (Masri, 1998), which Turner et al. (2012) interpreted as a transgressive wave ravinement flooding surface that reworked the previously exposed upper surface of the Umm Sahm

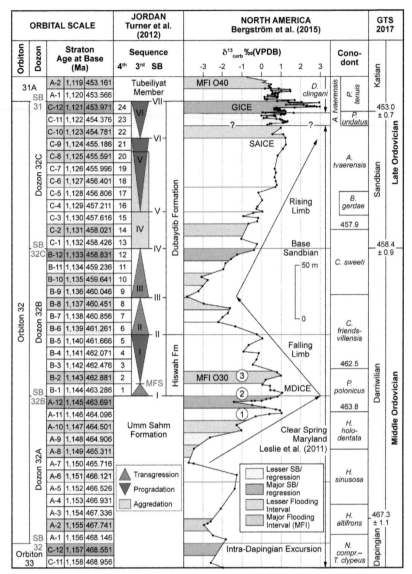

Figure 4 Orbital calibration (this study) of 24 Ordovician 4th-order sequences in Jordan (Turner et al., 2012) and δ¹³C curve in North America. *Reproduced with permission from Estonian Journal of Earth Sciences from Bergström, S.M., Saltzman, M.R., Leslie, S.A., Ferretti, A., Young S.A., 2015. Trans-atlantic application of the Baltic Middle and Upper Ordovician carbon isoptope zonation. Est. J. Earth Sci., 64 (1), 8–12.*

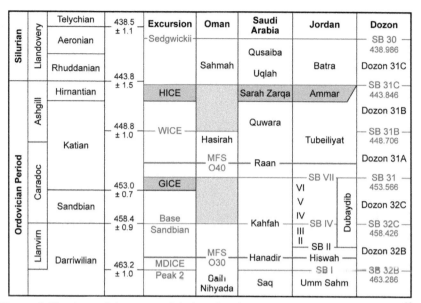

			Excursion	Oman	Saudi Arabia	Jordan	Dozon
Silurian	Llandovery	Telychian	438.5 ± 1.1	Sedgwickii			SB 30 — 438.986
		Aeronian			Qusaiba		
		Rhuddanian	443.8 ± 1.5	Sahmah	Uqlah	Batra	Dozon 31C
Ordovician Period	Ashgill	Hirnantian		HICE		Sarah Zarqa	Ammar	SB 31C — 443.846
		Katian	448.8 ± 1.0	WICE	Hasirah	Quwara	Tubeiliyat	Dozon 31B
							SB 31B — 448.706
	Caradoc			MFS O40	Raan		Dozon 31A
			453.0 ± 0.7	GICE		VI — SB VII	SB 31 — 453.566
		Sandbian	458.4 ± 0.9	Base Sandbian	Kahfah	V — IV — SB IV	Dubaydib	Dozon 32C
						III —	SB 32C — 458.426
	Llanvirn	Darriwilian	463.2 ± 1.0	MDICE	MFS O30	Hanadir	II — SB II	Dozon 32B
				Peak 2	Gaili Nihyada	Saq	Hiswah — SB I — Umm Sahm	SB 32B — 463.286

Figure 5 The interval between global mid-Ordovician–early Silurian positive $\delta^{13}C$ excursions varies from 4.2 to 5.4 myr, and averages 4.94 myr (see Figs. 4, 6 and 9 for age estimates from GTS, 2017 and TSCreator). In the orbital scale the excursions are tuned at 4.86 myr and interpreted as major advances by Gondwana ice sheets.

Formation, and Powell (1989) suggested Hiswah deposition was initiated by a rapid sea-level rise that blanketed large areas of the Arabian shelf in silt and mud.

The Hiswah Formation passes laterally to the Hanadir Member of the Qasim Formation in Saudi Arabia dated in the *D. murchisoni* Graptolite Biozone of the British Llanvirn Stage (Fig. 5; El-Khayal and Romano, 1988; Senalp and Al-Duaiji, 2001). Sharland et al. (2001) positioned MFS O30 in the Hiswah and Hanadir units in the Llanvirn, which in more recent geological time scales occurs in upper Darriwilian Stage. In the subsurface SB I passes to a regional unconformity beneath the Hiswah and Hanadir units, imaged by a high-amplitude seismic reflection (Andrews, 1991; McGillivray and Al-Husseini, 1992). The description of SB I indicates a rapid sea-level rise above a regional erosional unconformity capping terrestrial clastics, consistent with a possible post-glacial transgression at SB 32B.

The next series of tests involve comparing the 3rd-order sequences to the predictions of the scale. The Hiswah Formation consists of vertically stacked, coarsening-up sequences 1–5 forming Sequence I, capped by SB II (Fig. 4; Turner et al., 2012). If these 5 sequences are Stratons 32B-1 to 32B-5 then Sequence I lasted about 2.0 myr and represents the lower T-R sequence of Dozon 32B. Turner et al. (2012) tentatively placed the MFS of Sequence I at the base of 4th-order sequence 2 where the maximum coastal onlap is marked by finer-grained, graptolitic and brachiopod-bearing, outer-shelf silty shales. This position is consistent with the prediction that the 3rd-order MFS occurs in MFI Straton 32B-2 (462.881−462.476 Ma). The MFS recognized by Turner et al. (2012) is the only candidate for MFS O30 positioned in the Hiswah Formation and estimated at 465 Ma in GTS 1996 (Sharland et al., 2001) and 463 Ma in GTS 2008 (Forbes et al. (2010, their Fig. 12.1), consistent with base Straton 32B-2 at 462.881 Ma in this study (Figs. 4 and 5).

Above SB II if sequences 6−12 are correlated to Stratons 32B-6 to 32B-12 then they would build the predicted 2.8-myr sequence in upper Dozon 32B (Fig. 4). These seven transgressive sequences are separated at SB III at base sequence 9 (Turner et al., 2012), consistent with its correlation to lowstand Straton 32B-9 between MFI Stratons 32B-8 and 32B-10 (Fig. 4). SB IV unconformity in sequence 12 correlates to dozon-scale SB 32C (458.416 Ma), which apparently coincides with base Sandbian Stage estimated at 458.4 ± 0.9 Ma (Figs. 4 and 5, GTS, 2017).

If upper sequences 13−24 are correlated to Stratons 32C-1 to 32C-12 then they build Dozon 32C (Fig. 4). Turner et al. (2012) described their stacking pattern as highly complex and suggested alternative 3rd-order sequence architectures for sequences 1−9. The only clear data/model correlations are SB VI to lowstand Straton 32C-9, and SB VII to SB 31 (Fig. 4).

The comparisons of data/model sequences and their empirical/orbital age estimates indicate Jordan's sequences 1−24 are most likely Stratons 1,144−1,121. Although the results of the tests generally support the glacio-eustatic hypothesis of Turner et al. (2012) it is important to consider alternative interpretations such as far-field tectonism controlling the long-period cycles (Sharland et al., 2001) or orbital-forcing of aquifer-eustasy the 4th-order sequences (Wagreich et al., 2014). This objective is considered in the next section by examining the relationship between glaciations and Carbon-13 isotope (δ^{13}C) patterns.

3. GLACIO-EUSTASY AND CARBON-13 ISOTOPE

During the latest Ordovician ice sheets advanced into western Arabia and deposited glaciogenic sediments in incised channels several kilometers wide and 100s of meters deep in northern Saudi Arabia (Zarqa and Sarah

Formations, Vaslet, 1990; McGillivray and Al-Husseini, 1992; Fig. 5). In Jordan Hirnantian glaciogenic sediments have been mapped as the Ammar Formation at outcrop (Abed et al., 1993; Fig. 5) and the Trabeel Formation in subsurface (Andrews, 1991). In Oman the post-glacial transgression is represented by the Sahmah Formation (Forbes et al., 2010) above SB 31C (Figs. 3 and 5).

Brenchley et al. (2003) documented the relationship between the positive global δ^{13}C HICE Excursion (Hirnantian Isotope Carbon Excursion) and the glacio-eustatic signature of the Hirnantian Glaciation of Gondwana. They showed the δ^{18}O and δ^{13}C rising and falling limbs correlate to glacio-eustatic fall and rise, respectively, and the intermediate δ^{13}C plateau to the maximum lowstand and ice volume. By correlating biozones across HICE they showed the Hirnantian Stage is an unconformity-prone interval.

Turner et al. (2012, their Fig. 1) depicted the Hiswah-Dubaydib interval approximately between the Guttenberg Isotope Carbon Excursion (GICE) and the Middle Darriwilian Isotope Carbon Excursion (MDICE) and noted these positive excursions may also represent Gondwana glaciations (Fig. 4). Several other authors have interpreted MDICE and GICE as major glacio-eustatic unconformities and related them to late Middle—Late Ordovician glaciogenic sediments in Gondwana (Kaljo et al., 2003; Saltzman and Young, 2005; Tobin et al., 2005).

If MDICE and GICE represent the δ^{13}C signatures of Gondwana glaciations tuned by long-period orbital forcing then their predicted ages can be compared to empirical estimates. According to Goldman et al. (2007) base GICE and base *P. tenuis* Conodont Biozone occur near base Katian (Fig. 4, Bergström et al., 2015; Leslie et al., 2011). Base GICE is correlated to base lowstand Straton 32C-12 at 453.971 Ma implying base Katian is about 1.0 myr older than the GTS 2017 average of 453.0 ± 0.7 Ma (Fig. 4).

In the Clear Spring Section (Maryland, US; Leslie et al., 2011, Fig. 4) MDICE is characterized by equal-amplitude δ^{13}C Peaks 1—3, all occurring in the *P. polonicus* Conodont Biozone (Bergström et al., 2015; Leslie et al., 2011) approximately between 463.8 and 462.5 Ma (TSCreator, Fig. 4). Peak 2 is chosen as SB 32B (463.286 Ma) based on the correlation of the three peaks presented by Wu et al. (2016) showing it contains the maximum value in the Tingskullen Core (Wu et al., 2016) and Solberga-1 Cores (Lehnert et al., 2014) in Sweden, as well as the Mehikoorma-421 Core in Estonia (Ainsaar et al., 2010). Wu et al. (2016) also show an unconformity occurs between Peaks 1 and 3 in Sweden's Hällekis Quarry and Peaks 2 and 3 in South China's Maocaopu Section (Schmitz et al., 2010), consistent with correlating SB 32B to an unconformity near Peak 2.

Taking into account the average uncertainty of about ±1.0 myr or more in empirical calibrations in the Ordovician (GTS, 2017), and by analogy to HICE representing a glaciation, it seems reasonable to correlate top GICE and MDICE-Peak 2 to the starts of post-glacial transgressions at SB 31 and SB 32B (Figs. 4 and 5). Importantly these proposed correlations further support the hypothesis of Turner et al. (2012) that the sequences in Jordan are linked to 405-kyr glacio-eustatic cycles.

The interpretation of $\delta^{13}C$ patterns as glacio-eustatic proxies becomes more evident in the Katian Stage (Bergström et al., 2015, Fig. 6) when the $\delta^{13}C$ record is displayed alongside the orbital scale. Above the correlation between top GICE and SB 31 the first major Katian $\delta^{13}C$ falling limb correlates to MFI Straton 31A-2 (453.161−452.756 Ma) offering a likely position for MFS O40 estimated at about 453−452 Ma (Forbes et al., 2010; Sharland et al., 2001). MFS O40 occurs in the transgressive Raan Member in Saudi Arabia characterized by the *D. clingani* Graptolite Biozone (Vaslet, 1990), which corresponds to North America's *P. tenuis* and *B. confluens* Conodont Biozones and *D. spiniferus* Graptolite Biozone (Fig. 6; Goldman et al., 2007); the Raan Member probably correlates to the 2.0-myr T-R sequence in lower Dozon 31A between top GICE and mid-KOPE.

Another correlation occurs between positive Waynesville Excursion (WICE) and SB 31B (448.706 Ma) almost precisely near base *A. ordovicus* Conodont Biozone at 448.8 Ma (TSCreator). The mid-KOPE−WICE interval apparently contains the 2.8-myr sequence in upper Dozon 31A. The $\delta^{13}C$ pattern in the WICE−Elkhorn interval consists of five discrete bundles that form a negative $\delta^{13}C$ cycle and correlate to Stratons 31B-1 to 31B-5, highlighting another 2.0-myr T-R sequence. The Hirnantian $\delta^{13}C$ record in Anticosti Island (Quebec, Canada, Young et al., 2010) is partly represented by a hiatus (Brenchley et al., 2003; see their Fig. 14), and an interpretation is not attempted above Straton 31B-5.

4. PALEOZOIC GLACIATIONS AND PLATE TECTONICS

4.1 Ordovician−Silurian Snow Gun

Assuming all the major Ordovician positive $\delta^{13}C$ excursions are related to maximum advances by ice-sheets on Gondwana then one might ask why the Hirnantian glaciation was much more geographically extensive

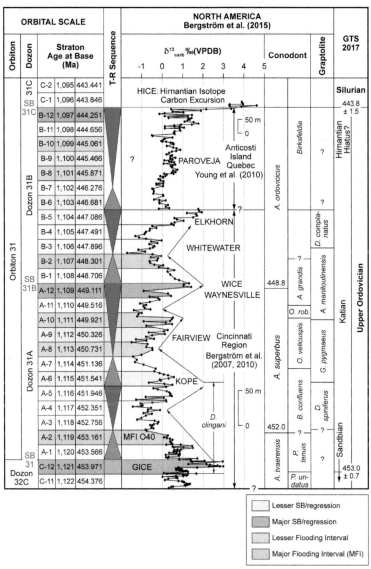

Figure 6 Orbital calibration (this study) of uppermost Sandbian—Hirnantian δ^{13}C curve in North America. The upper Hirnantian Stage is missing in the Anticosti Island Section (Brenchley et al., 2003), and a possible gap is indicated by a question mark in the *A. ordovicicus* conodont Zone, implying Stratons 31B-6 and 31B-12 are not sampled in the δ^{13}C curve. *Reproduced with permission from Estonian Journal of Earth Sciences from Bergström, S.M., Saltzman, M.R., Leslie, S.A., Ferretti, A., Young S.A., 2015. Transatlantic application of the Baltic Middle and Upper Ordovician carbon isotope zonation. Est. J. Earth Sci., 64 (1), 8—12.*

compared to older ones? The glacio-eustatic model of Matthews and Frohlich (2002) provides some insights that may answer this question. Their model relates orbital-forcing of insolation (solar irradiance) at high and low latitudes to the strength of thermohaline currents originating from low latitudes that rise along Antarctica's shelf. It assumes the rising waters drive warm moist air onto Antarctica's cold interior and modulate its ice cap via the snow gun effect (Fig. 7, Prentice and Matthews, 1991). These concepts should also apply to the Phanerozoic and older times because ocean-continent configurations conducive to building ice sheets existed.

In Fig. 8 Ordovician–Silurian plate-tectonic reconstructions (Cocks and Torsvik, 2006) are shown with arrows suggesting where deep thermohaline currents may have approached the cold coasts of Gondwana. The location of possible snow-gun regions are shown to start expanding in Mid-Ordovician, attain a maximum in Late Ordovician and then migrate to South America during the Silurian. The amplitude of the Hirnantian glacio-eustatic fall is estimated as 100 m (Brenchley et al., 2003), which is twice the 50 m estimate for the pre-Hirnantian falls interpreted in Jordan (Turner et al., 2012). These estimates suggest pre-Hirnantian ice sheets covered a smaller area compared to the Hirnantian ice sheets.

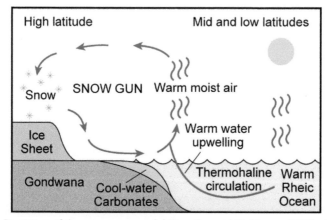

Figure 7 Depiction of the snow gun model (Prentice and Matthews, 1991): deep-water thermohaline currents originating in low-to mid-latitude Rheic Ocean rise along Gondwana's cold shelf and raise sea-surface and atmospheric temperatures. Warm moist air is transported to high-latitude ice-prone land causing snowfall to build an ice sheet. Cool-water carbonates are deposited at high latitudes during glacial advances.

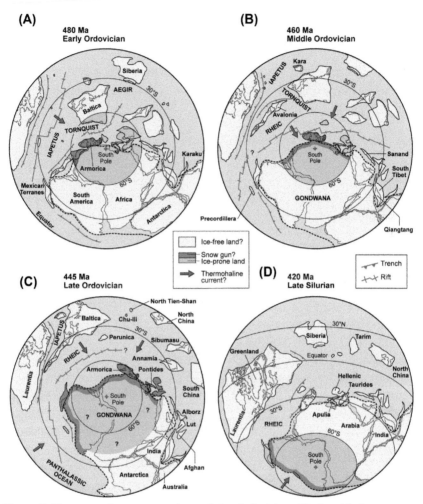

Figure 8 Plate-tectonic reconstructions of the Ordovician–Silurian. Likely snow-gun regions are sketched where suspected thermohaline currents may have intersected the coast of high-latitude ice-prone land in Gondwana. *Reproduced with permission from Cocks, L.R.M., Torsvik, T.H., 2006. European geography in a global context from the Vendian to the end of the Palaeozoic. In: Ghee, D.G, Stehenson, R.A. (Eds.), European Lithosphere Dynamics. Geological Society, London, Memoir 32, pp. 83–95.*

Pope and Steffan (2003) cited evidence for vigorous thermohaline circulation along southern and western Laurentia starting in late Mid-Ordovician and ending in Late Ordovician. They related this anomalous circulation event to the expansion of Gondwana glaciers, consistent with Rheic thermohaline currents invigorating the Gondwana snow gun. During the Silurian Gondwana's ice sheets migrated from Africa to South America as evident by Llandovery glaciogenic deposits in South America (Caputo et al., 2008; Diaz-Martinez and Grahn, 2007). This migration is consistent with the South Pole in Brazil instead of North Africa and the snow-gun region situated mainly along South America's coast next to the Panthalassic Ocean (Fig. 8).

4.2 Silurian Excursions

The greatest Silurian positive $\delta^{13}C$ excursion in the Canadian Arctic occurs in the lower part of the *S. sedgwicki* Graptolite Biozone in uppermost Aeronian (Melchin and Holmden, 2006, Fig. 9). Loydell (1998) correlated the Sedgwickii Excursion to the lowest Silurian sea level and a South American glaciation. He interpreted a sea-level rise in upper *sedgwickii* culminating in an MFS in earliest Telychian (*S. guerchi* Biozone). The orbital calibration correlates SB 30 (438.986 Ma) to the $\delta^{13}C$ maximum value in the Sedgwickii Excursion, and base MFI Straton 30A-2 (438.581 Ma) to the MFS at base Telychian (438.5 ± 1.1 Ma, GTS, 2017). It also correlates Arabian Plate mid-Aeronian MFS S10 estimated at 440 Ma (Sharland et al., 2001) to MFI Straton 31C-10 (Fig. 9).

Similar positive global excursions occur in late Silurian, such as the lower Sheinwoodian Ireviken Excursion (e.g., Cramer and Saltzman, 2007), upper Homerian Mulde Excursion (e.g., Jarochowska and Munnecke, 2016) and mid-Ludfordian Lau Excursion (e.g., Kozlowski, 2015). These excursions as well as younger Devonian and Carboniferous excursions can also be calibrated with the orbital scale.

4.3 Cambrian Ice Sheets

Like the Ordovician and Silurian the Cambrian $\delta^{13}C$ record is characterized by numerous global positive excursions and negative cycles (Kouchinsky et al., 2007, 2017; Peng et al., 2012) but was believed to be ice-free until Landing and MacGabhann (2010) proved otherwise. They identified a lower Cambrian sequence boundary in New Brunswick, Canada, which separates a shallowing-upward sandstone succession from the overlying terrestrial clastics. Lying sharply above the latter unit the basal 45 cm-thick bed in an 11 m-thick mudstone-dominated unit contains dropstones.

The sequence boundary is correlated across nearly 650 km in New Brunswick, which during the early Cambrian was part of the Avalonia Terrane (Landing, 2004; Landing and MacGabhann, 2010). Avalonia was situated at about 60°S in early Cambrian Gondwana and drifted across the

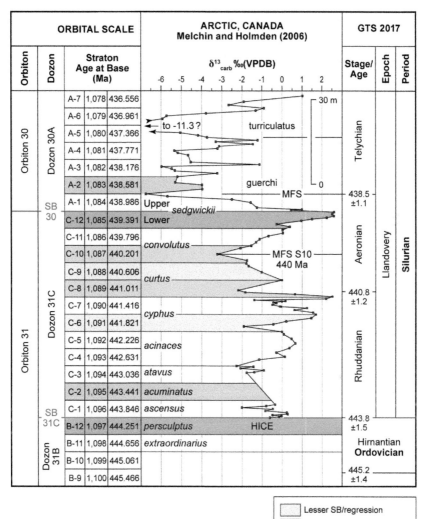

Figure 9 Orbital calibration (this study) of Rhuddanian–lowermost Telychian δ¹³C curve with graptolite zones in the Canadian Arctic. Positive Sedgwickii Excursion is correlated to the lowest glacio-eustatic lowstand in the Silurian (Loydell, 1998). Mid-Aeronian MFS S10 estimated at 440 Ma (Sharland et al., 2001) coincides with MFI Straton 31C-10, but the greatest highstand occurs at base Telychian *guerchi* Biozone (Loydell,1998) at base MFI Straton 30A-2. *Reproduced with GFF's permission from Melchin, M.J., Holmden, C., 2006. Carbon isotope chemostratigraphy of the Llandovery in Arctic Canada: Implications for global correlation and sea-level change. GFF, 128, 173–180.*

Rheic Ocean during the Ordovician (Fig. 8). Today it includes the eastern North America seaboard, southern Ireland, Wales, England, Belgium, the Netherlands and parts of northern Germany (Cocks et al., 1997; Cocks and Torsvik, 2006).

The evidence for this Cambrian glaciation has so far only been documented in one section and the dropstone bed is not accurately dated by biostratigraphy or geochronology (Landing and MacGabhann, 2010). It cannot therefore be correlated to a specific lower Cambrian positive δ^{13}C excursion from amongst numerous candidates. In contrast, the Paibian unconformity below the Furongian ('late Cambrian') Al Bashair Formation in Oman (Fig. 3, Forbes et al., 2010) can be correlated to a positive δ^{13}C excursion in North America but in this case the evidence for glaciation has yet to be found.

4.4 Cambrian SPICE and Ice

Base Paibian Stage represents base Furongian Series but not the traditional Middle/Upper Cambrian Boundary (Peng et al., 2012). It is not dated by high-precision radiometric techniques (Landing et al., 2015) and approximated at 497 Ma (GTS, 2012; Peng et al., 2012). Base Paibian is marked by a mass extinction and the start of the rising limb of global SPICE Excursion (Steptoean Positive Isotope Carbon Excursion) documented in North America's (Laurentia) Steptoean Stage (Saltzman et al., 1998; Saltzman and Thomas, 2012, Fig. 10). In North America's mid-continent, SPICE's rising and falling limbs correlate to a regional regression and transgression, respectively, and the intermediate δ^{13}C plateau to a lowstand (Fig. 10, Runkel et al., 1998; Saltzman et al., 2004). The regional SAUK II/III SB occurs at the top of the plateau and marks the lowest relative sea level (Runkel et al., 1998).

SB SAUK II/III is correlated to orbiton-scale SB 34 at 497.306 Ma (1.586 + 34 × 14.58), and the interpretation of relative sea level (Runkel et al., 1998) is comparable to the predictions for Stratons 35C-5 to 34A-4 (Fig. 10). Base Paibian occurs just below a significant unconformity evident by comparing SPICE in the Smithfield Canyon Section and Rhinehart Core (Figs. 11 and 12, Saltzman et al., 2004), where the plateau is altogether missing in the latter. In the orbital calibration lowstand Straton 32C-12 and the upper part of Straton 32C-11 are missing and an unconformity/hiatus of about 500 kyr occurs in early Paibian in the Rhinehart Core.

The early Paibian unconformity is most likely the one below the Al Bashair Formation in Oman. The Paibian Stage contains an MFS in middle

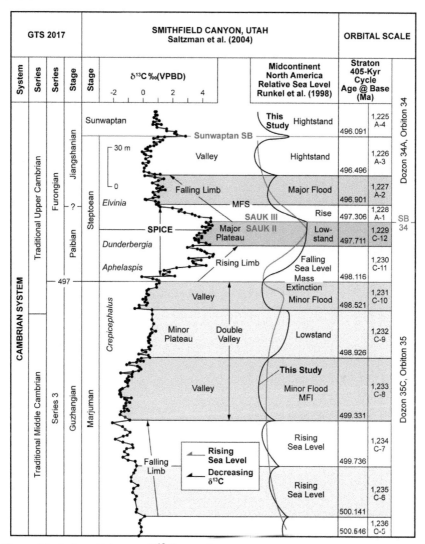

Figure 10 Carbon-13 isotope ($\delta^{13}C$) record spanning the transition between Cambrian Series 3 and Furongian Series in the Smithfield Canyon Section, Utah (U.S.), correlated to relative sea level in North America's mid-continent. The black sea-level curve of this study assumes the $\delta^{13}C$ record anti-correlates to glacio-eustasy, and is calibrated according to the predictions of the orbital scale. *Reproduced with SEPM's permission from Saltzman, M.R., Runkel, A.C., Cowan, C.A., Runnegar, B., Stewart, M.C., Palmer, A.R., 2004. The upper Cambrian SPICE (δ13C) event and the Sauk II-Sauk III regression: New evidence from Laurentian basins in Utah, Iowa and Newfoundland. J. Sediment. Res., 74, 366—377; blue curve, Runkel et al., 1998.*

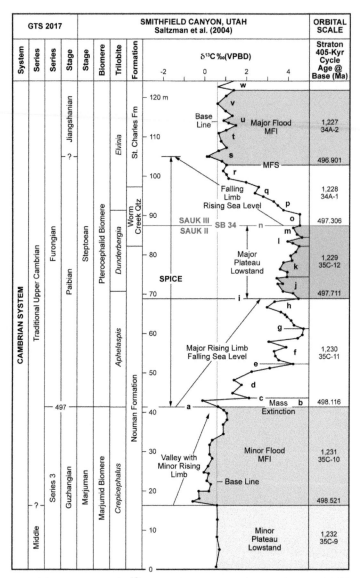

Figure 11 Trilobite zones and δ^{13}C record spanning SPICE in the Smithfield Canyon Section showing δ^{13}C markers and intervals 'a' to 'w'. *Reproduced with SEPM's permission from Saltzman, M.R., Runkel, A.C., Cowan, C.A., Runnegar, B., Stewart, M.C., Palmer, A.R., 2004. The upper Cambrian SPICE ($\delta13C$) event and the Sauk II-Sauk III regression: New evidence from Laurentian basins in Utah, Iowa and Newfoundland. J. Sediment. Res., 74, 366—377.*

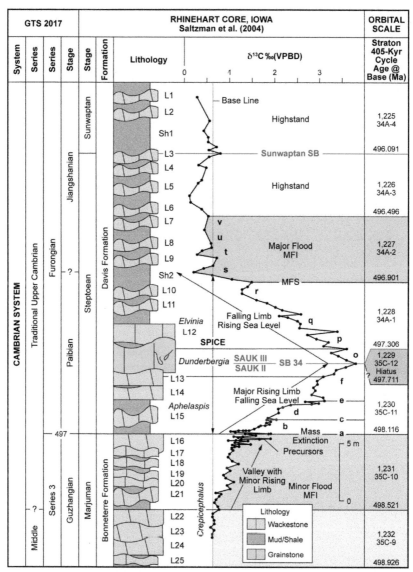

Figure 12 Trilobite zones and measured section spanning SPICE in the Rhinehart A-1 Core (Iowa, U.S., Runkel et al., 1998). Note Intervals 'h' to 'm' in the Smithfield Canyon Section (Fig. 11) are absent in the Reinhart Core, implying upper Straton 35C-11 and Straton 35C-12 are represented by a hiatus/unconformity. *Reproduced with SEPM's permission from Saltzman, M.R., Runkel, A.C., Cowan, C.A., Runnegar, B., Stewart, M.C., Palmer, A.R., 2004. The upper Cambrian SPICE (δ13C) event and the Sauk II-Sauk III regression: New evidence from Laurentian basins in Utah, Iowa and Newfoundland. J. Sediment. Res., 74, 366–377.*

Elvinia Trilobite Biozone (Runkel et al., 1998; Figs. 10—12). It is apparently older than the mid-Al Bashair MFS Cm30 (MFS1 of Droste, 1997) dated 501 Ma in mid-'late Cambrian' (Sharland et al., 2001) and revised to 492 Ma tentatively in upper Stage 9 (Forbes et al., 2010). The mid-*Elvinia* MFS occurs near the base of Stage 9, now named the Jianghshanian Stage, at about 496.9 Ma in the orbital scale.

The Paibian Stage apparently represents the SPICE glaciation between its onset at top Series 3 (top Guzhangian Stage) and its rapid melt-out in early Elvinia. The North American Steptoan Stage apparently ends at the positive Sunwaptan $\delta^{13}C$ Spike marking a sequence boundary (Fig. 10).

5. EARLY TRIASSIC GLACIO-EUSTASY

During the Late Carboniferous and Early Permian icehouse times, glaciers covered most of southern Arabia (Forbes et al., 2010; Sharland et al., 2001, and references therein). In Late Permian and throughout the Mesozoic and early Cenozoic the climate is generally assumed to have been greenhouse, and in particular the Early Triassic is interpreted as extremely hot and ice-free (Sun et al., 2012). By the Triassic Antarctica was near the South Pole (www.scotese.com, C.R. Scotese) and therefore a good candidate to host a polar ice cap. The orbital scale predicts orbiton-scale SB 17 occurred at 249.446 Ma ($1.586 + 17 \times 14.58$) in mid Early Triassic (251.90—247.2 Ma, GTS, 2017) but determining its precise position with respect to the Lower Triassic stages and substages is not straightforward.

5.1 Olenekian Gray Zone and SB 17

The Induan/Olenekian Boundary was initially positioned in the maximum value of the greatest Lower Triassic global $\delta^{13}C$ excursion in the Great Bank of Guizhou, southern China (Fig. 13; Payne et al., 2004; abbreviated DSX after the Dienerian/Smithian Substage Boundary). Based on biostratigraphic criteria however some authors positioned the stage boundary above Excursion DSX in the Guandao Section (Fig. 13, Lehrmann et al., 2015), or followed Payne et al. (2004) by positioning it in the maximum value (schematic in Burgess et al., 2014a,b; Iran, Horacek et al., 2007a). Other studies show the boundary below the excursion (Fig. 13, Jinya/Wali Section, South China, Galfetti et al., 2007; schematic in Romano et al., 2012), or depict the stage transition as a gray zone (United Arab Emirates, Clarkson et al., 2013; Oman, Richoz, 2006) and the base Olenekian

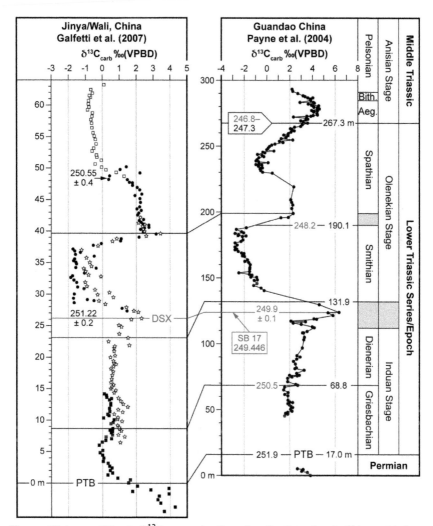

Figure 13 Lower Triassic δ¹³C curve in Guandao Section, South China with base Olenekian Stage positioned above Excursion DSX at 131.9 m (Lehrmann et al., 2015). In the Jinya/Wali Section, South China, Excursion DSX is dated 251.22 ± 0.2 Ma and base Olenekian is positioned below it (Galfetti et al., 2007). Other authors position the stage boundary in the maximum value in DSX or depict it as a gray zone (see text). The boundary was estimated at 249.5 Ma in GTS 2008–10, 251.2 Ma in GTS 2012–17, and 249.9 ± 0.1 Ma (Li et al., 2016, shown in blue). In this study Excursion DSX is correlated to SB 17 at 249.446 Ma. *From Payne, J.L., Lehrmann, D.J., Wei, J., Orchard, M.P., Schrag, D.P., Knoll A.H., 2004. Large perturbations of the carbon cycle during recovery from the end-Permian extinction. Science, 305, 506–509. Reprinted with permission from AAAS.*

Global Boundary Stratotype Section and Point (GSSP) has yet to be chosen by the International Commission on Stratigraphy (ICS).

Base Olenekian positioned in the maximum value of Excursion DSX was estimated at 249.5 Ma in GTS 2008−10 vintages thus offering a precise correlation to SB 17 (249.446 Ma), but the correlation seemed improbable after Galfetti et al. (2007, Fig. 13) obtained a ^{206}Pb/^{238}U 251.22 ± 0.22 Ma date from zircons recovered at the level of Excursion DSX in the Jinya/Wali Section. The correlation became more likely once again after Li et al. (2016, Fig. 13) dated base Olenekian at 249.9 ± 0.1 Ma based on a cyclostratigraphic analysis tied at the Permian/Triassic Boundary (PTB) at 251.90 Ma (Burgess et al., 2014a,b).

The 1.3 myr difference for the age of Excursion DSX between the 251.22 ± 0.22 Ma radiometric dating (Galfetti et al., 2007) and 249.9 ± 0.1 Ma cyclostratigraphic estimate (Li et al., 2016) exceeds their cited accuracies. This discrepancy implies the two empirical age estimates for Excursion DSX may carry uncertainties of the order of ±1.0 myr, and neither can be precisely compared to the predicted age of SB 17. The uncertainty of empirical estimates becomes even greater in the Spathian Substage (Fig. 13): the radiometric estimate of 250.55 ± 0.4 Ma in mid-Spathian (Jinya/Wali, Galfetti et al., 2007) is about 2.3 myr older than the 248.2 ± 0.1 Ma cyclostratigraphic estimate at base Spathian (Guandao, Li et al., 2016). This example illustrates some of the difficulties that are encountered when comparing empirical age estimates to deterministic ones predicted by the orbital scale.

5.2 Oman's Olenekian−Anisian Supersequence MSS

The application of the orbital scale offers a different approach for identifying SB 17 by calibrating Supersequence MSS of the Mahil Formation in Oman outcrops (Fig. 14, Pöppelreiter et al., 2011). Baud and Richoz (2013, and references therein) reviewed the Permian−Triassic transition in the Arabian Plate and the conflicting lithostratigraphic definitions of the Mahil Formation in Oman.

Supersequence MSS consists of three 3rd-order sequences (MS1−MS3), 8 cycle-sets (MCS1.1−MCS3.3) and 48 cycles (Fig. 14, Pöppelreiter et al., 2011) and is dated as Olenekian-early Anisian (Forbes et al., 2010, and reference therein). Base Supersequence MSS is a major sequence boundary

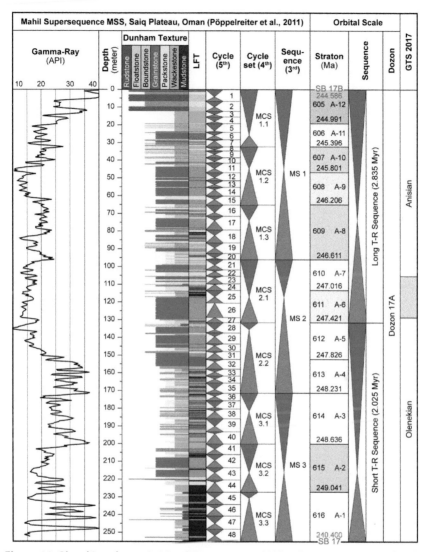

Figure 14 Olenekian—lower Anisian Supersequence MSS in Oman is interpreted in the Mahil Formation and is characterized by 48 5th-order cycles (Pöppelreiter et al., 2011). In this study the 5th-order cycles are interpreted as short-eccentricity 100-kyr cycles implying MSS represents 4.86-myr Dozon 17A. See Baud and Richoz (2013) for a review of the Permian—Triassic transition in the Arabian Plate and the lithostratigraphic definition of the Mahil Formation in Oman.

across Arabia and in Oman outcrops Rabu et al. (1986) described basalmost MSS to "commonly include decimeter-thick beds of dolomite with quartz and intra-formational breccia ... and a close succession of hardgrounds separating the dolomite beds and reflecting periodic emergence."

Al-Husseini (2015) estimated the duration of Supersequence MSS at about 4.8 myr by correlating the 48 5th-order cycles to 100-kyr short-eccentricity orbital cycles (Fig. 14). Four cycle-sets containing 4 or 5 cycles are interpreted as 4 stratons, and the other four containing 7 or 8 cycles represent 8 stratons. This example illustrates the durations of stratons can vary between about 300 and 500 kyr but average 405 kyr in 4.86-myr Dozon 17A. The orbital scale calibrates Supersequence MSS between SB 17 (249.446 Ma) and SB 17B (244.586 Ma) in Olenekian–early Anisian as dated by biostratigraphy.

5.3 Olenekian Gray Zone in the Italian Alps

The position of SB 17 can also be identified in other Tethyan realms including the extensively studied sections in the Italian Alps. In Fig. 15 the lithostratigraphic members of the Lower Triassic Werfen Formation and their depositional settings in the Italian Alps (Broglio Loriga et al., 1990) are correlated according to substages to the $\delta^{13}C$ record measured in the same region (Horacek et al., 2007b). The climate fluctuations are based on regional interpretations (Stefani et al., 2010, and references therein) and the analysis of $\delta^{18}O_{apatite}$ in South China (Sun et al., 2012). At outcrop in the Italian Alps the Olenekian gray zone coincides with the poorly exposed transition between the Seis and Gastropod Oolite members (e.g., Hofmann et al., 2014) and T-R sequences Sc2 and Sc3 (Gianolla et al., 1998).

The transitional interval between the two members was deposited during a rapid sea-level fall represented by regressive and lowstand Seis Member clastics, whereas the lower part of the Gastropod Oolite was deposited in intertidal to supratidal settings during the early transgression (Broglio Loriga et al., 1990; Fig. 15). This interval is an unconformity-prone lowstand, and in some sections it may be missing. Sun et al. (2012) interpreted the Dienerian as Cooling Event I and the Late Smithian as the warmest interval in the Early Triassic. These climate trends are based on $\delta^{18}O_{apatite}$ analysis and correlate to the rising and falling $\delta^{13}C$ limbs that intersect at Excursion DSX. Excursion DSX is correlated to the lowest relative sea level and coldest time and is therefore interpreted as the maximum glacio-eustatic lowstand matching SB 17 at 249.446 Ma.

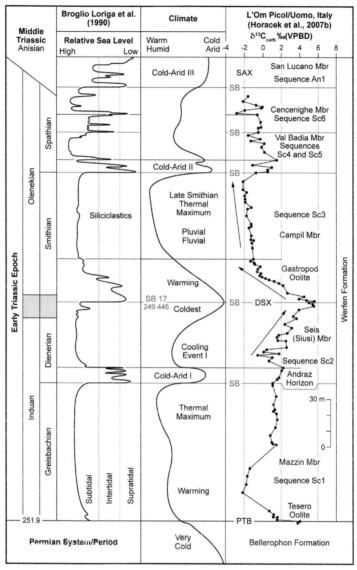

Figure 15 Members and depositional settings, T-R sequences Sc1 to Sc6 (Gianolla et al., 1998), and δ^{13}C curve (Horacek et al., 2007b) of the Lower Triassic Werfen Formation in the Italian Alps. Climate trends in the region based on Stefani et al. (2010) and cool/warm phases from South China (Sun et al., 2012). *Reproduced by permission from University of Padua from Broglio Loriga, C., Góczán, F., Haas, J., Lenner, K., Neri, C., Oravecz-Scheffer, A., Posenato, R., Szabo, I., Toth-Makk, A., 1990. The Lower Triassic sequences of the Dolomites (Italy) and Transdanubian Mid-Mountains (Hungary) and their correlation. Memorie di Scienze Geologiche, 42, 41—103.*

6. TUNING THE CENOMANIAN STAGE AND SEA LEVEL

6.1 Oman's Natih Reservoirs and Sequences

In Oman, the upper Albian, Cenomanian and lower Turonian stages are represented by the Natih Formation (Forbes et al., 2010). This highly fractured formation contains source rocks and stacked petroleum reservoirs in several giant oil fields (Terken, 1999), and has been characterized by various sequence-stratigraphic schemes at outcrop and subsurface. In Fig. 16 Natih Members G to A and Arabian 3rd-order sequences K120 to

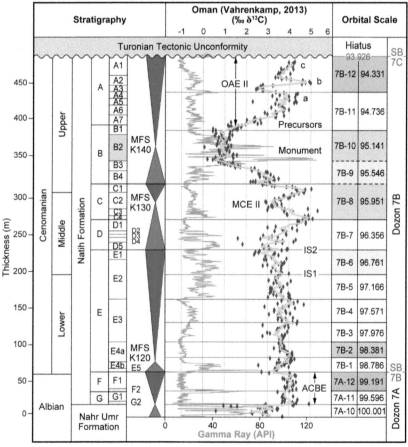

Figure 16 Oman's Natih Members G to A and Arabian 3rd-order sequences K120 to K140 dated by biostratigraphy (van Buchem et al., 2011) and correlated by Vahrenkamp (2013) to Natih subsurface units and δ^{13}C record. Correlation to δ^{13}C markers in southern England (Jarvis et al., 2006, Fig. 17) by Vahrenkamp (2013) is modified in this study in the interval between positive excursions OAE II (Oceanic Anoxic Event II) and ACBE (Albian Cenomanian Boundary Event).

K140 are shown as dated by biostratigraphy (van Buchem et al., 2011, and references therein) alongside the Natih subsurface units and δ^{13}C data (Vahrenkamp, 2013). The Natih E Member in the giant Fahud Field has also been characterized by T-R sequences ranging from 0.3—3 m thick 6th-order to 9—50 m thick 3rd-order, and their flow characteristics have been used to optimize the enhanced oil recovery strategy adopted by Petroleum Development Oman (Morettini et al., 2005). In the following discussion the Natih Formation is characterized by stratons and where evident 100-kyr 5th-order cycles in order to provide a deterministic interpretation of its sequences.

The **upper Cenomanian Boundary** is defined near the top of global positive δ^{13}C excursion OAE II (Oceanic Anoxic Event II) represented in the Culver Cliff Section, southern England (Jarvis et al., 2006; Fig. 17). In Oman OAE II occurs in the Natih Member A, and the stage boundary near-to or at the top of Sequence K140 (Vahrenkamp, 2013; van Buchem et al., 2011; Fig. 16). Above the Natih Formation a major Arabia-wide Turonian Tectonic Unconformity is related to regional uplift associated with the obduction of Tethyan ophiolites along the eastern and northern margins of the Arabian Plate (Boote et al., 1990; Droste and Van Steenwinkel, 2004; Filbrandt et al., 2006; Semail Ophiolite in Oman, Fig. 2). OAE II is here interpreted as a glacio-eustatic lowstand and an unconformity seen in the Trunch and Speeton Sections (Fig. 17). SB 7C (93.926 Ma) is positioned near OAE II-c close to the stage boundary dated between 93.9 and 94.1 Ma (Meyers et al., 2012; Eldrett et al., 2015), and merges with the Turonian Tectonic Unconformity (Fig. 16).

The **lower Cenomanian Boundary** occurs in the positive ACBE Excursion (Albian/Cenomanian Boundary Event) represented in England's Speeton Section (Jarvis et al., 2006, Fig. 17). Vahrenkamp (2013, Fig. 16) correlated the Albian Natih Members G and F to ACBE, and the thin Natih G2 and F2 siliciclastic subunits are here interpreted as glacio-eustatic lowstand deposits marking basal Stratons 7A-11 and 7A-12 (Fig. 16). SB 7B (98.786 Ma) is positioned at top Natih F in lowermost Cenomanian, as chosen for base Sequence K120 by van Buchem et al. (2011) and based on its description as a bored and iron-crusted hardground, but without evidence of subaerial exposure (Figs. 16 and 18).

Unlike the top Cenomanian, which is consistently dated at about 94.0 Ma the estimated age of the Albian/Cenomanian Boundary varies considerably. It was revised from 99.6 $\pm$ 0.9 Ma (GTS, 2008) to 100.5 $\pm$ 0.4 Ma (GTS, 2017) but in Japan estimates are 99.7 $\pm$ 0.3 Ma (ID-TIMS)

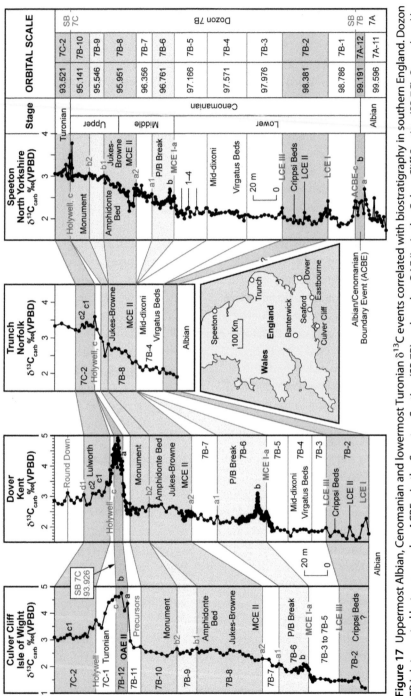

Figure 17 Uppermost Albian, Cenomanian and lowermost Turonian δ^{13}C events correlated with biostratigraphy in southern England. Dozon 7B is interpreted between marker ACBE-c in the Speeton Section (SB 7B) and marker OAE II-c in the Culver Cliff Section (SB 7C). *Reproduced by permission from Jarvis, I., Gale, A.S., Jenkyns, H.C., Pearce, M.A., 2006. Secular variation in Late Cretaceous carbon isotopes: A new δ^{13}C carbonate reference curve for the Cenomanian–Campanian (99.6–70.6 Ma). Geol. Mag., 143, 561–561.*

and 99.7 ± 1.3 Ma (LA-ICPMS) (Quidelleur et al., 2011), and 98.9 ± 1.1 Ma (^{40}Ar/^{39}Ar step-heating technique, Gaylor, 2013). The orbital calibration is consistent with the interpretation of Gale et al. (2002), which identified 12 correlative 4th-order sequences in southeast India and southwest Europe and considered them Cenomanian 405-kyr glacio-eustatic cycles. When 4.86 myr is added to 94.0 Ma at top Cenomanian its lower boundary is about 98.9 Ma.

Sequence K120 (Natih E Member) consists of six 4th-order sequences (Homewood et al., 2008, Fig. 18), which are here correlated to Stratons 7B-1 to 7B-6 with MFS K120 positioned in the organic-rich mudstones in MFI Straton 7B-2. **Sequence K130** (Natih D and C Members) at outcrop is subdivided into 5th-order sequences II-1 to II-7 (Homewood et al., 2008, Fig. 19). Sequences II-1 to II-4 are correlated to Straton 7B-7, and II-5 to II-7 to MFI Straton 7B-8 with upper Middle Cenomanian MFS K130 positioned in the latter. The signature of MFI Straton 7B-8 is recognized by the negative δ^{13}C Middle Cenomanian Event MCE II in England and in Natih Member C containing MFS K130 in Subunit C2 (Figs. 16 and 17).

Sequence K140 (Natih B and lower A Members) at outcrop is subdivided into sequences III-1 to III-12 (Homewood et al., 2008, Fig. 19). In the orbital interpretation 5th-order sequences III-1 to III-3 are correlated to lowstand Straton 7B-9 above the unconformity between the Natih C and B Members. In England lowstand Straton 7B-9 is interpreted in the δ^{13}C interval 'b1–b2' in the Speeton and Culver Cliff sections, and an unconformity in the Dover and Trunch sections (Fig. 17). The exceptionally thick Sequence III-4 containing several flooding surfaces is correlated to MFI Straton 7B-10. It is expressed as the negative δ^{13}C excursion in the Upper Cenomanian Monument Event in England and to the Natih Member B containing MFS K140 in organic-rich Subunit B2 (Figs. 16 and 17). The clinoforms III-5 to III-8 and III-9 to III-12 are interpreted as 5th-order 100-kyr sequences representing Stratons 7B-11 and 7B-12, respectively (Fig. 19). These shallow-marine sequences represent a lowstand that ends at SB 7C below Sequence IV.

Sequences IV and V at outcrop are attributed to a major transgression that is not evident in the subsurface (Figs. 16 and 19). These sequences are correlated to Lower Turonian Straton 7C-1 and MFI Straton 7C-2. The unconformity below Sequence IV is correlated to SB 7C and it is older than the Turonian Tectonic Unconformity.

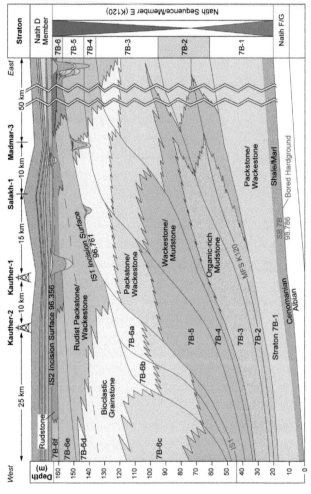

Figure 18 Natih Member E (Sequence K120, Fig. 16) interpreted in a 110-km traverse in North Oman as six 4th-order sequences (Homewood et al., 2008), here correlated to Stratons 7B-1 to 7B-6 with Lower Cenomanian MFS K120 in MFI Straton 7B-2. Incised channels IS1 and IS2 bound lowstand Straton 7B-6 and highlight glacio-eustatic fluctuations of 20 and 30 m, respectively.

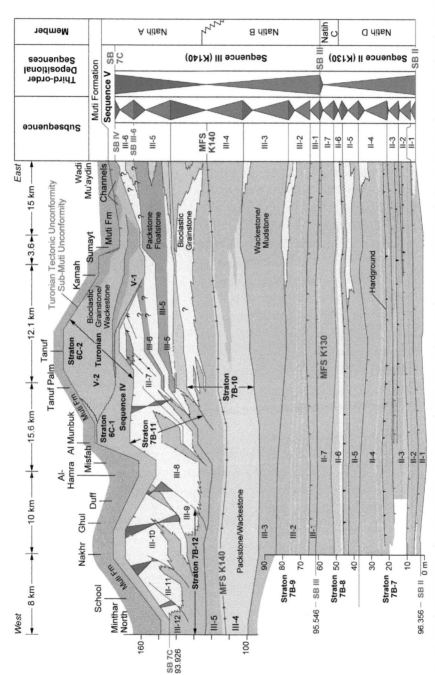

Figure 19 Natih D to A Members (Sequences K130 and K140, Fig. 16) in a 65-km traverse in North Oman outcrop (Homewood et al., 2008, and references therein) interpreted as Cenomanian Stratons 7B-7 to 7B-12 and Turonian Stratons 7C-1 and Turonian Stratons 7C-2. Turonian Tectonic Unconformity is associated with uplift and obduction of the Semail Ophiolite in eastern Oman (Fig. 2).

6.2 Cenomanian Glacio-Eustasy

The Cretaceous is generally assumed to be a greenhouse period with Antarctica either ice-free or hosting small ephemeral ice sheets (Miller et al., 2005). Gale et al. (2002) based on their fieldwork in southeast India and southwest Europe estimated Cenomanian glacio-eustatic fluctuations ranged between 2 and 20 m. In Oman two channels with depths of 20 and 30 m cut into the Natih E Member and are referred to Incision Surfaces IS1 and IS2 (Grélaud et al., 2010; Homewood et al., 2008; Fig. 18). In the orbital interpretation the older IS1 surface is correlated to base lowstand Straton 7B-6 and the positive Middle Cenomanian Event MCE I-a at base Middle Cenomanian thus placing both incision surfaces in lower Middle Cenomanian (Figs. 16 and 17).

Based on the height of the high-dip clinoforms in upper Sequence K120 (Natih Member E, Straton 7B-6, Fig. 18) and faunal content relative sea level was about 60 m and fluctuated by 20–30 m (Droste, 2010; van Buchem et al., 2011, and references therein). Antarctica's present-day ice cap holds the equivalent of 60 m of global sea level suggesting the 60 m sea level in Cenomanian Oman may essentially represent an ice-free Antarctica. The 20–30 m Cenomanian falls would then be attributed to short-lived ice caps of about one-half its present-day cap.

7. ANTARCTICA GLACIO-EUSTATIC SIGNATURE

During the Miocene the Arabian Plate was bounded to the west by the Red Sea and Gulf of Suez rift basins and to the north and east by the Tethyan Seaway connecting the Mediterranean and Arabian Seas (Fig. 2). The paleo-seaway was situated along the present-day Zagros-Taurus Suture Zone and formed the foredeep of the Tethys subduction zone (Fig. 2; e.g., Sharland et al., 2001). The oldest syn-rift flood in the Red Sea Basin occurred in mid-Aquitanian in restricted arid settings characterized by Egypt's Nukhul anhydrites and Saudi Arabia's Yanbu Salt (Al-Husseini et al., 2010; Hughes and Johnson, 2005, Fig. 20). Two younger restricted settings occurred in latest Burdigalian (e.g., Egypt's Rahmi/Markha and Saudi Arabia's An Numan anhydrites), and mid-Langhian–Serravalian (e.g., Egypt's Belayim anhydrites and South Gharib Salt, and Saudi Arabia's Kial anhydrites and Mansiyah Salt; Hughes and Johnson, 2005, Fig. 21).

The Red Sea evaporites are regionally extensive and occur between marls, shales and carbonates deposited in outer-neritic (100–200 m water

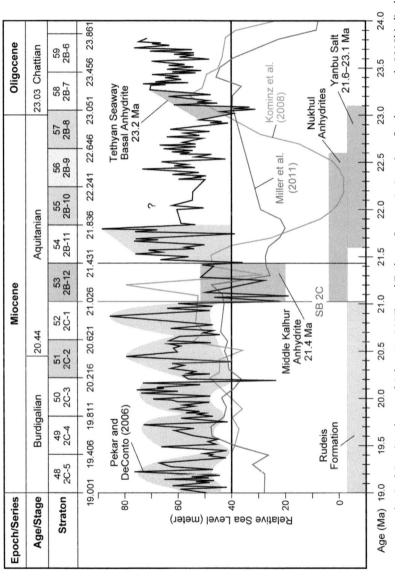

Figure 20 Evaporites in the Red Sea (Hughes and Johnson, 2005) and Tethyan Seaway in Iran (van Buchem et al., 2010b) displayed along relative sea-level curves (Kominz et al., 2008; Miller et al., 2011; reproduced with permission from K. Miller). Lowstand of about 30 m between 21.4 and 21.0 Ma (Pekar and DeConto, 2006) highlights Straton 2B-12 and attributed to expansion of Antarctica's ice cap.

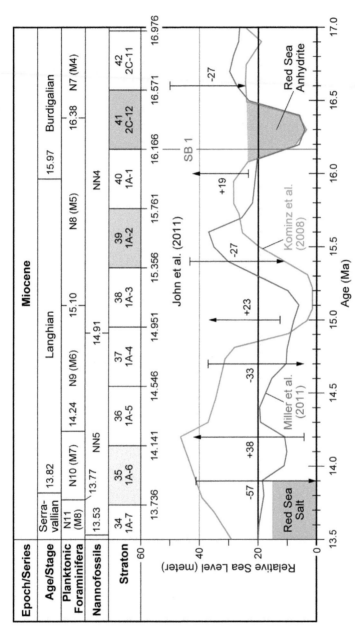

Figure 21 Lowstand between 16.5 and 16.0 Ma (John et al., 2011; Kominz et al., 2008; Miller et al., 2011; reproduced with permission from K. Miller) correlates to late Burdigalian anhydrites in the Red Sea and Gulf of Suez (Hughes and Johnson, 2005) and lowstand Straton 2C-12. The 30-m rise at 16.1–16.0 Ma correlates to SB1 (16.166 Ma). The major fall at about 13.9 Ma (John et al., 2011) occurs in mid lowstand Straton 1A-6 and may correlate to the precipitation of massive halite in the Red Sea.

depths) to deeper bathyal settings (e.g., Egypt's late Aquitanian—Burdigalian Rudies Formation and Langhian Shagar Member; e.g., Hughes and Johnson, 2005; Wescott et al., 1996; Youssef, 2011). The deep-marine clastics attain thicknesses of up to 1000 m and were transported by extensive river systems situated on mountain ranges flanking the basin and driven by pluvial periods (Al-Husseini et al., 2010; Al-Husseini, 2012). In the Tethyan Seaway, restricted settings are characterized by regional anhydrites in Iran and Iraq (Goff et al., 1995) dated by Strontium-isotope ratios in Iran (van Buchem et al., 2010b, their Fig. 18; Figs. 20 and 21): (1) Basal Anhydrite (23.2 Ma), (2) Middle Kalhur Anhydrite (21.4 Ma) and lower Gachsaran Anhydrite (18.5 Ma).

During the restricted periods open-marine flow was reduced or blocked between the Mediterranean and Arabian Seas in both the Red Sea and Tethyan Seaway (Fig. 2). During the lowstands water level in the basins may have even dropped below global level such that the subsequent flood may have filled desiccated basins. The final closure of the northern end of the Tethyan Seaway is marked by the Agha Jari continental clastics dated 12.3—12.8 Ma (Homke et al., 2004).

The restricted settings are interpreted as glacio-eustatic lowstands caused by the expansion of Antarctica's ice cap. During global lowstands four sills situated at Bab Al Mandeb Strait, northern Gulf of Suez, and both ends of the Tethyan Seaway limited or blocked open-marine circulation across both basins (Fig. 2). To confirm this interpretation early Miocene relative sea level is plotted from several papers (John et al., 2011; Kominz et al., 2008; Miller et al., 2011; Pekar and DeConto, 2006) and compared to the ages of the evaporites and stratons (Figs. 20 and 21).

The relative sea level curve of Pekar and DeConto (2006, Fig. 20) is based on high-resolution $\delta^{18}O$ data from ODP Site 1090 situated between South Africa and Antarctica and calibrated with the Astronomical Time Scale (ATS). It shows a eustatic drop centered at 23.0 Ma, which correlates to the Basal Anhydrite in Iran (23.2 Ma) and base Straton 2B-8. The subsequent rise correlates to the predicted MFI Straton 2B-8. A major drop occurs between 21.4 and 21.0, which correlates to Iran's Middle Kalhur Anhydrite (21.4 Ma) and lowstand Straton 2B-12 (21.431—21.026 Ma). The abrupt rise at 21.0 Ma correlates to SB 2B (21.026 Ma) and the late Burdigalian deep-marine flooding of the Red Sea (e.g., Egypt's Rudeis Formation).

The other two curves in Fig. 20 are based on back-stripping techniques and isotopic data in North America (Kominz et al., 2008; Miller et al., 2011).

The more recent curve by Miller et al. (2011) shows a falling sea-level trend in early Aquitanian, and a rise of about 30 m between 21.7 and 21.2 Ma. The falling trend may represent the Nukhul anhydrites and Yanbu Salt dated by strontium-isotope ratios between 23.1 and 21.6 Ma but these dates are not in stratigraphic order and not calibrated in ATS (Hughes and Johnson, 2005).

Both the relative sea-level curves of Kominz et al. (2008) and Miller et al. (2011) identify a 25 m, 400-kyr lowstand between 16.5 and 16.1 Ma (Fig. 21). Similarly John et al. (2011) based on interpreting seismic images and $\delta^{18}O$ records offshore Australia estimated a relative drop of about 27 m at 16.5 Ma and a rise of 19 m at 16.0 Ma. The 16.5–16.0 Ma lowstand correlates to lowstand Straton 2C-12 (16.571–16.166 Ma) and the latest Burdigalian anhydrites in the Red Sea. All three studies interpret a sea-level rise at 16.1 Ma, which is SB 1 (16.166 Ma, Fig. 1) and a highstand of about 20–30 m between 16.0 and 15.4 Ma, which correlates to post-glacial Stratons 1A-1 and 1A-2 and the early Langhian deep-marine flood in the Red Sea. The sea-level interpretation of ODP Site 1090 (Pekar and DeConto, 2006) ends at 16.0 Ma and differs from the other three.

John et al. (2011) also interpreted several more Langhian falls successively increasing from 37 and 57 m, which may correlate to the Belayim and Kial anhydrites and the massive South Gharib and Mansiyah salts in the Red Sea. The greatest fall of 59 m at 13.9 Ma in latest Langhian may represent Antarctica's ice cap reaching its present-day size. During the late Pliocene the Northern Hemisphere ice sheets started to advance and retreat after the complete closure of the Panama Seaway causing glacio-eustatic fluctuations to increase by another 70 to 100 m (see review in Al-Husseini, 2013).

8. CONCLUSIONS

This study presented examples of how Arabian sequences and global Carbon-13 isotope ($\delta^{13}C$) patterns can be independently dated with the orbital scale of glacio-eustasy (Fig. 1). Throughout the study the calibrations of stratons are cited to the nearest thousand years in order to remain synchronized with their correlative long-eccentricity 405-kyr cycles. Although the durations of stratons may range between 300 and 500 kyr as expressed in Oman's Supersequence MSS (Fig. 14) over a few million years they average

405 kyr suggesting the accuracy of the scale is about ±100 kyr. Confirming this accuracy is challenging because empirical age estimates in most Phanerozoic and older intervals carry much greater uncertainties as explained in the Early Triassic Olenekian gray zone (Fig. 13). The scale's accuracy is more convincingly demonstrated in the example from Jordan's Ordovician by comparing the predicted ages for SBs, MFSs, biozones and stages to those in GTS 2017 in a 10-myr interval (Fig. 4).

The orbital calibration in Jordan indicates three regional sequence boundaries correlate closely to maximum $\delta^{13}C$ values: (1) at Peak 2 in MDICE, (2) in a minor excursion at Base Sandbian, and (3) top GICE. The three sequence boundaries separate two groups of 12 stratons (4.86-myr dozons), which suggests positive excursions MDICE-Peak 2, Base Sandbian, GICE, WICE, HICE and Sedgwickii tune at 4.86 myr and represent the maximum spatial advances by ice sheets on Gondwana (Figs. 5, 6 and 9). This relationship suggests orbital-forcing of insolation modulates temperature gradients and thermohaline circulation causing the snow gun to trigger major glaciations every 4.86 myr. These major glaciations last about one million years and the locations and sizes of their ice sheets are determined by continent-ocean configurations (Figs. 7 and 8).

Numerous global $\delta^{13}C$ positive excursions and negative cycles occur in the Phanerozoic, Ediacaran and older periods (Saltzman and Thomas, 2012); Cambrian SPICE (Figs. 10−12), Early Triassic DSX (Figs. 13−15) and Cretaceous OAE II and ACBE (Figs. 16−19) are just a few examples of positive excursions. Determining whether $\delta^{13}C$ patterns are generally a glacio-eustatic proxy can be tested by repeating the steps outlined in the example from Jordan's Ordovician. Should this hypothesis prove to be true the outcome would be a global deterministically dated, chemostratigraphic framework tied by 405-kyr stratons and 3rd-order T-R sequences. Higher resolution Milankovitch sequences (20 to 100 kyr) could then be better resolved within this framework and used to calibrate biozones or characterize petroleum reservoirs such as Oman's Cenomanian Natih Formation (Figs. 16−19).

The interpreted 15−50 m eustatic fluctuations in Jordan's Ordovician (Turner et al., 2012) and the 20−30 m Cenomanian falls in Oman (Grélaud et al., 2010) are similar in magnitude and periodicity to those caused by Antarctica's dynamic ice cap during the Miocene (Figs. 20 and 21). In conclusion the results of this study support the view that high-latitude ice sheets occurred during greenhouse times.

ACKNOWLEDGMENTS

I am greatly indebted to late Professor Emeritus Robley K. Matthews for his guidance, and privileged to have coauthored his final contribution to the Earth Sciences in 2010: "Orbital-forcing glacio-eustasy: A sequence-stratigraphic time scale". Michael Montenari is thanked for inviting this contribution and Florian Maurer for reviewing the manuscript. I thank Arnold Egdane for designing the figures.

REFERENCES

Abed, A.M., Makhlouf, I.M., Amireh, B.S., Khalil, B., 1993. Upper Ordovician glacial deposits in southern Jordan. Episodes 6 (1 & 2), 316—328.

Ainsaar, L., Kaljo, D., Martma, T., Meidla, T., Männik, P., Nõlvak, J., Tinn, O., 2010. Middle and Upper Ordovician carbon isotope chemostratigraphy in Baltoscandia: a correlation tool and clues to environmental history. Palaeogeogr. Palaeoclimatol. Palaeoecol. 294, 189—201.

Al-Husseini, M.I., Mahmoud, M.D., Matthews, R.K., 2010. Miocene Kareem sequence, Gulf of Suez, Egypt. GeoArabia 15 (2), 175—204.

Al-Husseini, M.I., 2012. Late Oligocene—early Miocene Nukhul sequence, Gulf of Suez and Red Sea. GeoArabia 17 (1), 17—44.

Al-Husseini, M.I., 2013. Antarctica's glacio-eustatic signature in the Aptian and late Miocene—Holocene: implications for what drives sequence stratigraphy. GeoArabia 18 (1), 17—52.

Al-Husseini, M.I., 2015. Arabian orbital stratigraphy revisited — AROS 2015. GeoArabia 20 (4), 183—216.

Andrews, I.J., 1991. Palaeozoic lithostratigraphy in the subsurface of Jordan. Kingd. Jordan Nat. Resources Author. Bull. 2, 1—75.

Baud, A., Richoz, S., 2013. Permian—triassic transition and the Saiq/Mahil boundary in the Oman mountains: proposed correction for lithostratigraphic nomenclature. GeoArabia 18 (3), 87—98.

Bergström, S.M., Saltzman, M.R., Leslie, S.A., Ferretti, A., Young, S.A., 2015. Trans-atlantic application of the Baltic Middle and Upper Ordovician carbon isotope zonation. Est. J. Earth Sci. 64 (1), 8—12.

Boote, D.R.D., Mou, D., Waite, R.I., 1990. Structural evolution of the Suneinah foreland, Central Oman mountains. In: Robertson, A.H.F., Searle, M.P., Ries, A.C. (Eds.), The Geology and Tectonics of the Oman Region, vol. 49. Geological Society, London, Special Publication, pp. 397—418.

Brenchley, P.J., Carden, G.A., Hints, L., Kaljo, D., Marshall, J.D., Martma, T., Meldha, T., Nõlvak, J., 2003. High-resolution stable isotope stratigraphy of Upper Ordovician sequences: constraints on the timing of bioevents and environmental changes associated with mass extinctions and glaciation. GSA Bull. 115, 89—104.

Broglio Loriga, C., Góczán, F., Haas, J., Lenner, K., Neri, C., Oravecz-Scheffer, A., Posenato, R., Szabo, I., Toth-Makk, A., 1990. The lower Triassic sequences of the dolomites (Italy) and Transdanubian mid-Mountains (Hungary) and their correlation. Mem. Sci. Geol. 42, 41—103.

Burgess, S.D., Bowring, S., Shen, S-z., 2014a. High-precision timeline for Earth's most severe extinction. Proc. Natl. Acad. Sci. U.S.A. 111, 3316—3321.

Burgess, S.D., Bowring, S., Shen, S-z, 2014b. Correction. Proc. Natl. Acad. Sci. U.S.A. 111, 5060.

Caputo, M.V., Melo, J.H.G., Streel, M., Isbell, J.L., 2008. Late Devonian and early Carboniferous glacial records of south America. In: Fielding, C.R., Frank, T.D., Isbell, J.L. (Eds.), Resolving the Late Paleozoic ice Age in Time and Space, vol. 441. Geological Society of America, Special Paper, pp. 161–173.

Clarkson, M.O., Richoz, S., Wood, R.A., Maurer, F., Krystyn, L., McGurty, D.J., Astratti, D., 2013. A new high-resolution $\delta^{13}C$ record for the early Triassic: insights from the Arabian platform. Gondwana Res. 24, 233–242.

Cocks, L.R.M., Mckerrow, W.S., Van Staal, C.R., 1997. The margins of Avalonia. Geol. Mag. 134, 627–636.

Cocks, L.R.M., Torsvik, T.H., 2006. European geography in a global context from the Vendian to the end of the Palaeozoic. In: Ghee, D.G., Stehenson, R.A. (Eds.), European Lithosphere Dynamics, vol. 32. Geological Society, London, Memoir, pp. 83–95.

Cramer, B.D., Saltzman, M.R., 2007. Fluctuations in epeiric sea carbonate production during Silurian positive carbon isotope excursions: a review of proposed paleoceanographic models. Palaeogeogr. Palaeoclimatol. Palaeoecol. 245, 37–45.

Diaz-Martinez, E., Grahn, Y., 2007. Early Silurian glaciation along the western margin of Gondwana (Peru, Bolivia and northern Argentina): Palaeogeographic and geodynamic setting. Palaeogeogr. Palaeoclimatol. Palaeoecol. 245, 62–81.

Droste, H.J., 1997. Stratigraphy of the lower Paleozoic Haima Supergroup of Oman. GeoArabia 2 (4), 419–472.

Droste, H.J., 2010. High-resolution seismic stratigraphy of the Shu'aiba and Natih formations in the Sultanate of Oman: implications for Cretaceous epeiric carbonate platform systems. In: van Buchem, F.S.P., Gerdes, K.D., Esteban, M. (Eds.), Mesozoic and Cenozoic Carbonate Systems of the Mediterranean and the Middle East – Stratigraphic and Diagenetic Reference Models, vol. 329. Geological Society, London, Special Publication, pp. 145–162.

Droste, H.J., Van Steenwinkel, M., 2004. Stratal geometries and patterns of platform carbonates: the Cretaceous of Oman. In: Eberli, G., Massaferro, J.L., Sarg, J.F.R. (Eds.), Seismic Imaging of Carbonate Reservoirs and Systems, vol. 81. AAPG Memoir, pp. 185–206.

Eldrett, J.S., Ma, C., Bergman, S.C., Lutz, B., Gregory, F.J., Dodsworth, P., Phipps, M., Hardas, P., Minisini, D., Ozkan, A., Ramezani, J., Bowring, S.A., Kamo, S.L., Ferguson, K., Macaulay, C., Kelly, A.E., 2015. An astronomically calibrated stratigraphy of the Cenomanian, Turonian and earliest Coniacian from the Cretaceous western interior seaway, USA: implications for global chronostratigraphy. Cretac. Res. 56, 316–344.

El-Khayal, A.A., Romano, M., 1988. A revision of the upper part of the Saq Formation and Hanadir shale (lower ordovician) of Saudi Arabia. Geol. Mag. 125 (2), 161–174.

Filbrandt, J.B., Al-Dhahab, S., Al-Habsy, A., Harris, K., Keating, J., Al-Mahruqi, S., Ozkaya, S.I., Richard, P.D., Robertson, T., 2006. Kinematic interpretation and structural evolution of North Oman, Block 6, since the Late Cretaceous and implications for timing of hydrocarbon migration into Cretaceous reservoirs. GeoArabia 11 (1), 97–140.

Forbes, G., Jansen, H., Schreurs, J., 2010. Lexicon of Oman Subsurface Stratigraphy: Reference Guide to the Stratigraphy of Oman's Hydrocarbon Basins. GeoArabia Special Publication 5, Gulf PetroLink, Bahrain, p. 371.

Gale, A.S., Hardenbol, J., Kennedy, W.J., Young, J.R., Phansalkar, V., 2002. Global correlation of Cenomanian (Upper Cretaceous) sequences: evidence for Milankovitch control on sea level. Geology 30 (2), 291–294.

Galfetti, T., Bucher, H., Ovtcharova, M., Schaltegger, U., Brayard, A., Bruhwiler, T., Goudemand, N., Weissert, H., Hochuli, P.A., Cordey, F., Guodun, K.A., 2007. Timing of the Early Triassic carbon cycle perturbations inferred from new U–Pb ages and ammonoid biochronozones. Earth Planet. Sci. Lett. 258, 593–604.

Gaylor, J., 2013. $^{40}Ar/^{39}Ar$ Dating of the Late Cretaceous. Earth Sciences. PhD Thesis. Université Paris Sud-Paris XI, p. 275.

Gianolla, P., De Zanche, V., Mietto, P., 1998. Triassic sequence stratigraphy in the southern Alps (Northern Italy): definition of sequences and basin evolution. In: de Graciansky, P.-C., Hardenbol, J., Jacquin, T., Vail, P. (Eds.), Mesozoic and Cenozoic Sequence Stratigraphy of European Basins, vol. 60. SEPM Special Publication, pp. 719—747.

Goff, J.C., Jones, R.W., Horbury, A.D., 1995. Cenozoic Basin Evolution of the northern part of the Arabian Plate and its control on hydrocarbon habitat. In: GEO '94 Proceedings. Gulf PetroLink, Bahrain, pp. 402—412.

Goldman, D., Leslie, S.A., Nõlvak, J., Young, S.A., Bergström, M., Huff, W.D., 2007. The global Stratotype section and Point (GSSP) for the base of the Katian stage of the upper ordovician series at black Knob ridge, southeastern Oklahoma, USA. Episodes 30 (4), 258—270.

Gradstein, F.M., Ogg, J.G., 1996. A Phanerozoic time scale. Episodes 19, 3—5.

Grélaud, C., Razin, P., Homewood, P., 2010. Channelized systems in an inner carbonate platform setting: Differentiation between incisions and tidal channels (Natih Formation, Late Cretaceous, Oman). In: van Buchem, F.S.P., Gerdes, K.D., Esteban, M. (Eds.), Mesozoic and Cenozoic Carbonate Systems of the Mediterranean and the Middle East — Stratigraphic and Diagenetic Reference Models, vol. 329. Geological Society, London, Special Publications, pp. 163—186.

Haq, B.U., Al-Qahtani, A.M., 2005. Phanerozoic cycles of sea-level change on the Arabian Platform. GeoArabia 10 (2), 127—160.

Hinnov, L.A., Hilgen, F.J., 2012. Cyclostratigraphy and Astrochronology. The Geologic Time Scale. Elsevier, pp. 63—83.

Hofmann, R., Hautmann, M., Bucher, H., 2014. Recovery dynamics of benthic marine communities from the Lower Triassic Werfen Formation, northern Italy. Lethaia 48, 474—496.

Homewood, P., Razin, P., Grélaud, C., Droste, H.J., Vahrenkamp, V., Mettraux, M., Mattner, J., 2008. Outcrop sedimentology of the Natih Formation, northern Oman: a field guide to selected outcrops in the Adam Foothills and Al Jabal al Akhdar areas. GeoArabia 13 (3), 39—120.

Homke, S., Vergés, J., Garcés, M., Emami, H., Karpuz, R., 2004. Magnetostratigraphy of Miocene—Pliocene Zagros foreland deposits in the front of the Push-e Kush Arc (Lurestan Province, Iran). Earth Planet. Sci. Lett. 225, 397—410.

Horacek, M., Richoz, S., Brandner, R., Krystyn, L., Spötl, C., 2007a. Evidence for recurrent changes in Lower Triassic oceanic circulation of the Tethys: the δ^{13}C record from marine sections in Iran. Palaeogeogr. Palaeoclimatol. Palaeoecol. 252, 355—369.

Horacek, M., Brandner, R., Abart, R., 2007b. Carbon isotope record of the P/T Boundary and the Lower Triassic in the southern Alps: evidence for rapid changes in storage of organic carbon; correlation with magnetostratigraphy and biostratigraphy. Palaeogeogr. Palaeoclimatol. Palaeoecol. 252, 347—354.

Hughes, G.W., Johnson, R.S., 2005. Lithostratigraphy of the Red Sea region. GeoArabia 10 (3), 49—126.

Immenhauser, A., Matthews, R.K., 2004. Albian sea-level cycles in Oman: the 'Rosetta Stone' approach. GeoArabia 9 (3), 11—46.

Jarochowska, E., Munnecke, A., 2016. Late Wenlock carbon isotope excursions and associated conodont fauna in the Podlasie Depression, eastern Poland: a not-so-big crisis? Geol. J. 71 (5), 683—703.

Jarvis, I., Gale, A.S., Jenkyns, H.C., Pearce, M.A., 2006. Secular variation in Late Cretaceous carbon isotopes: a new δ^{13}C carbonate reference curve for the Cenomanian—Campanian (99.6—70.6 Ma). Geol. Mag. 143, 561—608.

John, C.M., Karner, G.D., Browning, E., Leckie, R.M., Mateo, Z., Carson, B., Lowery, C., 2011. Timing and magnitude of Miocene eustasy derived from the mixed siliciclastic-carbonate stratigraphic record of the northeastern Australian margin. Earth Planet. Sci. Lett. 304, 455—467.

Kaljo, D., Martma, T., Mannik, P., Viira, V., 2003. Implications of Gondwana glaciations in the Baltic late Ordovician and Silurian and a carbon isotopic test of environmental cyclicity. Bull. Sociéte Géologique France 174 (1), 59–66.

Kozlowski, W., 2015. Eolian dust influx and massive whitings during the Kozlowski/Lau Event: carbonate hypersaturation as a possible driver of the mid-Ludfordian Carbon Isotope Excursion. Bull. Geosci. 90 (4), 807–840.

Kominz, M.A., Browning, J.V., Miller, K.G., Sugarman, P.J., Mizintsevaw, S., Scotese, C.R., 2008. Late Cretaceous to Miocene sea-level estimates from the New Jersey and Delaware coastal plain coreholes: an error analysis. Basin Res. 20, 211–226.

Kouchinsky, A., Bengston, S., Pavlov, V., Runnegar, B., Tossander, P., Young, E., Ziegler, K., 2007. Carbon isotope stratigraphy of the Precambrian-Cambrian Sukharikha River section, northwester Siberian platform. Geol. Mag. 144, 609–618.

Kouchinsky, A., Bengtson, S., Landing, V., Steiner, M., Vendrasco, M., Ziegler, K., 2017. Terreneuvian stratigraphy and faunas from the Anabar uplift, Siberia. Acta Palaeontol. Pol. 62 (2), 311–440.

Landing, E., 2004. Precambrian–Cambrian boundary interval deposition and the marginal platform of the Avalon microcontinent. J. Geodyn. 37, 411–435.

Landing, E., MacGabhann, B.A., 2010. First evidence for Cambrian glaciation provided by sections in Avalonian new Brunswick and Ireland: Additional data for Avalon–Gondwana separation by the earliest Palaeozoic. Palaeogeogr. Palaeoclimatol. Palaeoecol. 285, 174–185.

Landing, E., Geyer, G., Buchwaldt, R., Bowring, S.A., 2015. Geochronology of the Cambrian: a precise middle Cambrian U–Pb zircon date from the German margin of west Gondwana. Geol. Mag. 152, 28–40.

Laskar, J., Robutal, P., Joutel, F., Gastineau, M., Correia, A., Levrard, B., 2004. A long term numerical solution for the insolation quantities of the Earth. Astron. Astrophys. 428 (1), 261–286.

Lehnert, O., Meinhold, G., Wu, R.C., Calner, M., Joachimski, M.M., 2014. δ^{13}C chemostratigraphy in the upper Tremadocian through lower Katian (Ordovician) carbonate succession of the Siljan district, central Sweden. Est. J. Earth Sci. 63 (4), 277–286.

Lehrmann, D.J., Ramezani, J., Bowring, S.A., Martin, M.W., Montgomery, P., Enos, P., Payne, J.L., Orchard, M.J., Wang, H., Wei, J., 2015. Timing of recovery from the end-Permian extinction: Geochronologic and biostratigraphic constraints from South China. Geology 34, 1053–1056.

Leslie, S.A., Saltzman, M.R., Bergström, S.M., Repetski, J.E., Howard, A., Seward, A.M., 2011. Conodont biostratigraphy and stable isotope stratigraphy across the Ordovician Knox/Beekmantown unconformity in the central Appalachians. In: Gutiérrez-Marco, J.-C., Rabano, I., Garcia-Bellido, D. (Eds.), Ordovician of the World, vol. 14. Publicaciones del Museo Geominero de Espana, pp. 301–308.

Li, M., Ogg, J.G., Zhang, Y., Huang, C., Hinnov, L., Chen, Z.-Q., Zou, Z., 2016. Astronomical-cycle scaling of the end-Permian extinction and the early Triassic Epoch of south China and Germany. Earth Planet. Sci. Lett. 441, 10–25.

Loydell, D.K., 1998. Early Silurian sea-level changes. Geol. Mag. 135, 447–471.

Masri, A., 1998. Batn Al Ghul (Jabal Al Harad) Sheet 3149 II. 1:50,000. Kingdom of Jordan Natural Resources Authority, Amman, Jordan.

Matthews, R.K., Frohlich, C., 2002. Maximum flooding surface and sequence boundaries: comparisons between observation and orbital forcing in the Cretaceous and Jurassic (65–190 Ma). GeoArabia 7 (3), 503–538.

Matthews, R.K., Al-Husseini, M.I., 2010. Orbital-forcing glacio-eustasy: a sequence-stratigraphic time scale. GeoArabia 15 (3), 155–167.

McGillivray, J.G., Al-Husseini, M.I., 1992. The Palaeozoic petroleum geology of central Arabia. AAPG (Am. Assoc. Pet. Geol.) Bull. 76, 1473–1490.

Melchin, M.J., Holmden, C., 2006. Carbon isotope chemostratigraphy of the Llandovery in Arctic Canada: implications for global correlation and sea-level change. GFF 128, 173—180.

Meyers, S.R., Siewert, S.E., Singer, B.S., Sageman, B.B., Condon, D.J., Obradovich, J.D., Jicha, B.R., Sawyer, D.A., 2012. Intercalibration of radioisotopic and astrochronologic time scales for the Cenomanian-Turonian boundary interval, Western Interior Basin, USA. Geology 401 (1), 7—10.

Miller, K.G., Wright, J.D., Browning, J.V., 2005. Visions of ice sheets in a greenhouse world. Mar. Geol. 217, 215—231.

Miller, K.G., Mountain, G.S., Wright, J.D., Browning, J.V., 2011. A 180-million-year record of sea level and ice volume variations from continental margin and deep-sea isotopic records. Oceanography 24 (2), 40—53.

Morettini, E., Thompson, A., Eberli, G., Rawnsley, K., Roeterdink, R., Asyee, W., Christman, P., Cortis, A., Foster, K., Hitchings, V., Kolkman, W., van Konijnenburg, J.-H., 2005. Combining high-resolution sequence stratigraphy and mechanical stratigraphy for improved reservoir characterization in the Fahud field of Oman. GeoArabia 10 (3), 17—44.

Ogg, J.G., Ogg, G.M., Gradstein, F.M., 2016. A Concise Geologic Time Scale. Elsevier, p. 240.

Ogg, J.G., Ogg, G.M., Gradstein, F.M., 2017. Time Scale Creator. Based on a Concise Geologic Time Scale 2016. Elsevier. https://engineering.purdue.edu/Stratigraphy/tscreator/.

Payne, J.L., Lehrmann, D.J., Wei, J., Orchard, M.P., Schrag, D.P., Knoll, A.H., 2004. Large perturbations of the carbon cycle during recovery from the end-Permian extinction. Science 305, 506—509.

Pekar, S.F., DeConto, R., 2006. High-resolution ice-volume estimates for the early Miocene: evidence for a dynamic ice sheet in Antarctica. Palaeogeogr. Palaeoclimatol. Palaeoecol. 231, 101—109.

Peng, S., Babcock, L.E., Cooper, R.A., 2012. The Cambrian period. In: Gradstein, F.M., Ogg, J.G., Schmitz, M., Ogg, G.M. (Eds.), The Geologic Time Scale 2012. Elsevier, pp. 437—488.

Pope, M.C., Steffan, J.B., 2003. Widespread, prolonged late Middle to Late Ordovician upwelling in North America: a proxy record of glaciation? Geology 31 (1), 63—66.

Pöppelreiter, M.C., Schneider, C.J., Obermaier, M., Forke, H.C., Koehrer, B., Aigner, T., 2011. Seal turns into reservoir: Sudair equivalents in outcrops, Al Jabal al-Akhdar, Sultanate of Oman. GeoArabia 16 (1), 69—108.

Powell, J.H., 1989. Stratigraphy and sedimentation of the Phanerozoic rocks in central and south Jordan, Part A: Ram and Khreim groups. King. Jordan Nat. Resources Author. Bull. 11, 72.

Prentice, M.L., Matthews, R.K., 1991. Tertiary ice sheet dynamics: the snow gun hypothesis. J. Geophys. Res. 96 (B4), 6811—6827.

Quidelleur, X., Paquette, J.L., Fiet, N., Takashima, R., Tiepolo, M., Desmares, D., Nishi, H., Grosheny, D., 2011. New U—Pb (ID-TIMS and LA-ICPMS) and $^{40}Ar/^{39}Ar$ geochronological constraints of the Cretaceous geologic time scale calibration from Hokkaido (Japan). Chem. Geol. 286 (3&4), 72—83.

Rabu, D., Béchennec, F., Beurrier, M., Hutin, G., 1986. Explanatory Notes to the Geological Map of the Nakhl Quadrangle, Sultanate of Oman. Geoscience Map, Scale 1: 100,000, Sheet NF 40—3E. Ministry of Petroleum and Minerals, Directorate General of Minerals, Sultanate of Oman, p. 42.

Richoz, S., 2006. Stratigraphie et variations isotopiques du carbone dans le Permien supérieur et le Trias inférieur de quelques localits de la Néotéthys (Turquie, Oman et Iran). Mémoires de Géologie (Université de Lausanne) 46, 283.

Romano, C., Goudemand, N., Vennemann, T.W., Ware, D., Schneebeli-Hermann, E., Hochuli, P.A., Brühwiler, T., Brinkmann, W., Bucher, H., 2012. Climatic and biotic upheavals following the end-Permian mass extinction. Nat. Geosci. 6, 57—60.

Runkel, A.C., McKay, R.M., Palmer, R., 1998. Origin of a classic cratonic sheet sandstone: stratigraphy across the Sauk II—Sauk III boundary in the upper Mississippi valley. GSA Bull. 110, 188—210.

Saltzman, M.R., Thomas, E., 2012. Carbon isotope stratigraphy. In: Gradstein, F.M., Ogg, J.G., Schmitz, M., Ogg, G.M. (Eds.), In The Geologic Time Scale 2012. Elsevier, pp. 207—232.

Saltzman, M.R., Young, S.A., 2005. Long-lived glaciation in the Late Ordovician? Isotopic and sequence-stratigraphic evidence from western Laurentia. Geology 33, 109—112.

Saltzman, M.R., Runnegar, B., Lohmann, K.C., 1998. Carbon isotope stratigraphy of upper Cambrian (Steptoean stage) sequences of the eastern Great basin: record of a global oceanographic event. GSA Bull. 110, 285—297.

Saltzman, M.R., Runkel, A.C., Cowan, C.A., Runnegar, B., Stewart, M.C., Palmer, A.R., 2004. The upper Cambrian SPICE (δ^{13}C) event and the Sauk II-Sauk III regression: new evidence from Laurentian basins in Utah, Iowa and Newfoundland. J. Sediment. Res. 74, 366—377.

Schmitz, B., Bergström, S.M., Wang, X.F., 2010. The middle Darriwilian (Ordovician) δ^{13}C excursion (MDICE) discovered in the Yangtze Platform succession in China: implications of its first recorded occurrences outside Baltoscandia. J. Geol. Soc. Lond. 167, 249—259.

Senalp, M., Al-Duaiji, A.A., 2001. Qasim Formation: Ordovician storm- and tide-dominated shallow-marine siliciclastic sequences, Central Saudi Arabia. GeoArabia 6 (2), 233—268.

Sharland, P.R., Archer, R., Casey, D.M., Davies, R.B., Hall, S.H., Heward, A.P., Horbury, A.D., Simmons, M.D., 2001. Arabian Plate Sequence Stratigraphy. GeoArabia Special Publication 2, Gulf PetroLink, Bahrain, p. 371.

Stefani, M., Furin, S., Gianolla, P., 2010. The changing climate framework and depositional dynamics of Triassic carbonate platforms from the Dolomites. Palaeogeogr. Palaeoclimatol. Palaeoecol. 290, 43—57.

Sun, Y.D., Joachimski, M.M., Wignall, P.B., Yan, C.B., Chen, Y.L., Jiang, H.S., Wang, L.N., Lai, X.L., 2012. Lethally hot temperatures during the Early Triassic greenhouse. Science 338, 366—370.

Terken, J.M.J., 1999. The Natih petroleum system of north Oman. GeoArabia 4 (2), 157—180.

Tobin, K.J., Bergström, S.M., De La Garza, P., 2005. A mid-Caradocian (453 Ma) drawdown in atmospheric pCO_2 without ice sheet development? Palaeogeogr. Palaeoclimatol. Palaeoecol. 226, 187—204.

Turner, B.R., Armstrong, H.A., Wilson, C.R., Makhlouf, I.M., 2012. High frequency eustatic sea-level changes during the middle to early late ordovician of southern Jordan: indirect evidence for a Darriwilian ice age in Gondwana. Sediment. Geol. 251—252, 34—48.

Vahrenkamp, V.C., 2013. Carbon-isotope signatures of Albian to Cenomanian (Cretaceous) shelf carbonates of the Natih Formation, Sultanate of Oman. GeoArabia 18 (3), 65—82.

van Buchem, F.S.P., Al-Husseini, M.I., Maurer, F., Droste, H.J., 2010a. Barremian—aptian Stratigraphy and Hydrocarbon Habitat of the Eastern Arabian Plate. In: GeoArabia Special Publication 4, vols. 1,2. Gulf PetroLink, Bahrain, p. 614.

van Buchem, F.S.P., Allan, T.L., Laursen, G.V., Lotfpour, M., Moallemi, A., Monibi, S., Motiei, H., Pickard, H.N.A., Tahmasbi, A.R., Vedrenne, V., Vincent, B., 2010b. Regional stratigraphic architecture and reservoir types of the Oligo-Miocene deposits in the Dezful Embayment (Asmari and Pabdeh formations) SW Iran. In: van Buchem, F.S.P., Gerdes, K.D., Esteban, M. (Eds.), Mesozoic and Cenozoic Carbonate Systems of the Mediterranean and the Middle East: Stratigraphic and Diagenetic Reference Models, vol. 329. Geological Society, London, Special Publications, pp. 219—263.

van Buchem, F.S.P., Simmons, M.D., Droste, H.J., Davies, R.B., 2011. Late Aptian to Turonian stratigraphy of the eastern Arabian Plate: depositional sequences and lithostratigraphic nomenclature. Pet. Geosci. 17, 211–222.

Vaslet, D., 1990. Upper Ordovician glacial deposits in Saudi Arabia. Episodes 13, 147–161.

Wagreich, M., Lein, R., Sames, B., 2014. Eustasy, its controlling factors and the limnoeustatic hypothesis — concepts inspired by Eduard Suess. Austr. J. Earth Sci. 107 (1), 115–131.

Wescott, W.A., Krebs, W.N., Dolson, J.C., Karamat, S.A., Nummedal, D., 1996. Rift basin sequence stratigraphy: some examples from the Gulf of Suez. GeoArabia 1 (2), 343–358.

Wu, R., Calner, M., Lehnert, O., 2016. Integrated conodont biostratigraphy and carbon isotope chemostratigraphy in the Lower—Middle Ordovician of southern Sweden reveals a complete record of the MDICE. Geol. Mag. 154 (2), 334–353.

Young, S.A., Saltzman, M.R., Ausich, W.I., Desrochers, A., Kaljo, D., 2010. Did changes in atmospheric CO_2 coincide with latest Ordovician glacial-interglacial cycles? Palaeogeogr. Palaeoclimatol. Palaeoecol. 296, 376–388.

Youssef, A., 2011. Early—middle Miocene Suez syn-rift-Basin, Egypt: a sequence stratigraphy framework. GeoArabia 16 (1), 113–134.

CHAPTER SIX

Muschelkalk Ramp Cycles Revisited

Annette E. Götz*,1 and Ákos Török§

*University of Portsmouth, School of Earth and Environmental Sciences, Portsmouth, United Kingdom
§Budapest University of Technology and Economics, Department of Engineering Geology and Geotechnics, Budapest, Hungary
1Corresponding author: E-mail: annette.goetz@port.ac.uk

Contents

Abstract

Anisian Muschelkalk ramp cycles are well documented from the northwestern Tethys shelf and its northern Peri-Tethys Basin by detailed studies in Germany, the Netherlands, Poland, and Hungary. High-resolution correlation based on cyclostratigraphy has been attempted in recent years, but is still hampered by precise age control. The recent approach of Tethys–Peri-Tethys correlation of the eccentricity-cycle-scaled Anisian carbonate series integrating global climate signatures improves the so far established small-scale cycle correlation using high-frequency eustatic signatures in the Milankovitch frequency band. Cyclostratigraphic calibration of Lower Muschelkalk ramp deposits of different palaeogeographic settings by astronomical cycle-tuning of gamma-ray, isotopic and magnetostratigraphic data and integration of global climate signatures is the next step towards refined correlation.

1. INTRODUCTION

The Anisian is a crucial time interval in Earth's history to understand the prolonged biotic recovery phase of the most severe mass extinction at the end of the Permian and its actual time span (Song et al., 2011; Chen and Benton, 2012; Benton, 2015; Foster and Sebe, 2017; Nützel et al., 2018).

Stratigraphy & Timescales, Volume 3
ISSN 2468-5178
https://doi.org/10.1016/bs.sats.2018.08.003

In the marine realm, carbonate production in shallow epeiric seas is still reduced, Tethyan reefal communities were represented by low diversity faunas (Pruss and Bottjer, 2005; Velledits et al., 2011; Martindale et al., 2017), and adjacent epicontinental seas lacked reefal buildups or are characterized by monospecific bivalve patch reefs and low diversity faunal assemblages (Szulc, 2000, 2007). On the other hand, the Anisian marks a time of climate amelioration with middle Anisian warming event (Holz, 2015; Trotter et al., 2015; Li et al., 2018; Götz et al., 2018a), and two prominent transgressive phases in the Bithynian and Pelsonian (Götz and Feist-Burkhardt, 2012; Haq, 2018), the latter interpreted as a phase of global warming (Retallack, 2013; Götz et al., 2018a). These marine pulses are documented in the development of carbonate ramp systems along the western Tethys shelf in Poland, Slovakia, Hungary, Bulgaria, and the Dolomites (e.g., Michalík et al., 1992; Haas et al., 1995; Philip et al., 1996; Török, 1998; Budai and Vörös, 2006; Jaglarz and Szulc, 2003; Götz et al., 2003; Rychliński and Szulc, 2005; Götz and Török, 2008; Stefani et al., 2010; Chatalov, 2013, 2018; Matysik, 2016) and characteristic diachronous facies successions in the northern Peri-Tethys Basin described from Germany and Poland (Szulc, 2000). Highest diversity of conodonts, foraminifers, and phytoplankton (Martini et al., 1996; Götz and Gast, 2010; Götz and Feist-Burkhardt, 2012), ammonoid diversification (Vörös, 2014) as well as Tethyan immigration by ammonoids and crinoids into the Peri-Tethys (Klug et al., 2005) via the southern gates (Fig. 1) mark the Anisian transgressive phases. Timing and correlation of the ramp deposits of these two different palaeogeographic settings at high resolution is still in a state of flux (Feist-Burkhardt et al., 2008; Götz and Török, 2008; Ajdanlijsky et al., 2004, 2018; Chatalov, 2018). However, the integration of previous cyclostratigraphic interpretations (Kramm, 1997; Kędzierski, 2000; Rameil et al., 2000; Götz, 2004) into the recently refined global chronostratigraphic scheme of the Anisian (Li et al., 2018) including prominent climate signatures (Götz et al., 2018a) allows for a more precise correlation of Tethyan and Peri-Tethyan Muschelkalk cycles.

2. PREVIOUS WORK

Cyclostratigraphic interpretations of Anisian Muschelkalk deposits of the Peri-Tethyan realm (the so-called "Germanic Basin") date back to the early works of Fiege (1938), Jubitz (1954, 1958), Schüller (1967), and

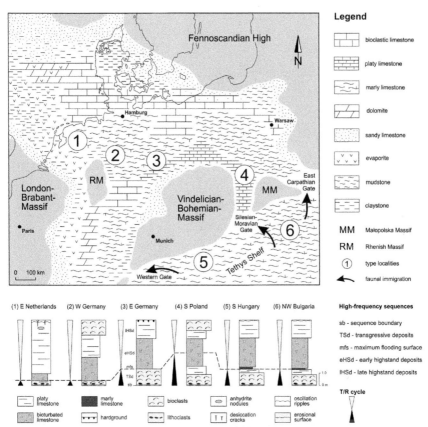

Figure 1 Palaeogeography of the northwestern Tethys shelf and Peri-Tethys Basin (central and southeast Europe) during Pelsonian times and development of small-scale cycles interpreted as high-frequency sequences, the basic stratigraphic building blocks of the Anisian Lower Muschelkalk sequence, in the different palaeogeographic settings; compiled after Pöppelreiter (2002), Götz (1996), Rameil et al. (2000), Kędzierski (2000), Götz and Török (2008), and Götz et al. (2018b).

Schulz (1972) in Germany who recognized a small-scale cyclic pattern of facies (Fig. 2A) that they interpreted to be related to changes in water depth. Bio- and lithoclastic wacke–packstones and grainstones (Fig. 2A and B) and flaser-bedded mudstones, so-called "Wellenkalke" (Fig. 2C), represent the dominant lithotypes (Götz, 1996). First attempts at intrabasinal correlation based on Schüller's descriptions and interpretation of metre-scale cycles are published by Kolb (1976), Merz (1987) and Kramm (1997). Major progress in Muschelkalk stratigraphy was made when Szulc (2000) provided a revised sequence stratigraphic interpretation for the Peri-Tethys Basin

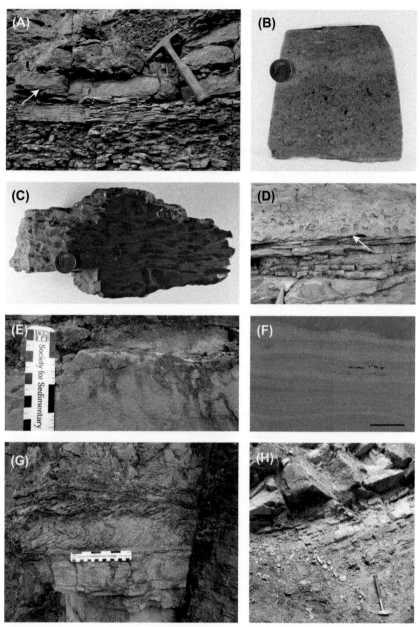

Figure 2 Characteristic Muschelkalk lithotypes and cycle elements. (A) lithoclastic limestone (transgressive deposits) with erosional base and hardground (arrow) overlying
nodular and flaser-bedded limestone (highstand deposits), Steudnitz quarry (Germany).
(B) bioclastic limestone (transgressive deposits) with bivalve shells and crinoid columnal
segments, Steudnitz quarry (Germany). (C) flaser-bedded limestone ("Wellenkalk" = wavy limestone), Steudnitz quarry (Germany). (D) thin (2.5 cm) tempestite layer
(arrow) overlying flaser-bedded limestone ("Wellenkalk"), Lapis road cut (S Hungary). (E)
hardground, Opletnya section (NW Bulgaria). (F) laminated dolomite (scale bar 1 cm), F-
33 borehole (65.90 mbs), Bükkösd (S Hungary). (G) sigmoidal deformation structure,
Opletnya section (NW Bulgaria). (H) platy dolomite, Sás völgy section (S Hungary).

defining major sequence boundaries in relation to the Tethyan sequences, and by contrasting this with and refining the first sequence interpretation of Aigner and Bachmann (1992). It was Kędzierski (2000) who then published a first regional correlation of Anisian small-scale cycles of the Germanic Basin from 90 to 100 m thick Lower Muschelkalk sections in eastern Germany and Poland along a W-E transect to emphasis their potential for high-resolution stratigraphy, however, without the integration of cyclic successions of the western basin to understand the causes of lateral changes and the complex hierarchical stacking pattern. An integrated sedimentological-palynological study of the type section of the Anisian Lower Muschelkalk (Steudnitz, E Germany; Fig. 3) in the central basin part by Rameil et al. (2000) provided further evidence of the hierarchical cyclic stacking pattern within the Milankovitch frequency band, and was later supported by chemostratigraphy (Conradi et al., 2007). Götz (2004) then highlighted the spatial and stratigraphic variations of cyclic patterns along a NW–SE transect, integrating data from Lower Muschelkalk sections in the Netherlands (Pöppelreiter, 2002), central Germany (Götz, 1996), and Poland (Kędzierski, 2000). These metre-scale cycles where interpreted as 100-ka cycles (Götz, 2004) and used for cyclostratigraphic calibration of the Lower Muschelkalk of the Germanic Middle Triassic (Menning et al., 2006). Anisian ramp deposits of the proximal northwestern Tethys shelf, described from S Hungary, also reveal a prominent hierarchical cyclicity (Götz et al., 2003), and characteristic metre-scale facies successions were interpreted as having formed through high-frequency sea-level changes in tune with orbital cycles (Götz and Török, 2008). Futhermore, these authors provided a first Tethyan–Peri-Tethyan correlation of the Anisian deposits including isotopic and biostratigraphic data.

3. TOWARDS REFINED CORRELATION

There are three main points that have to be taken into account when aiming for refined correlation of Anisian Muschelkalk ramp cycles in different palaeogeographic settings of central and southeast Europe: (1) Anisian syndepositional tectonism, (2) Anisian ramp morphologies, and (3) refined Anisian age control. Previous attempts of a cyclic interpretation of Anisian Muschelkalk carbonates of the Peri-Tethys Basin neglected syndepositional tectonism. Kędzierski (2000) interpreted small-scale facies successions of the Polish Muschelkalk as incomplete symmetric to complete

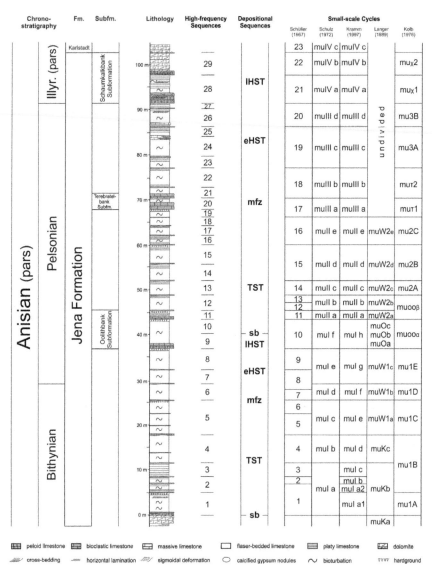

Figure 3 The type section Steudnitz (E Germany, Thuringia) of the Anisian Lower Muschelkalk, its cyclo- and sequence stratigraphic interpretation (modified from Rameil et al., 2000) and contrasting cyclostratigraphic schemes previously developed for Lower Muschelkalk deposits in different parts of the western Peri-Tethys Basin: central Germany (Schüller, 1967; Schulz, 1972; Kramm, 1997), NW Germany (Langer, 1989), NE Germany (Kolb, 1976). Abbreviations used: *Fm.*, Formation, *Subfm.*, Subformation.

asymmetric transgressive—regressive cycles or semi-cycles, related to short-term sea-level fluctuations. Thicknesses and total number of cycles strongly differ from those in the central basin (Rameil et al., 2000). However, the cycle and semi-cycle formation characteristic of the southeastern seaway (Silesian-Moravian Gate; Fig. 1), connecting the intracratonic Peri-Tethys Basin and the open Tethys shelf, may simply record syndepositional tectonic pulses, to be expected in a peripheral position towards the Tethys spreading centre (Feist-Burkhardt et al., 2008). It was recognized that the reactivation of Variscan structures must have strongly affected sedimentation in the southeastern gate areas of the basin located in S Poland (Szulc, 2000). Seismically induced synsedimentary deformations described from Anisian Muschelkalk carbonates (Szulc, 1993; Föhlisch and Voigt, 2001) where interpreted to be linked to ancestral Variscan faults and lineaments with the most prominent signatures documented from the Polish Holy Cross Mountains through Upper and Lower Silesia to eastern Germany (Szulc, 2000; Feist-Burkhardt et al., 2008). Transtensional tectonics along the southern Peri-Tethys Basin margin with strike-slip movements are the dominant feature of crustal motion during the Anisian. In the Tethyan realm, intense tectonic activity affected only the southern Alpine basins whereas the northern Alpine and Carpathian basins as part of the proximal shelf area were relatively passive areas. Thus, one expects a different record of small-scale cyclicity in the Anisian of S Poland representing the Peri-Tethys—Tethys seaway, contrasting with that of the central basin (Germany) and proximal shelf (S Hungary) with a very similar cyclicity (Götz and Török, 2008).

Cyclic patterns strongly relate to the distinct ramp morphology. Homoclinal ramps of the Peri-Tethys Basin and proximal Tethys shelf show a striking similarity in small-scale facies successions (Götz and Török, 2008, Fig. 4). In contrast, the Anisian carbonates of the southeastern seaway seem to reflect a much more complex ramp morphology (Matysik, 2016) with protected lagoons, patchy shoals, migrating shoal bars and channels, rimmed or distally steepened with first reefal buildups in the Pelsonian (Szulc, 2007), most likely reflecting variations in accommodation space related to syndepositional tectonism. In this region small-scale facies successions document incomplete cycles or semi-cycles with prominent load structures and slumps (Kędzierski, 2000).

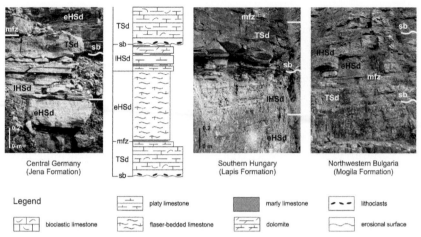

Figure 4 Small-scale cycles interpreted as high-frequency sequences in Lower Muschelkalk ramp successions of Germany, Hungary and Bulgaria. Striking similarity in small-scale facies successions is related to their deposition on homoclinal ramps characteristic of the Peri-Tethys Basin and proximal Tethys shelf. Abbreviations used: *eHSd*, early highstand deposits, *lHSd*, late highstand deposits, *mfz*, maximum flooding zone, *sb*, sequence boundary, *TSd*, transgressive deposits.

Synsedimentary deformations were also reported from the central Peri-Tethys Basin in outcrops of E Germany (Götz, 1996; Rüffer, 1996; Voigt and Linnemann, 1996) and from Tethyan shelf deposits of Slovakia, S Hungary and NW Bulgaria (Míchalik, 1997; Török, 1997, 2000; Chatalov, 2001), interpreted as seismically induced structures. However, recently Ajdanlijsky et al. (2018) noted that sigmoidal and other types of small-scale synsedimentary deformation are a striking feature of highstand deposits in metre-scale elementary sequences of Anisian ramp deposits in NW Bulgaria (Fig. 2G). Their repeated presence in the same level of the elementary sequences is interpreted to indicate a cyclic sedimentary rather than a palaeoseismic control. It is concluded that they developed during a stage of decreasing accommodation space when the weakly lithified mainly mud- and wackestones became unstable because of shallow channel erosion and slumped over a very short distance, forming thin lens- and/or wedge-like disturbed bodies. In the southern Hungarian Muschelkalk ramp system of the Mecsek Mountains synsedimentary deformation seems to be also related to mud-dominated highstand deposits, similar to the occurrence within the cycles of NW Bulgaria. Again, the ramp morphology seems to have played a crucial role in how sedimentary cycles are recorded and how synsedimentary deformation can originate from very different processes. In this regard, the Anisian Muschelkalk seems to contain a multifaceted ramp inventory of sedimentary structures not yet fully understood in terms of ramp dynamics related to different palaeogeographic settings and auto- *vs.* allocyclic processes.

The Anisian timescale has been refined in recent years (Bachmann and Kozur, 2004; Menning et al., 2006; Kozur and Bachmann, 2008; Mundil et al., 2010; Ogg, 2012). However, Hinnov and Ogg (2007) pointed out that the correlation of the eccentricity-cycle-scaled Lower Triassic continental and early Middle Triassic shallow marine series of the Germanic Basin (Menning et al., 2006) to marine stratigraphy is still uncertain and it was concluded by Tanner (2010) that Triassic cyclostratigraphic studies remain far from the goal of developing a reliable, astronomically-calibrated timescale.

Recently, Nitsch et al. (2018) reviewed the cyclostratigraphy of the Middle Triassic of the Germanic Basin, contrasting the different schemes developed for Lower Muschelkalk successions in different parts of the basin (Fig. 3). It is noted that variations in cycle numbers documented by different authors and from different localities, respectively, might partly be attributed to the use of event beds such as tempestites as cycle boundaries which are not necessarily documented as distinct beds in all basin parts. On the other hand, single event deposits are hardly recognizable in condensed successions of the maximum flooding interval. The authors questioned the regional correlation of metre-scale cycles based on small-scale lithofacies patterns and prominent bounding surfaces and concluded that basin-wide correlation is most accurate using medium-scale cyclicity in the range of tens of metres. Indeed, gamma-ray log signatures detected from inner ramp deposits of the Netherlands and NW Germany (Borkhataria et al., 2006) clearly display medium-scale transgressive—regressive cycles traceable over a distance of more than 200 km. However, outcrop studies enable a higher resolution and in numerous previous studies, metre-scale cycles of the Peri-Tethys Basin were correlated basin wide from the western inner ramp of the Netherlands to the outer ramp of eastern Germany by erosional surfaces and transgressive lag deposits followed by peak abundance of marine phytoplankton indicating maximum flooding (Götz, 1996, 2002, 2004; Götz and Feist-Burkhardt, 1999, 2000; Rameil et al., 2000; Pöppelreiter, 2002;

Pöppelreiter et al., 2005). It was concluded that the small-scale cycles may represent the short orbital eccentricity cycle of 100,000 years and stacked small-scale sequences forming sets of 3 to 4 sequences may be interpreted as reflecting the eustatic signal related to the 400,000 year eccentricity cycle (Rameil et al., 2000). This pattern is less pronounced in the Polish Lower Muschelkalk while showing a prominent consistency in the Hungarian part of the northwestern Tethys shelf (Götz and Török, 2008), where also numerous tempestites (Fig. 2D) are recorded (Török, 1993), clearly distinguishable from transgressive deposits of small-scale cycles (Fig. 2A and B). Again, the ramp morphology is the key point to how event beds such as tempestites and sea-level fluctuations of different orders are recorded in characteristic facies successions. Chatalov (2013) highlighted the similarity of the Anisian ramp system of S Hungary to the homoclinal ramp setting of NW Bulgaria and a recent study of Ajdanlijsky et al. (2018) showed that the Anisian ramp initialization stage is well documented by a hierarchical cyclicity with metre-scale elementary sequences as basic building blocks of the ramp succession. They commonly start with high-energy facies representing tidally influenced oolitic and/or bioclastic bars formed during transgression, when the previously very shallow, intertidal or supratidal environment was flooded. The corresponding transgressive surface is well developed and commonly erodes into the underlying sediment. The increased water depth and diminished current energy led to the abandonment of the high-energy bars and low-energy mud- and wackestones then predominate. The rapid shift from high-energy to low-energy deposits as well as reduced sedimentation rate are interpreted to reflect maximum flooding. Highstand deposits of metre-scale elementary sequences are mud-dominated, part of them dolomitized, with evaporite pseudomorphs and/or tepees. The boundaries of elementary sequences cannot always be placed at a discrete bedding surface and rather define thin sequence-boundary zones (Strasser et al., 1999). Below the following transgressive surface, thin marly lowstand deposits may occur. In some cases, however, a prominent transgressive surface directly overlies the sequence boundary of the elementary sequence, implying very low accommodation. During the Pelsonian third-order maximum flooding phase, small-scale cycles are dominated by bioturbated, marly limestones and the high-frequency maximum flooding surface is marked by a thin condensed layer of laminated marl showing peak abundance of marine acritarchs (Götz et al., 2018b). Within third-order sequence boundary zones, small-scale cycles are characterized by platy, dolomitic limestones (Fig. 2H) as recently also recorded in drill cores from the Anisian Muschelkalk in S Hungary (Fig. 2F). The striking similarity of small-scale cycles detected in mid and outer ramp deposits of central Germany, southern Hungary and northwestern Bulgaria (Fig. 4) is indicative of their allocyclic formation and might thus in future studies allow for tracing metre-scale ramp cycles along the northwestern Tethys shelf into the Peri-Tethys Basin over several hundreds of kilometres (Fig. 1).

To date, available information on the duration of the Anisian varies more than 1 Myr ranging from 6.7 Myr (Kozur and Bachmann, 2008), 6.5 Myr (Hagdorn et al., 2016), 6.4 Myr (Menning et al., 2006), 5.6 Myr (Ogg, 2012), and 5.3 Myr (Ogg et al., 2016) to 5.2 Myr (Mundil et al., 2010), while the duration of the Lower Muschelkalk ramp system based on Milankovitch cycles is calculated at 2.1 Myr (Bachmann and Kozur, 2004; Kozur and Bachmann, 2008; Hagdorn et al., 2016). The duration of the Anisian calculated by Menning et al. (2006) is based on using 100 kyr Milankovitch cycles with a relatively constant number of 104 throughout the Germanic Basin, and by applying the same approach Kozur and Bachmann (2008) provided a first estimation for the marine Lower Muschelkalk in the Germanic Basin. Nitsch et al. (2018) compared the calculations using the long (400 kyr) and short (100 kyr) eccentricity cycles, concluding that the mean of the cycle duration may significantly differ from the Milankovitch signatures and thus a best fit approach as a first step towards time estimation based on cyclic patterns is suggested. A duration of 1.5 to 2.1 Myr for the Anisian Lower Muschelkalk is discussed. Furthermore, these authors advocated a uniform order of cycles to be applied for time estimation of individual cycle duration in the Germanic Triassic — in continental and marine settings. However, this approach is highly questionable with regard to strong randomness in the stratigraphic record as outlined by Schlager (2010). The argumentation for an ordered hierarchy of cycles is based on the observation of superposition of cycles e.g. in outcrops. However, superposition of shorter and longer trends is also a characteristic of random processes such as Brownian walk and without time series analysis or other quantitative techniques, indicating that a particular pattern arises from the superposition of cycles of different periods, it is no argument for an ordered hierarchy of cycles. Without doubt, distinguishing order versus randomness is particularly important for small-scale cycles of lower rank, i.e. in the Milankovitch frequency band.

With regard to a high-precision global Anisian timescale, Li et al. (2018) provided new data from a key reference section in South China. Astronomical tuning of gamma-ray and magnetic susceptibility data from the studied Guandao section based on interpreted 405 kyr long orbital eccentricity cycles provided a high-resolution astronomical time indicating a 5.3 Myr duration for the Anisian stage with the Olenekian—Anisian stage boundary at 246.8 ± 0.1 Ma and the Anisian—Ladinian stage boundary at 241.5 ± 0.1 Ma. These authors concluded that the middle Anisian humid phase appears to have been a global event that lasted from 244.5 Ma to 244 Ma. This humid phase and the middle Anisian warming coincided with a maximum flooding surface at Guandao. Furthermore, it was found that the sea-level changes at Guandao in the eastern Tethys generally correlate with sea-level fluctuations in the western Tethyan and Boreal regions, supporting their eustatic origin.

A humid middle Anisian (Pelsonian) phase within the general arid to semi-arid Anisian climate was deduced from floral assemblages and palaeosoil occurrence in the northwestern Tethyan realm (Kustatscher et al., 2010; Haas et al., 2012). Maximum flooding during the Pelsonian is documented in condensed ramp deposits marked by firm- and hardgrounds (Fig. 5) described from Muschelkalk successions of the northwestern Tethys shelf and northern Peri-Tethys Basin (Török, 1998; Rameil et al., 2000; Pöppelreiter, 2002; Götz and Török, 2008; Götz and Gast, 2010; Götz and Feist-Burkhardt, 2012; Stefani et al., 2010; Chatalov, 2013; Matysik, 2016) and thus provides further evidence of middle Anisian global warming related to a global CO_2 crisis in the Triassic (Retallack, 2013; Götz et al., 2018a).

The Pelsonian maximum flooding zone spans a time interval of about 400-kyr in the Peri-Tethys Basin, i.e. four 100-kyr cycles (Rameil et al., 2000), matching the duration of a 0.5 Myr middle Anisian humid phase as recently calculated by Li et al. (2018). A major Pelsonian flooding event was also identified across the entire Arabian Plate by biostratigraphic calibration to the *Balatonites balatonicus* ammonoid zone of the Tethyan standard (Davies and Simmons, 2018). In the northwestern Tethyan and Peri-Tethyan realm, conodonts and palynomorphs provide the biostratigraphic framework to underpin the dating of this humid phase (Götz and Feist-Burkhardt, 1999; Narkiewicz, 1999; Götz et al., 2003; Kovács and Rálisch-Felgenhauer, 2005; Götz and Gast, 2010; Kustatscher et al., 2010; Götz et al., 2018b; c) expressed in a major flooding event in both palaeogeographic settings. A recent study

Figure 5 Hardgrounds and crinoidal pack-/grainstones of the Pelsonian Terebratel Beds (Terebratelbank Subformation; see Fig. 3) represent the most pronounced flooding phase within the Anisian Lower Muschelkalk of the Peri-Tethys Basin. Road cut along the motorway A4, west of Eisenach (E Germany, Thuringia).

of conodont material from sections in Germany, Hungary and Bulgaria provides a new oxygen isotope ($\delta^{18}O$) record from conodont apatite (Fig. 6), revealing a much more dynamic pattern of warming phases in the Anisian of the northwestern Tethyan realm (Götz et al., 2018a). Two warming events in the Bithynian and Pelsonian are inferred, previously recognized as maximum flooding phases in the northwestern Tethyan and Peri-Tethyan realm (Götz and Török, 2008) and will provide a basis for future Tethyanwide and global correlations. The identification and utilization of climate signatures as basin-wide, regional and global correlation tools are thus proposed to also overcome the constraints of tectonic signatures obscuring the eustatic beat in different palaeogeographic settings.

4. CONCLUSIONS

Anisian Muschelkalk ramp cycles represent high-frequency eustatic signatures in the Milankovitch frequency band with the 100,000 years eccentricity signal being most pronounced. Their regional occurrence is a

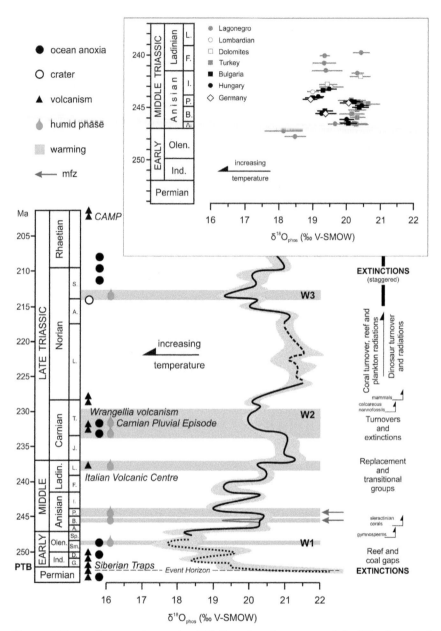

Figure 6 First order warming cycles (W1-3) and major events of the Triassic period (modified from Trotter et al., 2015); the inserted graph illustrates oxygen isotope compositions of conodont apatite ($\delta^{18}O_{phos}$) for the Anisian stage plotting data from Turkey, the Dolomites, and the Lagonegro and Lombardian basins (Trotter et al., 2015) and from Bulgaria (Balkan Mts.), Hungary (Mecsek Mts.), and the Lower Muschelkalk type

characteristic feature of shallow ramp systems of the Tethyan and Peri-Tethyan realm and has been the subject of numerous previous studies. They have also been used as the basis for cyclostratigraphic calibration of the Anisian Lower Muschelkalk of the Peri-Tethys Basin. The duration of the Anisian and Lower Muschelkalk, respectively, calculated based on 100 kyr Milankovitch cycles with a relatively constant number throughout the Germanic Basin, is to date seen as the most practicable way towards a high-resolution stratigraphy.

The Anisian climate, being much more dynamic as previously stated, with warming episodes of different orders recently recorded by new oxygen isotope data, needs to be studied in more detail. Including new available high-precision data from the eastern Tethys, capturing a Pelsonian humid phase of 0.5 Myr (244.5−244 Ma) that is documented by a major Pelsonian transgressive event of approximately 0.4 Myr in the western Tethyan realm, provides strong evidence for a globally synchronous signature and fortifies the Tethys−Peri-Tethys correlation approach of the eccentricity-cycle-scaled Anisian carbonate series integrating global climate signatures. Still, astronomical cycle-tuning of gamma-ray, isotopic and magnetostratigraphic data is required aiming at an astronomically-calibrated timescale of so far mainly biostratigraphically-constrained cyclic stratigraphy.

ACKNOWLEDGMENTS

Numerous field campaigns and workshops within the framework of DFG-funded research projects (FE 435/3−1, 3−2, GO 761/1−1) and discussions with colleagues and students over the past fifteen years enabled this compilation on the Lower Muschelkalk cyclostratigraphy. The thorough review by Maurice E. Tucker (Bristol) is gratefully acknowledged.

section Steudnitz in E Germany (Götz et al., 2018a). Two warming events in the Bithynian and Pelsonian are inferred, previously recognized as maximum flooding phases (mfz) in the northwestern Tethyan and Peri-Tethyan realm (Götz and Török, 2008). These events are highlighted in the main figure with *red arrows*. The red curve illustrates the new data (Götz et al., 2018a) in contrast to the less pronounced signal documented in previously published data (Trotter et al., 2015). Abbreviations used: *CAMP*, Central Atlantic Magmatic Province; *PTB*, Permian-Triassic Boundary. Triassic substages (Lucas, 2010): *A.*, Aegean; *A.*, Alaunian; *B.*, Bithynian; *D.*, Dienerian; *F.*, Fassanian; *G.*, Griesbachian; *I.*, Illyrian; *J.*, Julian; *L.*, Lacian; *L.*, Longobardian; *P.*, Pelsonian; *S.*, Sevatian; *Sm.*, Smithian; *Sp.*, Spathian; *T.*, Tuvalian.

REFERENCES

Aigner, T., Bachmann, G.H., 1992. Sequence stratigraphic framework of the German Triassic. Sediment. Geol. 80, 115—135.

Ajdanlijsky, G., Strasser, A., Tronkov, D., 2004. Route VII. Cyclicity in the Lower Triassic series between Opletnya railway station and Sfrazhen hamlet. Geological routes in the northern part of Iskar Gorge. In: Sinnyovsky, D. (Ed.), Guide of Field Geological Training. Vanio Nedkov Publishing House, Sofia, pp. 90—101.

Ajdanlijsky, G., Götz, A.E., Strasser, A., 2018. The Early to Middle Triassic continental-marine transition of NW Bulgaria: sedimentology, palynology and sequence stratigraphy. Geol. Carpathica 69 (2), 129—148.

Bachmann, G.H., Kozur, H.W., 2004. The Germanic Triassic: correlations with the international chronostratigraphic scale, numerical ages and Milankovitch cyclicity. Hallesches Jahrb. Geowiss. 26, 17—62.

Benton, M.J., 2015. When Life Nearly Died: The Greatest Mass Extinction of All Time. Thames & Hudson, London, p. 352.

Borkhataria, R., Aigner, T., Pipping, K.J.C.P., 2006. An unusual, muddy, epeiric carbonate reservoir: the Lower Muschelkalk (Middle Triassic) of the Netherlands. AAPG (Am. Assoc. Pet. Geol.) Bull. 90 (1), 61—89.

Budai, T., Vörös, A., 2006. Middle Triassic platform and basin evolution of the southern Bakony Mountains (Transdanubian Range, Hungary). Riv. Ital. Paleontol. Stratigr. 112, 359—371.

Chatalov, A., 2001. Deformational structures in the Iskar Carbonate Group (Lower—Upper Triassic) from the western Balkanides. Geologica Balcanica 30 (3/4), 43—57.

Chatalov, A., 2013. A Triassic homoclinal ramp from the Western Tethyan realm, Western Balkanides, Bulgaria: integrated insight with special emphasis on the Anisian outer to inner ramp facies transition. Palaeogeogr. Palaeoclimatol. Palaeoecol. 386, 34—58.

Chatalov, A., 2018. Global, regional and local controls on the development of a Triassic carbonate ramp system, Western Balkanides, Bulgaria. Geol. Mag. 155, 641—673.

Chen, Z.Q., Benton, M.J., 2012. The timing and pattern of biotic recovery following the end-Permian mass extinction. Nat. Geosci. 5, 375—383.

Conradi, F., Götz, A.E., Rameil, N., McCabe, R., 2007. Integrating chemostratigraphy and palynofacies into sequence stratigraphic models: a case study of the Lower Muschelkalk (Anisian) of the Germanic Basin. Geophys. Res. Abstr. 9. EGU2007-A-01763.

Davies, R.B., Simmons, M.D., 2018. Chapter 4: Triassic sequence stratigraphy of the Arabian Plate. In: Poppelreiter, M.C. (Ed.), Lower Triassic to Middle Jurassic Sequence of the Arabian Plate. EAGE, Houten, pp. 101—162.

Feist-Burkhardt, S., Götz, A.E., Szulc, J., (coordinators), Borkhataria, R., Geluk, M., Haas, J., Hornung, J., Jordan, P., Kempf, O., Michalík, J., Nawrocki, J., Reinhardt, L., Ricken, W., Röhling, H.-G., Rüffer, T., Török, Á., Zühlke, R., 2008. Triassic. In: McCann, T. (Ed.), The Geology of Central Europe, vol. 2. Geological Society London, pp. 749—821.

Fiege, K., 1938. Die Epirogenese des Unteren Muschelkalkes in Nordwestdeutschland. Zentralblatt für Mineralogie, Geologie und Paläontologie 1938 B, 143—170.

Föhlisch, K., Voigt, T., 2001. Synsedimentary deformation in the Lower Muschelkalk of the Germanic Basin. Special Publ. Int. Assoc. Sedimentol. 31, 279—297.

Foster, W.J., Sebe, K., 2017. Recovery and diversification of marine communities following the late Permian mass extinction event in the western Palaeotethys. Glob. Planet. Change 155, 165—177.

Götz, A.E., 1996. Fazies und Sequenzanalyse der Oolithbänke (Unterer Muschelkalk, Trias) Mitteldeutschlands und angrenzende Gebiete. Geol. Jahrb. Hess. 124, 67—86.

Götz, A.E., 2002. Hochauflösende Stratigraphie im Unteren Muschelkalk (Mitteltrias, Anis) des Germanischen Beckens. Schriftenreihe der Dt. Geol. Ges. 15, 101—107.

Götz, A.E., 2004. Zyklen und Sequenzen im Unteren Muschelkalk des Germanischen Beckens. Hallesches Jahrb. Geowiss. 18, 91—98.

Götz, A.E., Feist-Burkhardt, S., 1999. Sequenzstratigraphische Interpretation der Kleinzyklen im Unteren Muschelkalk (Mitteltrias, Germanisches Becken). Zbl. Geol. Paläontol., Teil I (1997) 7/9, 1205—1219.

Götz, A.E., Feist-Burkhardt, S., 2000. Palynofacies and sequence analysis of the Lower Muschelkalk (Middle Triassic, German basin). Zbl. Geol. Paläontol., Teil I (1998) 9/10, 877—891.

Götz, A.E., Török, Á., 2008. Correlation of Tethyan and Peri-Tethyan long-term and high-frequency eustatic signals (Anisian, Middle Triassic). Geol. Carpathica 59 (4), 307—317.

Götz, A.E., Gast, S., 2010. Basin evolution of the Anisian Peri-Tethys: implications from conodont assemblages of Lower Muschelkalk key sections (central Europe). Ger. J. Geol. 161, 39—49.

Götz, A.E., Feist-Burkhardt, S., 2012. Phytoplankton associations of the Anisian Peri-Tethys Basin (central Europe): evidence of basin evolution and palaeoenvironmental change. Palaeogeogr. Palaeoclimatol. Palaeoecol. 338, 151—158.

Götz, A.E., Török, Á., Ajdanlijsky, G., 2018a. Anisian climate change inferred from a new $\delta^{18}O$ record from conodont apatite. In: Abstracts XXI International Congress, Carpathian Balkan Geological Association.

Götz, A.E., Ajdanlijsky, G., Strasser, A., 2018b. Palynology of a Middle Triassic (Anisian) ramp system (NW Bulgaria): towards a refined age control and depositional model. In: Abstracts XXI International Congress, Carpathian Balkan Geological Association.

Götz, A.E., Luppold, F.W., Hagdorn, H., 2018c. Biostratigraphie und Zonierung der Conodonten des Muschelkalks. In: Deutsche Stratigraphische Kommission (Ed.), Stratigraphie von Deutschland XIII. Muschelkalk, Schriftenreihe der deutschen Gesellschaft für Geowissenschaften, vol. 91 [in press].

Götz, A.E., Török, Á., Feist-Burkhardt, S., Konrád, G., 2003. Palynofacies patterns of Middle Triassic ramp deposits (Mecsek Mts., S Hungary): a powerful tool for high-resolution sequence stratigraphy. Mitt. Ges. Geol. Bergbaustud. Österr. 46, 77—90.

Haas, J., Kovács, S., Török, Á., 1995. Early Alpine shelf evolution in the Hungarian segment of the Tethys margin. Acta Geol. Hung. 38, 95—110.

Haas, J., Budai, T., Raucsik, B., 2012. Climatic controls on sedimentary environments in the Triassic of the Transdanubian Range (western Hungary). Palaeogeogr. Palaeoclimatol. Palaeoecol. 353—355, 31—44.

Hagdorn, H., Simon, T., Dittrich, D., Friedlein, V., Geyer, G., Kramm, E., Nitsch, E., 2016. Muschelkalk. In: STG 2016 (German Stratigraphic Commission, ed.; editing, coordination and layout: Menning, M., Hendrich, A.), Stratigraphic Table of Germany 2016. Potsdam (German Research Centre for Geosciences), in German, (1) Table plain 100 x 141 cm, (2) Table folded A4.

Haq, B.U., 2018. Triassic Ocean and Sea Levels Re-examined [in press].

Hinnov, L.A., Ogg, J.G., 2007. Cyclostratigraphy and the astronomical time scale. Stratigraphy 4, 239—251 (New York).

Holz, M., 2015. Mesozoic paleogeography and paleoclimates — a discussion of the diverse greenhouse and hothouse conditions of an alien world. J. S. Am. Earth Sci. 61, 91—107.

Jaglarz, P., Szulc, J., 2003. Middle Triassic evolution of the Tatricum sedimentary basin: an attempt of sequence stratigraphy to the Wierchowa Unit in the Polish Tatra Mountains. Ann. Soc. Geol. Pol. 73, 169—182.

Jubitz, K.-B., 1954. Zur praktischen Anwendung der feinstratigraphischen und kleintektonischen Methode. Freiberger Forschungsh. C 9, 80—112.

Jubitz, K.-B., 1958. Zur feinstratigraphisch-geochemischen Horizontierungsmethodik in Kalksedimenten (Trias). Geologie 7, 863—923.

Kędzierski, J., 2000. Sequenzstratigraphie des Muschelkalks im östlichen Teil des Germanischen Beckens (unpubl. PhD). University Halle-Wittenberg, Halle (Saale), 116 p.

Klug, C., Schatz, W., Korn, D., Reisdorf, A.G., 2005. Morphological fluctuations of ammonoid assemblages from the Muschelkalk (Middle Triassic) of the Germanic Basin — indicators of their ecology, extinctions, and immigrations. Palaeogeogr. Palaeoclimatol. Palaeoecol. 221, 7—34.

Kolb, U., 1976. Lithofazielle und geologische Untersuchungen der Wellenkalkfolge des Subherzynen Beckens. Freiberger Forschungsh. C 316, 41—70.

Kovács, S., Rálisch-Felgenhauer, E., 2005. Middle Anisian (Pelsonian) platform conodonts from the Triassic of the Mecsek Mts (South Hungary) — their taxonomy and stratigraphic significance. Acta Geol. Hung. 48, 69—105.

Kozur, H.W., Bachmann, G.H., 2008. Updated correlation of the Germanic Triassic with the Tethyan scale and assigned numeric ages. Berichte Geol. Bundes-Anstalt 76, 53—58.

Kramm, E., 1997. Stratigraphie des Unteren Muschelkalks im Germanischen Becken. Geologica et Paleontologica 31, 215—234.

Kustatscher, E., van Konijnenburg-van Cittert, J.H.A., Roghi, G., 2010. Macrofloras and palynomorphs as possible proxies for palaeoclimatic and palaeoecological studies: a case study from the Pelsonian (Middle Triassic) of Kühwiesenkopf/Monte Prà della Vacca (Olang Dolomites, N-Italy). Palaeogeogr. Palaeoclimatol. Palaeoecol. 290, 71—80.

Langer, A., 1989. Lithostratigraphische, technologische und geochemische Untersuchungen im Muschelkalk des Osnabrücker Berglandes. Mitteilungen aus dem geologischen Institut der Universität Hannover 29, 1—114.

Li, M., Huang, C., Hinnov, L., Chen, W., Ogg, J., Tian, W., 2018. Astrochronology of the Anisian stage (Middle Triassic) at the Guandao reference section, South China. Earth Planet. Sci. Lett. 482, 591—606.

Lucas, S.G., 2010. The Triassic chronostratigraphic scale: history and status. In: Lucas, S.G. (Ed.), The Triassic Timescale, Geological Society, London, Special Publications, vol. 334, pp. 17—39.

Martindale, R.C., Foster, W.J., Velledits, F., 2017. The survival, recovery, and diversification of metazoan reef ecosystems following the end-Permian mass extinction event. Palaeogeogr. Palaeoclimatol. Palaeoecol. [in press].

Martini, R., Peybernès, B., Zaninetti, L., Fréchngues, M., 1996. Découverte de foraminifères dans les intervalles transgressifs de deux séquences de dépôt anisiennes (Muschelkalk) du Bassin de la Weser (Hesse, Allemagne du Nord). Geobios 24 (5), 505—511.

Matysik, M., 2016. Facies types and depositional environments of a morphologically diverse carbonate platform: a case study from the Muschelkalk (Middle Triassic) of Upper Silesia, Southern Poland. Ann. Soc. Geol. Pol. 86, 119—164.

Menning, M., Gast, R., Hagdorn, H., Käding, K.-C., Simon, T., Szurlies, M., Nitsch, E., 2006. Zeitskala für Perm und Trias in der Stratigraphischen Tabelle von Deutschland 2002, zyklostratigraphische Kalibrierung der höheren Dyas und Germanischen Trias und das Alter der Stufen Roadium bis Rhaetium 2005. Newslett. Stratigr. 41, 173—210.

Merz, G., 1987. Zur Petrographie, Stratigraphie, Paläogeographie und Hydrologie des Muschelkalks (Trias) im Thüringer Becken. Z. für Geol. Wiss. 15, 457—473.

Michalík, J., Masaryk, P., Lintnerová, O., Papšová, J., Jendrejáková, O., Reháková, D., 1992. Sedimentology and facies of a storm-dominated Middle Triassic carbonate ramp (Vysoká Formation, Malé Karpaty Mts., Western Carpathians). Geol. Carpathica 43 (4), 213—230.

Michalík, J., 1997. Tsunamites in a storm-dominated Anisian carbonate ramp (Vysoká Formation, Malé Karpaty Mts., Western Carpathians). Geol. Carpathica 48 (4), 221—229.

Mundil, R., Pálfy, J., Renne, P.R., Brack, P., 2010. The Triassic timescale: new constraints and a review of geochronological data. In: Lucas, S.G. (Ed.), The Triassic Timescale, Geological Society, London, Special Publications, vol. 334, pp. 41—60.

Narkiewicz, K., 1999. Conodont biostratigraphy of the Muschelkalk (Middle Triassic) in the central part of the Polish lowland. Geol. Q. 43, 313—328.

Nitsch, E., Kramm, E., Simon, T., 2018. Zyklostratigraphie des Muschelkalks. In: Deutsche Stratigraphische Kommission (Ed.), Stratigraphie von Deutschland XIII. Muschelkalk, Schriftenreihe der deutschen Gesellschaft für Geowissenschaften, vol. 91 [in press].

Nützel, A., Kaim, A., Grădinaru, E., 2018. Middle Triassic (Anisian, Bithynian) gastropods from North Dobrogea (Romania) and their significance for gastropod recovery from the end-Permian mass extinction event. Papers in Palaeontology 1—36.

Ogg, J.G., 2012. Triassic. In: Ogg, J.G., Schmitz, M.D., Ogg, G.M. (Eds.), The Geologic Time Scale 2012. Elsevier, Amsterdam, pp. 681—730.

Ogg, J.G., Ogg, G.M., Gradstein, F.M., 2016. The Concise Geological Time Scale 2016, 1-234. Elsevier, Amsterdam.

Philip, J., Masse, J., Camoin, G., 1996. Tethyan carbonate platforms. In: Nairn, A.E.M., Ricou, L.-E., Vrielynck, B., Dercourt, J. (Eds.), The Ocean Basins and Margins, The Tethys Ocean, vol. 8. Plenum Press, New York-London, pp. 239—266.

Pöppelreiter, M., 2002. Facies, cyclicity and reservoir properties of the Lower Muschelkalk (Middle Triassic) in the NE Netherlands. Facies 46, 119—132.

Pöppelreiter, M., Borkhataria, R., Aigner, T., Pipping, K., 2005. Production from Muschelkalk carbonates (Triassic, NE Netherlands): unique play or overlooked opportunity? In: Doré, A.G., Vining, B.A. (Eds.), Petroleum Geology: North-West Europe and Global Perspectives — Proceedings of the 6th Petroleum Conference. Geological Society London, pp. 299—315.

Pruss, S.B., Bottjer, D.J., 2005. The reorganization of reef communities following the end-Permian mass extinction. Comptes Rendus Palevol 4, 553—568.

Rameil, N., Götz, A.E., Feist-Burkhardt, S., 2000. High-resolution sequence interpretation of epeiric shelf carbonates by means of palynofacies analysis: an example from the Germanic Triassic (Lower Muschelkalk, Anisian) of East Thuringia, Germany. Facies 43, 123—144.

Retallack, G.J., 2013. Permian and Triassic greenhouse crises. Gondwana Res. 24, 90—103.

Rüffer, T., 1996. Seismite im Unteren Muschelkalk westlich von Halle. Hallesches Jahrb. Geowiss. 18, 119—130.

Rychliński, T., Szulc, J., 2005. Facies and sedimentary environments of the Upper Scythian—Carnian succession from the Belanské Tatry Mts., Slovakia. Ann. Soc. Geol. Pol. 75, 155—169.

Schlager, W., 2010. Ordered hierarchy versus scale invariance in sequence stratigraphy. Int. J. Earth Sci. (Geol. Rundschau) 99, 139—151.

Schüller, M., 1967. Petrographie und Feinstratigraphie des Unteren Muschelkalks in Südniedersachsen und Nordhessen. Sediment. Geol. 1, 353—401.

Schulz, M.-G., 1972. Feinstratigraphie und Zyklengliederung des Unteren Muschelkalks in N-Hessen. Mittl. Geol. Paläont. Inst. Univ. Hambg. 41, 133—170.

Song, H., Wignall, P.B., Chen, Z.Q., Tong, J., Bond, D.P.G., Lai, X., Zhao, X., Jiang, H., Yan, C., Niu, Z., Chen, J., Yang, H., Wang, Y., 2011. Recovery tempo and pattern of marine ecosystems after the end-Permian mass extinction. Geology 39, 739—742.

Stefani, M., Furin, S., Gianolla, P., 2010. The changing climate framework and depositional dynamics of the Triassic carbonate platforms from the Dolomites. Palaeogeogr. Palaeoclimatol. Palaeoecol. 290, 43—57.

Strasser, A., Pittet, B., Hillgärtner, H., Pasquier, J.-B., 1999. Depositional sequences in shallow carbonate-dominated sedimentary systems: concepts for a high-resolution analysis. Sediment. Geol. 128, 201—221.

Szulc, J., 1993. Early Alpine tectonics and lithofacies succession in the Silesian part of the Muschelkalk basin. A Synopsis. In: Hagdorn, H., Seilacher, A. (Eds.), Muschelkalk. Schöntaler Symposium 1991. Goldschneck-Verlag, Werner K. Weidert, Korb, pp. 19—28.

Szulc, J., 2000. Middle Triassic evolution of the northern Peri-Tethys area as influenced by early opening of the Tethys ocean. Ann. Soc. Geol. Pol. 70, 1—48.

Szulc, J., 2007. Sponge-microbial stromatolites and coral-sponge reef recovery in the Triassic of the western Tethys Domain. In: Lucas, S.G., Spielmann, J.A. (Eds.), The Global Triassic, New Mexico Museum of Natural History and Science Bulletin, vol. 41, p. 402.

Tanner, L.H., 2010. Cyclostratigraphic record of the Triassic: a critical examination. In: Lucas, S.G. (Ed.), The Triassic Timescale, Geological Society, London, Special Publications, vol. 334, pp. 119—138.

Török, Á., 1993. Storm influenced sedimentation in the Hungarian Muschelkalk. In: Hagdorn, H., Seilacher, A. (Eds.), Muschelkalk. Schöntaler Symposium 1991. Goldschneck-Verlag, Werner K. Weidert, Korb, pp. 133—142.

Török, Á., 1997. Triassic ramp evolution in southern Hungary and its similarities to the Germano-type Triassic. Acta Geol. Hung. 40, 367—390.

Török, Á., 1998. Controls on development of Mid-Triassic ramps: examples from southern Hungary. In: Wright, V.P., Burchette, T.P. (Eds.), Carbonate Ramps, Geol. Soc. London, Spec. Publ., vol. 149, pp. 339—367.

Török, Á., 2000. Muschelkalk carbonates in southern Hungary: an overview and comparison to German Muschelkalk. In: Bachmann, G.H., Lerche, I. (Eds.), Epicontinental Triassic. Zentralblatt für Geologie und Paläontologie, Teil I, 1998(9—10), pp. 1085—1103.

Trotter, J.A., Williams, I.S., Nicora, A., Mazza, M., Rigo, M., 2015. Long-term cycles of Triassic climate change: a new $\delta^{18}O$ record from conodont apatite. Earth Planet. Sci. Lett. 415, 165—174.

Velledits, F., Pero, C., Blau, J., Senowbari-Daryan, B., Kovács, S., Piros, O., Pocsai, T., Szugyi-Simon, H., Dumitrica, P., Palfy, J., 2011. The oldest Triassic platform margin reef from the Alpine-Carpathian Region (Aggtelek, NE Hungary): platform evolution, reefal biota and biostratigraphic framework. Riv. Ital. Paleontol. Stratigr. 117, 221—268.

Vörös, A., 2014. Ammonoid diversification in the Middle Triassic: examples from the Tethys (eastern Lombardy, Balaton Highland) and the Pacific (Nevada). Centr. Europ. Geol. 57 (4), 319—343.

Voigt, T., Linnemann, U., 1996. Resedimentation im Unteren Muschelkalk — das Profil am Jenzig bei Jena. Beiträge zur Geologie von Thüringen. Neue Folge 3, 153—167.

CHAPTER SEVEN

The Down-dip Preferential Sequence Record of Orbital Cycles in Greenhouse Carbonate Ramps: Examples From the Jurassic of the Iberian Basin (NE Spain)

Beatriz Bádenas[1] and Marcos Aurell
Departamento de Ciencias de la Tierra-IUCA, Universidad de Zaragoza, Zaragoza, Spain
[1]Corresponding author: E-mail: bbadenas@unizar.es

Contents

Stratigraphy & Timescales, Volume 3
ISSN 2468-5178
https://doi.org/10.1016/bs.sats.2018.07.002

Abstract

High-frequency sequences of different scales were recorded in the distinct carbonate ramps developed in the Iberian Basin (Spain) in the Jurassic times. This work reviews and compares the sedimentary record of these sequences in four Iberian carbonate ramps: (1) inner-to proximal outer ramp areas of a non skeletal-dominated homoclinal ramp (upper Sinemurian-lowermost Pliensbachian); (2) middle-to proximal outer ramp areas of a skeletal-dominated homoclinal ramp (upper Pliensbachian); (3) shallow to relatively deep domain of a microbial/siliceous sponge-dominated distally steepened ramp (Bajocian); and (4) inner-to open ramp domains of a homoclinal ramp with coral-microbial reefs (upper Kimmeridgian). The comparative review reveals that the sequences were recorded differently from shallow to deep ramp areas. In shallow ramp areas there is a preferential record of eccentricity-related sequences, whereas high-frequency (precession) sequences are ubiquitous in relatively deep ramp settings. This preferential preservation was likely controlled by the interplay between accomodation changes and internal processes controlling accummulation above and below fair-weather and storm wave base levels, rather than by climate (warm *vs.* cold greenhouse climate modes) and type of carbonate production. Comparison with cyclostratigraphic analysis performed in similar settings indicates that this variable down-dip sequence record seems to be a common feature in greenhouse carbonate ramps: eccentricity-related, meter-thick facies sequences are dominant in shallow ramp areas whereas the deep outer ramp-basin is characterized by preservation of precession or sub-Milankovitch cycles. It is emphasized that the middle to outer ramp sedimentary domain (located around the storm wave base) has the greatest potential for cyclostratigraphic analysis, because this domain is more likely to record both eccentricity- and/or precession-related sequences.

1. INTRODUCTION

Cyclostratigraphy is a subdiscipline of Stratigraphy that deals with the characterization, correlation, and interpretation of periodic or nearly periodic cyclic variations in the stratigraphic record, most commonly those linked to insolation changes induced by orbital Milankovitch cycles, and with their application in geochronology to refine the geological time-scale (e.g., Hilgen et al., 2004; Strasser et al., 2006). Decoding the environmental changes linked to the orbitally induced insolation changes may be difficult due to the complexity of the atmospheric, oceanic, sedimentary, and biological systems. Furthermore, the environmental changes may have different amplitude and frequency and may be out of phase, meaning that they do not translate one-to-one into the sedimentary record (e.g., Strasser et al., 2006). In the sedimentary systems, the orbital signal can also be absent or distorted by interruptions caused by "abnormal" processes (e.g., non-deposition, erosion, event-bed deposition) or by variations of recording processes (e.g., accumulation rates, diagenesis) that often depend of the nature of the sedimentary environment, thus complicating the identification of cycles (Weedon, 2003). Nevertheless, because the cyclostratigraphic data contain information about both "normal" and "abnormal" environmental variations and the processes that produce the records themselves (Weedon, 2003), Cyclostratigraphy represents a useful technique to improve understanding of depositional systems (e.g., Herbert, 1994), usually with a higher time resolution than other stratigraphic methods.

Every sedimentary system reacts differently to a given set of global, regional, or local factors, so that it is impossible to propose general rules for the formation of the cyclostratigraphic record (Einsele et al., 1991). This is particularly true when analyzing the orbital imprint in carbonate marine depositional systems, because of the strong influence of orbitally induced insolation changes on sedimentation and accommodation. The insolation changes have a complex direct and indirect influence on water temperature and chemistry, freshwater and nutrient input through rivers, circulation patterns, etc., and thereby on carbonate production, transport and accumulation (Pomar and Hallock, 2008; Westphal et al., 2010). In addition, the orbitally induced insolation changes produce changes on eustatic sea level and then on accommodation, through glacio-eustasy, thermo-eustasy or aquifer-eustasy (e.g., Lambeck et al., 2002; Sames et al., 2016), leading to the generation of hierarchically stacked high-frequency sequences of different scales (e.g., Mitchum and Van Wagoner, 1991; Strasser et al., 1999; Fischer et al., 2004).

High-frequency carbonate sequences originated in peritidal to shallow marine environments are not generally an unequivocal evidence of orbital-driven sea-level changes, as they can also be generated by internal/autocyclic mechanisms (e.g., lateral migration of sedimentary bodies or changes in carbonate production rate and sediment transport direction: Ginsburg, 1971; Pratt et al., 1992; Burgess, 2006; Yang et al., 2014) or by other external/allocyclic mechanisms (i.e., high-frequency accomodation changes due to synsedimentary fault movements: e.g., Bosence et al., 2009). Laterally equivalent high-frequency sequences have also been recognized in deep outer platform limestone-marl/clay successions. Their interpretation suggests climate-driven changes in clay input, in pelagic production, or in exported carbonate due to high-frequency sea-level changes controlling phases of shallow water carbonate productivity and exporting capability (e.g., Einsele and Ricken, 1991; Colombié and Strasser, 2003; Munnecke and Westphal, 2004; Boulila et al., 2010). Therefore, even in the presence of Milankovitch orbital forcing, the features and stacking of high-frequency sequences in carbonate marine depositional systems do not exclusively reflect the orbital influence in accommodation and sediment type, but also the coeval imprint of other internal and external mechanisms controlling production and accumulation of sediments and accommodation (e.g., Tresch and Strasser, 2011; Laya et al., 2013).

The high-frequency sequences represent the "building blocks" of the cyclostratigraphic record and contain key information for both sedimento-logical and cyclostratigraphic analysis of the sedimentary record. Carbonate sedimentologists study the high-frequency sequences to understand how and where carbonate sediments have been accumulated in response to internal and external mechanisms (e.g., Strasser and Védrine, 2009; Bádenas et al., 2004; Colombié and Strasser, 2005; Brandano et al., 2015). Cyclostratigraphers search for criteria to identify an orbital control, including wide lateral continuity of sequences, persistent regularity in their hierarchical stacking pattern, and correspondence of the estimated pe-riodicities to those of the orbital cycles (e.g., Weedon, 2003; D'Argenio et al., 2004). Using this combined sedimentological-cyclostratigraphical point of view, the present work describes and compares the record of orbitally driven high-frequency sequences in four distinct carbonate ramp depositional systems in the greenhouse Jurassic of the Iberian Basin (NE Spain) (Fig. 1): (1) upper Sinemurian-lowermost Pliensbachian non skeletal-dominated homoclinal ramp; (2) upper Pliensbachian skel-etal-dominated homoclinal ramp; (3) Bajocian microbial/siliceous sponge-dominated distally steepened ramp; and (4) upper Kimmeridgian homoclinal ramp with coral reefs. The selected case studies correspond to particular carbonate ramps located at relative low latitude and influenced

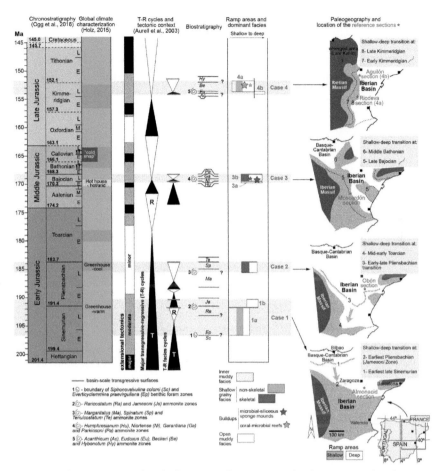

Figure 1 Chronostratigraphical, climatic, sedimentary and palaeogeographical setting of case studies 1–4. The distribution of key ammonite and benthic foraminifera biozones in the logged intervals is indicated (see legend in the lower-left part). In the paleogeographical maps (adapted form Aurell et al., 2003), the location of the reference sections is indicated.

by different climate conditions (warm or cold), regional tectonic context and long-term trend in relative sea level. The main aim of the comparison is to illustrate how the interaction of different internal and external factors controlled the preservation potential of orbitally controlled high-frequency sequences from shallow to deep ramp areas, thus providing some criteria to identify the orbital control in similar greenhouse carbonate ramps, and in particular, those concerning the lateral continuity of sequences.

2. METHODS AND TERMINOLOGY

The data exposed here provide a synthetic view of previous studies performed by the authors and co-workers, after detailed bed-by-bed logging and sampling in key outcrops of the carbonate ramp successions. The facies and high-frequency sequences were studied in several logs (few tens of meters to some kilometers apart) and physically correlated or mapped when possible. For the purpose of conciseness, the main results are synthetized using key reference stratigraphic sections (Fig. 1).

2.1 Ramp Domains

Three of the studied greenhouse ramps (cases 1, 2 and 4) are homoclinal ramps (*sensu* Read, 1982), which are subdivided into inner, middle and outer ramp domains (Fig. 2A). In the inner ramp, low energy (tidal flats and/or lagoons) and high-energy areas (barrier-islands, sand blankets, shoals) occur. The middle ramp locates between the fair-weather wave and the storm wave bases. In the proximal middle ramp grainy storm deposits and muddy facies intercalate. This is an area of relative high-energy compared to distal middle ramp, because of the major influence of the storm-related waves and unidirectional flows that resedimented coarse grains and lime mud from shallow areas. The storm-related unidirectional flows decrease down dip from the distal middle ramp to the outer ramp and muddy facies dominate.

By contrast, the Bajocian ramp (case 3; Fig. 2B) is a distally steepened ramp with a slope break located around the storm wave base, which clearly subdivides the ramp into a shallow, relative high-energy domain dominated by grainy facies, and a deep, low-energy muddy domain. For comparison purposes, the term shallow ramp is used here for both the inner to proximal middle ramp areas of the homoclinal ramp and the shallow ramp domain of the distally steepened ramp, which are characterized by constant or episodic wave and current reworking. The term deep ramp is applied to the open, low-energy ramp areas, including the outer ramp well below the storm wave base in the homoclinal ramps and the deep ramp domain of the distally steepened ramp. For the middle to outer ramp transitional area of the homoclinal ramps, the term shallow to deep ramp transition (abbreviated shallow/deep ramp) is used.

2.2 High-Frequency and Low-Frequency Sequences

The concept of high-frequency sequences of different scales (i.e., 4th–6th-order sequences of Vail et al., 1991) is used here from a descriptive point of

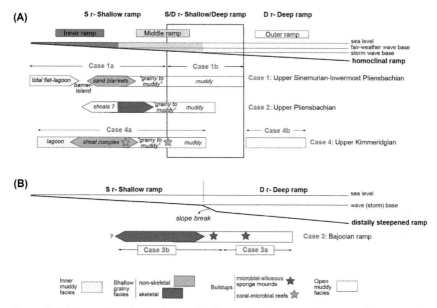

Figure 2 (A) Ramp domains in the studied homoclinal ramps of cases studies 1, 2 and 4. (B) Ramp domains in the studied distally steepened ramp of case study 3. For comparison purposes the terms shallow ramp, deep ramp, and shallow to deep ramp transition (abbreviated shallow/deep ramp) are used. See explanation in text.

view, which implies a hierarchy of sequences in which smaller sequences stack in longer sequences, independently of their possible order or time duration.

For each case study, high-frequency sequences (HFSs) are defined as the elementary building blocks or smaller stratal packages bounded by discontinuity surfaces (i.e., omission or erosion) identifiable in the field. In shallow water successions, where carbonate productivity is high, the HFSs usually have a shallowing-upward facies trend and their boundaries are accompanied by a superposition of facies that does not conform the Walther's law (e.g., Jones and Desrochers, 1992). However, because other vertical facies trends can also be present (deepening-upward, deepening-shallowing, aggradational; e.g., Spencer and Tucker, 2007; Bosence et al., 2009) and the sharp facies change at the sequence boundary may be absent (e.g., stacked aggradational HFSs), the main criterion used here for the identification of HFSs is the recognition of sequence boundaries rather than facies trends. In particular, in the shallow ramp successions studied, the HFS boundaries correspond to sharp bedding surfaces, usually Fe-rich and/or bioturbated, and locally erosive and delimiting unconformable beds. Subsequently, facies trends within the HFSs were also used to define them and describe their internal facies stacking as deepening-upward, shallowing-upward, deepening-shallowing, or aggradational. Key stratigraphic surfaces and long-term vertical facies trends allowed grouping HFSs into lower-frequency sequences (LFSs).

In shallow/deep and deep ramp environments characterized by lime mudstone-dominated successions, the HFSs correspond to groups of limestone beds (called here bundles of beds, from a descriptive point of view), which are bounded by sharp, usually Fe-rich and/or bioturbated, bedding surfaces. Vertical facies changes within bundles are subtle or even absent due to a greater homogeneity of facies in deep ramp areas compared to shallow ramp. Key stratigraphic surfaces and changes in bedding geometry allowed recognizing lower-frequency sequences (LFSs), including sets of bundles and lots of sets.

2.3 Time Control, Hierarchical Stacking Pattern and Estimated Periodicities

Sequence stratigraphy (i.e., identification of basin-scale transgressive-regressive, T-R, facies cycles), biostratigraphy (mainly ammonites) and locally chemostratigraphy (strontium isotopes: case 2) were used to constrain the time duration of the studied successions, by using the synthesis of geological time scale (GTS) by Ogg et al. (2016) as a reference (Fig. 1). Nevertheless, it is important to notice the updating of the GTS over the last years. The comparison of the time duration of some chronostratigraphic intervals reported in this work (Table 1) shows significant discrepancies between the Hardenbol et al. (1998) chart and the GTS 2004 and GTS 2012 (Gradstein et al., 2004, 2012). In the recent synthesis by Ogg et al. (2016) used here, there has been a major modification in the duration of the Middle and Late Jurassic intervals (i.e., late Bajocian: case 3, and late Kimmeridgian: case 4) compared to the GTS 2004 (Table 1).

Table 1 Comparison of the Time Duration Assigned in the Geological Time Charts to Some Selected Intervals Studied in the Present Work. The Gray Shading Indicates the Proposal Considered in the Present Work

	Hardenbol et al. (1998)	Gradstein et al. (2004)	Gradstein et al. (2012)	Ogg et al. (2016)
Late Kimmeridgian	1.9 My (152.6–150.7)	4 My (154.8–150.8)	2.6 My (154.7–152.1)	
Late Bajocian	3.7 My (172.9–169.2)	2.7 My (170.4–167.7)	1.1 My (169.4–168.3)	
Late Pliensbachian	1.9 My (191.5–189.6)	4 My (187.0–183.0)	4.8 My (187.5–182.7)	4.5 My (188.2–183.7)
Late Sinemurian	3.5 My (198.8–195.3)	5.2 My (194.8–189.6)	4.5 My (195.3–190.8)	4.1 My (195.5–191.4)

Even if the time span of the studied successions is not always well constrained, the average duration of the sequences can roughly be estimated by considering the time duration of the studied successions and the number of sequences, and by analyzing their hierarchical stacking pattern (e.g., Strasser et al., 2006). For relatively deep ramp successions, spectral analysis of stratigraphic data (magnetic susceptibility: case 1; bed thickness: case 4) has been used as complementary tool to decipher orbital cycles and possible equivalence with the recorded sequences. The hierarchy of sequences and cycles and the ratios between their corresponding frequencies, can be compared to those proposed for Milankovitch cycles in Pre-Pleistocene times (e.g., Berger et al., 1992; Laskar et al., 2004). However, because of the uncertainties concerning both the ancient Milankovitch cycle periods (e.g., Waltham, 2015) and the time duration of the studied intervals, the approach used here is to consider an average duration of $\sim$400, $\sim$100 and $\sim$20 ky for long eccentricity, short eccentricity and precession cycles, respectively, and therefore an average 1—4—20 ratio of sequences linked to these orbital cycles (Strasser et al., 2006). Obliquity ($\sim$40 ky) cycles have not been considered since they have more of an effect at higher latitudes, and have weaker or less apparent influence during ice-free periods than the other Milankovitch cyles (e.g., Zachos et al., 2001).

3. STRATIGRAPHIC AND PALAEOGEOGRAPHIC BACKGROUND

During the Jurassic, shallow epeiric seas covered wide areas of western Europe including the eastern part of the Iberian Plate (e.g., Dercourt et al., 1993). A large part of the central and western Iberian Plate was an uplifted high, the so-called Iberian Massif, and an intracratonic basin, the Iberian Basin, located eastwards (Fig. 1). General greenhouse conditions (Holz, 2015) and paleolatitude around 25—35° (Osete et al., 2011) favored widespread deposition of coastal to marine carbonates in waters depths of up to *ca.* 100 m (Aurell et al., 2003, 2010; Gómez and Goy, 2005).

The Jurassic successions of the Iberian Basin are bounded by major angular and erosive unconformities developed around the Triassic-Jurassic and Jurassic-Cretaceous boundaries, linked to major phases of extensional movements led by the westward extension of the Tethys Ocean and the opening of the Central Atlantic and Bay of Biscay (Salas et al., 2001; Aurell et al., 2003, 2016). Reactivation of normal faults also controlled sedimentation

throughout the Jurassic and resulted in different palaeogeographic distribu-
tions of shallow to open platform domains (Fig. 1): during the Early Jurassic
and the early-Middle Jurassic, the platforms were open to the north (Boreal
affinity; case 1—3), whereas during the late-Middle Jurassic and the Late
Jurassic, the Iberian platforms were open to the Tethyan Ocean (case 4).

Despite the tectonic imprint, widespread transgressive and regressive
events suggest a certain influence of regional or global sea-level changes
on the Jurassic facies distribution in the basin (Aurell et al., 2003, 2010).
Three major T-R cycles with transgressive peaks related to global sea-level
rises are identified (Fig. 1). In particular, the studied carbonate ramps
developed either during the trangressive hemicycle of the major Lower
Jurassic T-R cycle (cases 1 and 2), around the transgressive peak of the
Middle Jurassic T-R cycle (case 3), or during the regressive hemicycle of
the Upper Jurassic T-R cycle (case 4). Details on the chronostratigraphic
framework, particular climate conditions and tectonic context of the studied
cases are offered below.

4. UPPER SINEMURIAN-LOWERMOST PLIENSBACHIAN SEQUENCE RECORD

4.1 Stratigraphy and Sedimentary Environments

The upper Sinemurian-lowermost Pliensbachian ramp successions are
exposed along a 12 km long continuous outcrop located near Almonacid
de la Cuba village (Figs. 1 and 3). The upper Sinemurian shallow ramp limestones belong to the uppermost transgressive and regressive parts of the Het-
tangian-Sinemurian T-R facies cycle (case 1a; Bádenas et al., 2010). This
succession is ∼35 m in thickness and has an uncertain duration of
∼4 My, from the boundary between *Siphonovalvulina colomi* and *Everticy-
clammina praevirguliana* Zones (Boudagher-Fadel and Bosence, 2007;
Bosence et al., 2009) up to a basin-scale transgressive surface partly related
to extensional tectonics, which occurred at the uppermost *Raricostatum*
Zone (Fig. 1; Aurell et al., 2003). The lowermost Pliensbachian limestones
represent deposition in the shallow/deep ramp. The succession is 57 m in
thickness and was deposited during the transgressive hemicycle of the lower
Pliensbachian T-R facies cycle (case 1b; Sequero et al., 2017). Its upper
boundary is a basin-scale transgressive surface at the upper part of the *Jamesoni* Zone, likely related to a sea-level rise also recorded in other Tethyan and
Boreal successions (Aurell et al., 2003). The time duration of this succession
is ∼1.5 My.

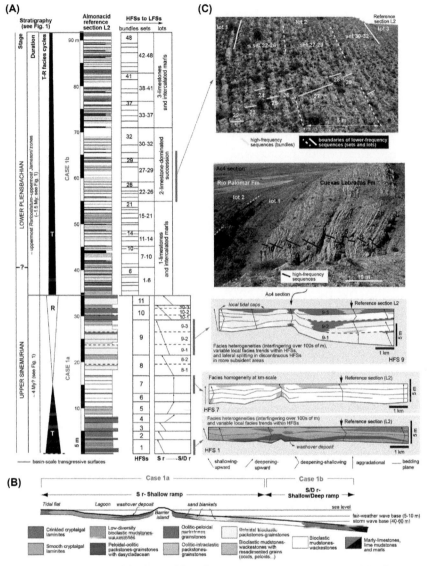

Figure 3 (A) Facies and high- and low-frequency sequences (HFSs, LFSs) of case 1 (upper Sinemurian-lowermost Pliensbachian) in the reference section of Almonacid. (B) Synthetic facies model showing the distribution of the main facies belts. The occurrence of cases 1a and 1b in the different domains of the carbonate ramp is indicated. (C) Selected field views showing the distribution of HFSs and LFSs. The three insets below show the lateral and vertical facies distribution in three selected HFSs in case 1a (HFSs 1, 7 and 9).

General greenhouse climate conditions and a long-term Hettangian-earliest Pliensbachian warming trend in the western Tethys (e.g., van de Schootbrugge et al., 2005; Dera et al., 2011) favored carbonate sedimentation in the Iberian Basin. As a whole, the upper Sinemurian-lowermost Pliensbachian successions represent deposition in a low-angle carbonate ramp dominated by non-skeletal grains (Fig. 3B). The shallow ramp area included tidal flats with cryptalgal laminites and muddy protected lagoons developed landward of a barrier–island system. Shallow high-energy areas encompass a wide spectrum of carbonate sand blankets with different proportions of ooids, peloids, intraclasts and oncoids and pass down dip to bioturbated bioclastic muddy sediments with tempestites (Fig. 3B). Deposition of carbonate muds in the shallow/deep ramp was related to benthic production (mainly brachiopods and bivalves), resedimentation from shallow domains, and possible pelagic production.

4.2 Shallow Ramp Sequences

Analysis and correlation of twelve sections in the upper Sinemurian shallow ramp limestones (case 1a) revealed the presence of 11 HFSs, each few m-thick (1−6.5 m), which are continuous along the 12 km-long outcrop (Fig. 3A and C). These sequences are bounded by transgressive surfaces and have variable facies heterogeneities and sedimentary trends depending on the environment of formation (Bádenas et al., 2010). Shallowing-upward and deepening-shallowing HFSs occur in protected lagoon–tidal flat areas (HFSs 1−3) and in the high-energy domain (HFSs 8−11). They show significant facies heterogeneities (interfingering over 100s of m), controlled mainly by lateral migration of a mosaic of facies over an irregular topography (Fig. 3C). Deepening-upward and aggradational sequences (HFSs 4−7) with facies homogeneity at km-scale were generated in the distal areas of the shallow ramp.

Some of the HFSs (HFSs 8−10) split laterally into 2 or 3 shallowing-upward discontinuous HFSs that are traceable only for 3−5 km in thickened areas (Fig. 3C), reflecting the infill of wedge-shaped accommodation space resulting from local differential subsidence, which is also confirmed by the local presence of tilted beds. The discontinuous nature of these HFSs can only be detected by correlation because in 1D log they have a similar field expression to continuous HFSs. Based on these observations, a basin-scale model with variable record of HFSs depending on the rates of subsidence (higher number of HFSs in areas with greater subsidence) was proposed

(Bádenas et al., 2010), later supported by analysis of HFSs in separated areas of the basin (Aurell and Bádenas, 2015a). Therefore, the estimated duration of HFSs in individual 1D logs could not be used in support of climatically induced (Milankovitch) sea-level changes because of the lateral variability in the number of HFSs (i.e., from 11 to 16 in ~4 My), with a very uncertain time duration (~350–250 ky?) (Table 2).

At a larger scale, in different Sinemurian plate margins from western Tethys, Bosence et al. (2009) proposed a dominant tectonic control on relative sea-level changes, with synsedimentary fault movement over-riding the reduced rates of climatically induced sea-level changes in the

Table 2 Summary of the Main Features of Orbitally Driven Sequences in the Studied Jurassic Carbonate Ramps of the Iberian Basin (Black: Shallow Ramp; Dark Gray: Shallow/Deep Ramp; Light Gray: Deep Ramp) and their Average Duration Close to that of Precession (P), Short Eccentricity (e) and Long Eccentricity (E) Cycles. Accumulation (Preservation) Rates are not Corrected from Decompaction, Because Several Evidences Point out to Relatively Rapid Lithification and Limited Compaction: Homogeneous Thickness of HFSs and Internal Beds Including Muddy and Grainy Facies Changing Laterally (case 1a: i.e., HFS 1 in Fig. 3; Case 4a); Undeformed Trace Fossils and Presence of Delicate Fossils (e.g., Thin-Shelled Bivalves; Spicule Sponges) in Chaotic, Even Vertical, Dispositions (Cases 1b, 2 and 4b); and Microbial-Encrusted Grainy Facies with Stromatactis Cavities (Case 3b) and Muddy Facies with Depositional Inclination (Case 3a: i.e., HFSs 5–10 in Fig. 5). Note Low Accumulation Rates in the Shallow Ramp Compared to the Deep Ramp

Number of sequences / hierarchy		Thickness of sequences	Duration	Referential orbital cycles			Accumulation (preservation) rate
				P (~20 ky)	e (~100 ky)	E (~400 ky)	
Case 4: upper Kimmeridgian homoclinal ramp with coral reefs							
4a	26 HFSs	1 LFS–4.5 HFSs	1–7 m	~2 My		~80 ky		~5 cm/ky
	6 LFSs		12–20 m				~340 ky	
4b	28 bundles	1 set–4.5 bundles	mean 1.4 m	~0.5 My	~18 ky			~10 cm/ky
	2+ (4) sets		5–8 m			~85 ky		
Case 3: Bajocian microbial/siliceous sponge-dominated distally steepened ramp							
3b	11 HFSs		1–5 m	~1.1 My		~100 ky		~3 cm/ky
3a	10 HFSs			<0.3 My	~30 ky			~10 cm/ky
Case 2: upper Pliensbachian skeletal-dominated homoclinal ramp							
	16 HFSs	1 LFS–4 HFSs	mean 1 m	~1.6 My		~100 ky		~1 cm/ky
	4 LFSs		3–5 m				~400 ky	
Case 1: upper Sinemurian-lower Pliensbachian non-skeletal dominated homoclinal ramp							
1b	48 bundles	1 lot–4.8 sets–16 bundles	mean 1.2 m	~1.5 My	~30 ky			~4 cm/ky
	10 sets		mean 5.5 m			~150 ky		
	3 lots		mean 19.4 m				~500 ky	
1a	11 continuous HFSs or 16 (continuous and discontinuous) HFSs		1–6.5 m	~4 My?		~350–250 ky? tectonic influence on possible eccentricity signal		~1 cm/ky

Early Jurassic greenhouse conditions. However, the tectonic imprint does not exclude the climate-driven eustatism (e.g., Tucker and Garland, 2010). In the ramp succession studied here, the climate-controlled (eccentricity?) sequences could be preserved potentially in areas of greater subsidence, whereas they would merge laterally to form "tectonic-dominated" sequences in areas of low subsidence. In this sense, the sequences recorded in areas of greater subsidence would fit the concept of "islands of order" (Schlager, 2010) in a random (tectonic controlled) sequence record, where one deterministic driver (climate) dominates.

4.3 Shallow/Deep Ramp Sequences

The lowermost Pliensbachian shallow/deep ramp lime mudstone-dominated successions (case 1b) are arranged in HFSs to LFSs (Fig. 3B and C): 48 bundles of beds (1.2 m in mean thickness and including a variable number of beds), 10 sets (5.5 m, formed by 3–7 bundles) and 3 lithological lots (19.4 m, including 3 or 4 sets). Detailed facies analysis reflect subtle vertical facies changes and indicates that most of the bundles correspond to shallowing-upward sequences, whereas sets and lots are mostly deepening-shallowing sequences (Sequero et al., 2017). The ratio of lots, sets and bundles (1–4.8–16) and their average duration ($\sim$500, $\sim$150 and $\sim$30 ky in $\sim$1.5 My time interval) is close to that of long eccentricity, short eccentricity and precession cycles, respectively. Sets (short eccentricity cycles) and lots (long eccentricity cycles) would mainly reflect high-frequency sea-level changes controlling phases of shallow-water carbonate productivity and exporting capability. Long eccentricity-related sequences have also been recorded in time-equivalent shallow ramp successions in southern areas of the Iberian Basin (Cortés et al., 2009). Bundles are related to precession driven seasonality variations controlling changes in shallow-water carbonate productivity and exported carbonate.

Spectral analysis of magnetic susceptibility (MS associated with magnetite) also indicates the presence of two orders of cyclicity equivalent to the field sequences: precession cycles with thicknesses of 1.1 and 1.3 m (similar to bundle thickness) and long eccentricity cycles of 19.5 m (similar to lot thickness) (Sequero et al., 2017). MS changes would be related to changes in continental erosion and fluvial/aeolian/marine transport of detrital sediment (e.g., Ellwood et al., 2006; Dechamps et al., 2015), which can be driven by Milankovitch cycles (e.g., Boulila et al., 2008; Wu et al., 2012).

5. UPPER PLIENSBACHIAN SEQUENCE RECORD

5.1 Stratigraphy and Sedimentary Environments

The upper Pliensbachian bioclastic limestones (case 2: Figs. 1 and 4) correspond to the regressive deposits of the upper Pliensbachian T-R facies cycle (Aurell et al., 2003) and were analyzed by Val et al. (2017) in two, 15 km apart, sections. The reference section at Obón is ∼17 m in thickness and spans from the upper *Margaritatus* Zone to the top of the *Spinatum* Zone, which is a basin-scale trangressive surface (Fig. 1; Aurell et al., 2003; Gómez and Goy, 2005). The duration of the succession, also calibrated using Sr isotopes, is ∼1.6 My (Val et al., 2017; assignment based on Gradstein et al., 2012; Ruhl et al., 2016).

Deposition took place in the transition between the shallow ramp and the shallow/deep ramp (i.e., middle ramp) of a storm-dominated carbonate ramp, characterized by the accumulation of skeletal debris of bivalves, brachiopods and echinoderms grading distally to progressively muddier sediments (Fig. 4B). Predominance of these skeletal-dominated carbonates was probably linked to the late Pliensbachian cooling interlude (Fig. 1), which is recorded by an oxygen-isotope positive trend in Obón (Val et al., 2017) and in many deep marine successions of western Europe (e.g., Dera et al., 2011). In the shallow ramp area below the fair-weather wave base, the storm-induced waves and offshore-directed return flows caused the accumulation of amalgamated and proximal tempestites composed of para-autochthonous (entire and disarticulated) fossils in a packstone matrix of small and rounded skeletal grains resedimented from relatively shallower areas. Matrix-supported facies including diluted tempestites dominated offshore, and lime mudstones, lime-mudstones and marls deposited in the shallow/deep ramp.

5.2 Shallow and Shallow/Deep Ramp Sequences

Sixteen HFSs (1 m in mean thickness) and 4 LFSs (3—5 m thick) have been recognized and correlated in the Obón-San Pedro area (Fig. 4). The HFSs are bounded by prominent sharp bedding surfaces of erosion or non-sedimentation (Fig. 4C), and have a very variable number of beds, some of which also include intra-bed erosional surfaces (i.e., base of amalgamated and proximal tempestites). Vertical facies trends in HFSs are usually shallowing-upward or aggradational in proximal areas, and deepening-upward or deepening-shallowing in distal areas (Fig. 4D). LFSs include 4 HFSs and are deepening-shallowing upward or shallowing-upward sequences bounded by prominent bedding surfaces (Fig. 3C). The lower

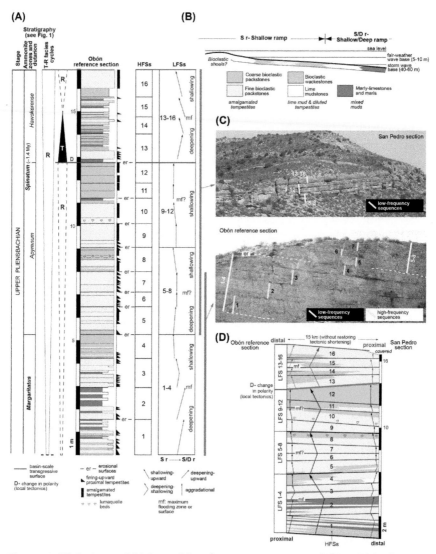

Figure 4 (A) Facies and high- and low-frequency sequences (HFSs, LFSs) of case 2 (upper Pliensbachian) in the reference section of Obón. (B) Synthetic facies model showing the distribution of the main facies belts. (C) Selected field views showing the distribution of HFSs and LFSs. (D) Summary of the vertical and lateral facies distribution along the 15 km-long transect reconstructed between Obón and San Pedro sections.

three LFSs (1–4, 5–8 and 9–12) reflect a long-term shallowing-upward facies trend, but the upper LFS 13–16 records a sharp deepening and change in facies polarity, reflecting the modulation by local tectonics of the long-term regressive trend (Fig. 4A and D; Val et al., 2017).

LFS 1–4 locates within the upper *Margaritatus* Zone, and of the upper three LFSs belong to the *Spinatum* Zone (~1.4 My). The estimated duration of the entire succession (~1.6 My) and the observed hierarchy of sequences (1 LFS including 4 HFS), point out LFSs and HFSs fit well with long- and short eccentricity cycles, respectively (Table 2). Sets of sequences defined in coeval successions in the Iberian Basin (Comas-Rengifo et al., 1999) are similar in number and thickness to LFSs defined in Obón-San Pedro. In addition, Ruhl et al. (2016) proposed the existence of three long eccentricity cycles for the *Spinatum* Zone in Cardigan Bay and Cleveland basins (UK), which fits the three LFSs identified in the Iberian Basin within this zone.

2In the context of a storm-dominated carbonate ramp and cool greenhouse climate (with the probable formation of polar ice caps: Price, 1999), the recorded sequences were probably linked to climate-induced sea-level oscillations controlling carbonate benthic production and resedimentation (Val et al., 2017), the boundaries of HFSs and LFSs being linked to stages of sharp decrease in both carbonate production and resedimentation. The link between storm frequency and short- and long eccentricity cycles has been proposed by Long (2007) for Ordovician-Early Silurian successions, and interpreted as reflecting eustatic sea level changes, climate-induced differences in sediment flux from shallow-water carbonate factories, or orbitally-induced shifts in the intensity and position of storm belts.

6. BAJOCIAN SEQUENCE RECORD

6.1 Stratigraphy and Sedimentary Environments

The outcrop near the village of Moscardón exposes the Bajocian microbial/siliceous sponge dominated facies in the transitional area between the shallow ramp domain and a relatively deep NW–SE-trending subsiding area (case 3: Figs. 1 and 5; Aurell et al., 2003; Gómez and Fernández-López, 2006). The outcrop allows a precise reconstruction of the facies architecture in two ~1 km-long depositional-dip oriented exposures (Aurell and Bádenas, 2015b). The succession is up to 70 m thick and its time frame is provided by a precise ammonite biostratigraphy (Gómez and Fernández-López, 2006). The lower part of the succession represents the upper part of the Aalenian-lower Bajocian T-R facies cycle (Baj-1 Sequence; case 3a) in the middle and upper part of the *Humphresianum* (~0.3 My) Zone. The upper

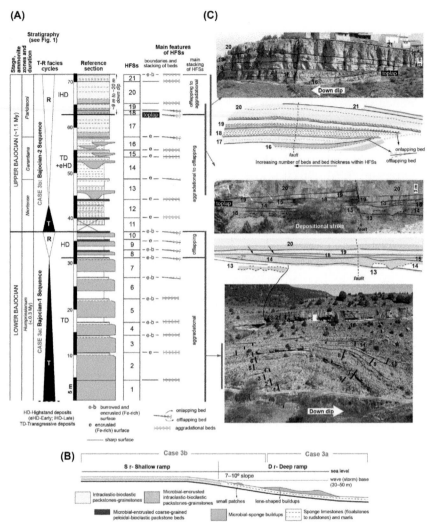

Figure 5 (A) Facies and high-frequency sequences (HFSs) of case 3 (Bajocian) in the reference section of Moscardón. (B) Synthetic facies model showing the distribution of the main facies belts. The occurrence of cases 3a and 3b in the different domains of the carbonate ramp is indicated. (C) Selected field views showing the distribution of HFSs.

part of the succession corresponds to the upper Bajocian T-R facies cycle (Baj-2 Sequence; case 3b), with an estimated duration of ~1.1 My.

General hot and arid conditions predominated during the early Bajocian (Holz, 2015, Fig. 1), and in particular in the western Tethys, where high temperatures favored reef growth (Dromart et al. 1996; Leinfelder et al., 2002; Brigaud et al., 2009); however, during the late

Bajocian there was a decreasing trend in western Tethys seawater temperatures (about 5—8°C: Brigaud et al., 2009). Nevertheless, high concentration of nutrients supplied from terrestrial sources rather than temperature has been suggested as the major factor favoring the dominance of microbial/siliceous sponge facies in the Iberian Basin (Aurell and Bádenas, 2015b) and in other Tethyan areas (e.g., Olivier et al., 2008). The Iberian Bajocian carbonate ramp was a distally steepened ramp with a low-angle slope located around the storm wave base (Figs. 1 and 5B). The shallow ramp was the locus of deposition of intraclastic-bioclastic packstones-grainstones including sponge debris, which underwent episodic stabilization by microbial crusts. The ramp slope and deep ramp included a wide variety of facies, such as meter-sized microbial-siliceous sponges patches, and lens-shaped mounds with *Hexactinellida* and *Lithistida* sponges, and marls and burrowed sponge-rich limestones with bivalves, brachiopods, echinoderms, gastropods and ammonites, also episodically encrusted by microbial crusts.

6.2 Shallow and Deep Ramp Sequences

The Bajocian succession includes 21 HFSs (1—5 m in thickness; Fig. 5). HFSs are bounded by sharp bedding surfaces, usually Fe-rich, bioturbated and microbial-encrusted, and have variable stacking pattern depending on their location within the long-term sequences. HFSs 1—10 (Baj-1 Sequence) are formed by deep ramp microbial-sponge boundstones with marls/sponge limestones at their lower part, and microbial-encrusted peloidal-bioclastic packstones on top. HFSs 1—7 (trangressive deposits) are stacked in an aggradational arrangement to form ~ 30 m-thick and 50—100 m-wide microbial-sponge mounds; in contrast, HFSs 8—10 (regressive deposits), have offlap geometries and significant lateral variations in thickness, with thick marls/sponge limestones almost filling the intermound depressions (Fig. 5C).

HFSs 11 and 12 (trangressive deposits of Baj-2 Sequence) consist of marls/sponge limestones with microbial-encrusted packstone caps and display an aggradational stacking pattern. By contrast, HFSs 13—17 (early regressive deposits) display an aggradational to offlap geometry; laterally they show a facies transition in the slope area, from shallow intraclastic-bioclastic packstones-grainstones to deep marls/sponge limestones with sponge patches and lens-shaped mounds. The contact with HFSs 18—21 (late regressive deposits) is a toplap surface that truncates some of the underlaying HFSs (Fig. 5C) and grades down dip into a conformable contact. HFSs 18—21 are composed of shallow grain-supported facies and are bounded by sharp erosive surfaces, locally microbial-encrusted and delimiting unconformable beds. These HFSs have an offlapping to aggrada-tional stacking pattern and progressively thicken down dip, the bed number and thickness also increasing down dip (Fig. 5C).

The recorded HFSs suggest the existence of high-frequency sea-level changes superimposed on the long-term Baj-1 and Baj-2 T-R sequences (Aurell and Bádenas, 2015b). This is supported by the fact that the internal stacking pattern of beds within HFSs reproduces at a small scale the stacking of HFSs within the long-term sequences (Fig. 5A). However, the imprint of the high-frequency sea-level changes was variable across the shallow and deeper areas. Deep ramp HFSs 1–10 ($\sim$ <0.3 My in total duration) are within the range of precession cycles, whereas HFSs 11–21 ($\sim$1.1 My) can be tentatively assigned to short eccentricity cycles.

The HFS boundaries reflect stages of low sedimentary rates that would be linked to deepening events. In the shallow ramp area subjected to storm wave action, episodic stages of quiet waters due to the rise of storm wave base would allow microbial stabilization and/or cementation of the grainy sediment. In the deep ramp area, cessation of mound growth and microbial stabilization, Fe-enrichment and bioturbation at the HFS boundaries, would reflect deepening events decreasing sediment supply necessary for the mound organisms to baffle and bind and/or deterioration of living conditions for the organisms, such as variations of nutrient con-centration and/or oxygen fluctuations (e.g., Leinfelder and Schmid, 2000; Hebbeln and Samankassou, 2015). These variations could be linked to possible oscillations of the pycnocline/nutricline, i.e., the area where the suspension-feeding macrofauna, including siliceous sponges, thrive (e.g., Stanton, 2006; Pomar et al., 2012; Pomar and Haq, 2016). Sea-level fluctuations driven by orbital (eccentricity) cycles have been proposed as the main factor controlling growth stages of sponge-microbial reefs in relatively deep (mid-outer ramp) areas in the Late Jurassic in Germany (Pawellek and Aigner, 2003a).

7. UPPER KIMMERIDGIAN SEQUENCE RECORD

7.1 Stratigraphy and Sedimentary Environments

A wide homoclinal carbonate ramp, including coral-microbial buildups, covered the eastern part of Iberia during the late Kimmeridgian (case 4: Fig. 1). Of particular interest for the analysis of shallow ramp areas are the outcrops located in Sierra de Albarracín (Aurell and Bádenas, 2004; Bádenas and Aurell, 2010; Alnazghah et al., 2013; San Miguel et al., 2017). In this review, the

100 m-thick Riodeva section is used to illustrate this shallow ramp domain (case study 4a; Figs. 1 and 6A; Bádenas and Aurell, 2010). This succession represents the regressive deposits of the upper Kimmeridgian T-R facies cycle and has a duration of ~ 2 My according to Ogg et al. (2016), although the biostratigraphic data do not allow a precise definition of the early/late Kimmeridgian and Kimmeridgian/Tithonian boundaries (Fig. 1; Fezer, 1988; Nose, 1995). A 43 m-thick succession within the Aguilón section (case 4b: Figs. 1 and 6B) has been selected to illustrate the deep ramp areas (Bádenas et al., 2003, 2005). It displays the transgressive deposits of the T-R facies cycle, probably within the middle part of the late Kimmeridgian, around the *Eudoxus* zone (~0.5 My).

Globally cool greenhouse conditions predominated during the Kimmeridgian (Fig. 1; Holz, 2015), and in the western Tethys coral-microbial reef growth was optimized due to relatively warm temperatures (Leinfelder et al., 2002; Dera et al., 2011). The studied ramp was a storm-dominated ramp opened to the Tethys Ocean, with a windward orientation with respect to winter storms and hurricanes (Figs. 1 and 6C; Bádenas and Aurell, 2001, 2008). The shallow ramp included lagoon and high-energy (shoal and foreshoal) areas with a wide range of non-skeletal facies (Fig. 6C). Coral-microbial buildups with patch and pinnacle morphology of variable size (up to 16 m high) grew mainly in shoal and foreshoal areas. The shallow/ deep ramp was the loci of deposition of carbonate muds containing tempestites decreasing offshore and scarce skeletal grains (mainly lituolids, bivalves, sepulids, gastropods, and solitary corals), and lime mudstones dominated in the deep ramp. Quantification of carbonate production indicates that part of the carbonate mud was supplied from the shallow high-productivity areas (Aurell et al., 1998; Boylan et al., 2002).

7.2 Shallow Ramp Sequences

The Riodeva section illustrates HFSs and LFSs mostly developed in shallow ramp (Fig. 6A). Twenty-six meter-thick HFSs with variable vertical facies trends (deepening-shallowing, shallowing-upward and aggradational) can be differentiated (Figs. 6A and 7A—C). HFSs are bounded by sharp bedding planes linked to transgressive surfaces. Long-term facies trends allow the recognition of six deepening-shallowing or shallowing-upward LFSs, each including 4—5 HFSs. Lateral tracing of LFS boundaries and facies correlation over a broad area (20 × 20 km, Fig. 5D: Bádenas and Aurell, 2010) reflects the growth of coral-microbial buildups during deepening, followed by progradation of shallow facies.

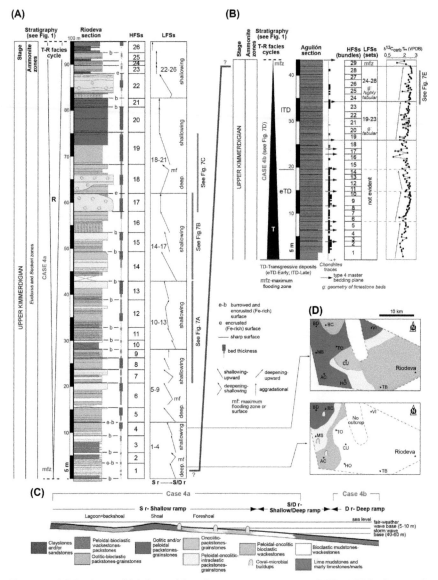

Figure 6 (A) Facies and high- and low-frequency sequences (HFSs, LFSs) of case 4b (upper Kimmeridgian) in the reference section of Riodeva. (B) Facies and high- and low-frequency sequences (HFSs, LFSs) of case 4b (upper Kimmeridgian) in the reference section of Aguilón. (C) Synthetic facies model showing the distribution of the main facies belts. The occurrence of cases 4a and 4b in the different domains of the carbonate ramp is indicated. (D) Facies distribution in LFS 1–4 over a broad area (20 × 20 km).

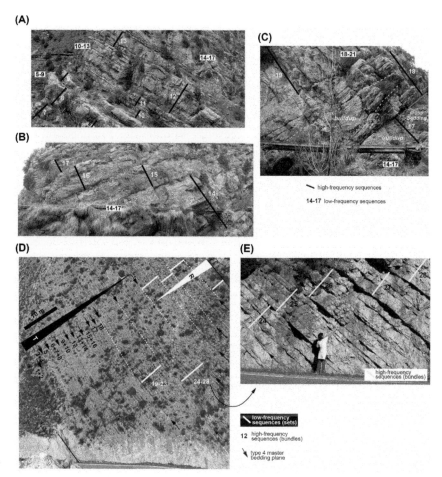

Figure 7 Selected field views showing the distribution of high- and low-frequency sequences in case 4a (A, B and C) and case 4b (D and E). See stratigraphic location in Fig. 6.

The observed LFSs were related to sea-level fluctuations 5—10 m in amplitude, formed in tune with eccentricity cycles (Aurell and Bádenas, 2004; Bádenas and Aurell, 2010). Local evidences of subaerial exposure have been recorded on top of some LFSs in more proximal areas of the ramp (Aurell and Bádenas, 2004), reflecting local record of sea-level fall after the subsequent transgressive surface of the next LFS. HFSs 1—26 were developed during a ~2 My interval and therefore would roughly fit the short eccentricity cycles. Grouping of 4—5 HFSs into LFSs suggests the imprint of the long eccentricity cycles.

7.3 Deep Ramp Sequences

The deep ramp lime mudstone successions in Aguilón represent the transgressive deposits of the upper Kimmeridgian T-R facies cycle, the progressive long-term increase in depositional depth being defined by a decreasing
trend in skeletal content and *Chondrites* traces (Fig. 6B; Bádenas et al., 2003,
2005). HFSs (bundles of up to 10 beds) and LFSs (sets of 5 bundles) can be
differentiated in the Aguilón succession. Bundles are bounded by sharp
bedding planes and have a mean thickness of 1.4 m; sets are bounded by
type 4 master bedding planes (following the terminology of Schwarzacher
and Fischer, 1982), which are usually associated with thin marly beds and
changes in the geometry of limestone beds (Bádenas et al., 2003). In the
selected trangressive succession, there are 28 bundles (Fig. 7B, E). The random
distribution of type 4 bedding planes in the lower part of the succession
(early transgressive deposits) does not allow for a clear definition of sets,
but they are clearly identified in the upper part (late transgressive deposits)
and in the regressive deposits of the succession (Fig. 7B and D; Bádenas
et al., 2003).

Bundles and sets were related to precession cycles and short eccentricity
cycles, respectively, which were also detected by spectral analysis of bed
thickness (Bádenas et al., 2003). Considering the duration of ∼0.5 My
for the studied trangressive succession, the 28 bundles correlate well with
precession cycles and the sets of 4−5 bundles with short eccentricity cycles.
Sets were interpreted as reflecting cyclic changes in shallow carbonate productivity and export from shallow reef regions caused by high-frequency
sea-level fluctuations. Carbon isotope trends recorded at set-scale
(Fig. 6B) reflect these shifts in carbonate export (Bádenas et al., 2003,
2005). Bundles are related to variations in exported carbonate due to
changes in shallow-water carbonate productivity probably linked to
precession driven seasonality variations. Both bundle and set boundaries
correspond to omission surfaces related to periods of reduced shallow-water
carbonate production and export.

8. DISCUSSION: THE VARIABLE DOWN-DIP RECORD OF SEQUENCES IN THE IBERIAN JURASSIC RAMPS

The Jurassic carbonate ramp successions in the Iberian Basin show a
distinct stacking of orbitally controlled sequences. The age calibration of
the sequences and their distribution along the ramp domains reflect

there is a variable record of these orbitally controlled sequences from shallow-to relatively deep ramp areas (Table 2 and Fig. 8). As discussed below, this variable record is thought to be mainly related to different accumulation rates (or more strictly, preservation rates: Strasser and Samankassou, 2003), which result from the interaction of the intrinsic processes of sediment production and redistribution and the long- and short-term changes in accommodation.

8.1 The Sedimentary Record of Orbital Cycles in Shallow Ramp

In shallow ramp domains, short- and long eccentricity cycles caused sea-level variations that generated HFSs and LFSs, respectively (mostly subtidal cycles: *sensu* Osleger, 1991). Shallowing-upward, deepening-shallowing, aggradational or deepening-upward HFSs and LFSs formed in the upper Pliensbachian and upper Kimmeridgian homoclinal ramp successions (Fig. 8A). Short eccentricity-related HFSs sequences related to sea-level variations and fluctuations of storm wave base are recorded in the shallow area of the distally steepened Bajocian ramp. Long eccentricity-related sequences cannot be defined here because of the short time span of this shallow-water succession (i.e., only 5 HFSs: Fig. 4A). In the "tectonic dominated" record of the upper Sinemurian ramp, the potential preservation of eccentricity-related peritidal and subtidal HFSs in areas of greater subsidence cannot be rouled out.

The shallow ramp successions show low accummulation rates ($< \sim 5$ cm/ky; Fig. 8A), if compared with those compiled in carbonate platforms for the Phanerozoic over time spans similar to those of the studied successions, i.e., ~ 10 cm/ky in $\sim 1-5$ My time intervals (Kemp and Sadler, 2014). These authors indicate that at $\sim 1-5$ My time intervals (similar to those of the successions studied here) the primary control on calculated accummulation rates is the long-term accommodation space (subsidence). In the Iberian Jurassic ramps, low accummulation rates are likely related to the combined effect of a long-term still stand and loss in accomodation space (i.e., predominat regressive context of the studied shallow ramp successions: Fig. 8A) and the internal processes of sediment production and redistribution due to deposition above in areas of constant or episodic wave and current reworking. The predominance of subtidal sequences and their layer-cake arrangement (i.e., lateral continuity at km-scale; e.g., upper Pliensbachian and upper Kimmeridgian ramps) fits well the modelized stratigraphy of layered cycles in ramps with erosion performed by Read et al. (1991).

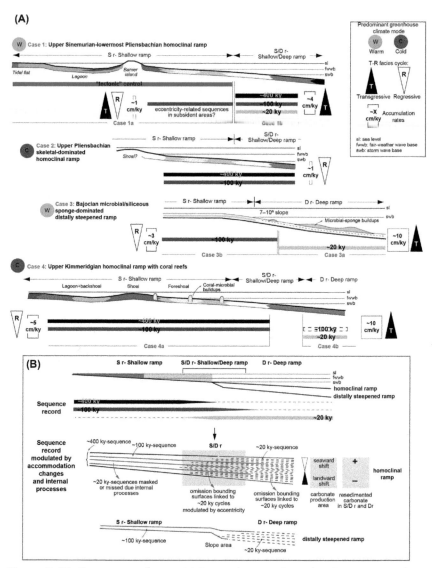

Figure 8 (A) Occurrence of eccentricity- and precession-related sequences along the different domains of the Iberian Jurassic carbonate ramps (see facies legends in Figs. 3–6). (B) Summary of the distribution of eccentricity- and precession-related sequences in the Iberian Jurassic carbonate ramps and model of preservation potential of short eccentricity- and precession-related sequences within a long eccentricity-related sequence depending on the interaction between sediment accumulation-accommodation. See explanation in text.

The thicknesses of eccentricity-related sequences in the different Jurassic shallow ramp areas show overlapping figures. The short eccentricity-related HFSs of the "greenhouse warm" Bajocian shallow ramp have a similar thickness range (1—5 m) to those in the "greenhouse cold" upper Kimmeridgian shallow ramp (1—7 m) (Table 2). It is interesting also to notice that in the "greenhouse cold" upper Pliensbachian shallow ramp the long eccentricity-related LFSs are of similar thickness (3—5 m) than the short eccentricity-related HFSs in the Bajocian and upper Kimmeridgian ramps. Differences in amplitude of the orbitally induced sea-level oscillations has been discussed by Husinec and Read (2007) to explain the distinct sequence record in shallow areas of barrier-type platforms during greenhouse cold conditions (precession-eccentricity driven or even sub-Milankovitch sequences) versus greenhouse warm conditions (probably driven by obliquity and short-term and long-term eccentricity). However in the Jurassic cases studied here, the shallow ramp area record only short- and long eccentricity-related sequences, independently of the climate conditions. Therefore, the observed sequence record and sequence thickness can be more directly linked with long-term accommodation and internal factors (production and redistribution), rather than with possible differences in amplitude and imprint of the orbitally induced sea-level oscillations during greenhouse cold *vs.* greenhouse warm conditions. The lateral down-dip increase in thickness of short eccentricity-related HFSs (and their internal beds) in the shallow ramp of the distally steepened Bajocian ramp (Fig. 4B and C) is a direct evidence of the control exterted by available accommodation. The overlapping figures of thicknesses of long eccentricity-related LFSs in the upper Pliensbachian and short eccentricity-related HFSs in the Bajocian and upper Kimmeridgian ramps can be partly related to differences in carbonate production and accumulation, with significant lower accumulation rates in the skeletal (bivalves, brachiopods and echinoderms) upper Pliensbachian storm-dominated ramp (Table 2).

In the shallow ramp, the precession-related sequences are masked or missed (i.e., "missed beats": Goldhammer et al., 1990; Osleger, 1991), due mainly to internal processes. In the low-energy, innermost portion of the ramp (lagoon and/or tidal flats), due to the existence of morphological barriers (shoals, barrier islands), the irregular topography and the shallow depth, autocyclic processes (storms, changes in carbonate productivity, minor drops in relative sea level) would override or mask the imprint of small-scale precession-related changes in carbonate production or possible precession-related sea-level variations (e.g., Tresch and Strasser, 2011). In high-energy shallow ramp areas, precession-related sequences were not preserved due to non deposition or erosion by constant or episodic wave and current reworking. This effect is likely also important in the distally steepened Bajocian ramp, due to storm reworking at the flat area above the low–angle slope.

8.2 The Sedimentary Record of Orbital Cycles in Shallow/Deep Ramp

The precession-related sequences are masked or missed in the shallow ramp area, but they can be preserved in the shallow/deep ramp dominated by muddy sediments. In particular, precession-related sequences are recorded as bundles of limestone beds reflecting variations in exported carbonate due to changes in shallow–water carbonate productivity, during transgressive stages (e.g., lowermost Pliensbachian ramp), but not during regressive stages (e.g., upper Pliensbachian and upper Kimmeridgian ramps; Fig. 8A). The imprint of precession cycles has also been recorded in distal middle ramp transgressive deposits of the upper Kimmeridgian of the Iberian Basin (Colombié et al., 2014).

The observed variability in the sequence record in shallow/deep ramp areas can be linked to the combined effect of accommodation changes (long term and/or short term) and processes involved in sediment accummulation, in particular variations in exported carbonate. Fig. 8B illustrates the preservation potential of short eccentricity- and preccesion-related sequences within a long eccentricity-related sequence, depending on the interaction between sediment accumulation-accommodation. During stages of accommodation gain, the landward shift of the carbonate production area would lead to a reduction in the carbonate exported to distal middle ramp, and therefore the episodes of reduced carbonate production and exportation linked to both eccentricity and precession cycles would be represented by well defined omission surfaces (i.e., sequence boundaries). During stages of accommodation loss, the opposite occurs: due to the increase in resedimentation coeval with the seaward shift of the carbonate factory, only those omission surfaces linked to precession cycles modulated by eccentricity would be recorded. In addition, erosion due to deposition around the storm wave base could also control the existence of some "missed preccesional beats" in these areas.

8.3 The Sedimentary Record of Orbital Cycles in Deep Ramp

In the deeper areas well below storm wave base (i.e., distally steepened Bajocian ramp and upper Kimmeridgian homoclinal ramp) the accummulation (preservation) rates are high (~10 cm/ky) compared to the shallow-water domains (Table 2). This is probably related to the long-term transgressive context of these successions and the deposition below storm wave base, which allow a more complete stratigraphic record to be preserved (Fig. 8A). The sequence record is dominated by precession-related sequences (bundles of beds in the upper Kimmeridgian ramp, HFSs in the Bajocian ramp), and short eccentricity-related sequences (sets of bundles) are only locally recorded. Preccesion-related changes in seasonality caused variations in shallow carbonate production and exportation that generated bundles of limestone beds (upper Kimmeridgian), or pycnocline/nuctricline oscillations that controlled stages of microbal-sponge growth (Bajocian).

The upper Kimmeridgian outer ramp succession (Fig. 6) illustrates the effect of accommodation changes on the preservation potential of preccesion- and short eccentricity-related bundles and sets in outer ramp areas. Here, the episodes of reduced carbonate production and exportation linked to precession cycles would have a greater effect on the formation of omission surfaces (i.e., boundaries of bundles) than in shallow-deep ramp areas, so that all "preccesional beats" would be preserved (Fig. 8B). Numerous prominent omission surfaces recorded at the early trangressive deposits (Fig. 6) are coherent with the landward shift of the carbonate (exporting) area. During late transgression and regression (Bádenas et al., 2003), a more ordered signal of prominent omission surfaces grouping bundles of limestone beds occurs (i.e., short eccentricity-related sets of bundles), therefore reflecting a more direct imprint of precession cycles modulated by eccentricity, similar to that observed in distal middle ramp areas during regressions (Fig. 8D).

In the distally steepened Bajocian ramp, precession-related pycnocline/nuctricline oscillations controlled microbal-sponge growth (HFS 1–10 in Fig. 5; Fig. 8B). In this slope area, the preccesion-related sequences most likely have a complex aggradational, onlapping or offlaping stacking pattern, in response to lower-frequency pycnocline/nuctricline variations, as indicated by the stratigraphy observed in the field (Fig. 4).

9. A REVIEW OF THE SEQUENCE RECORD IN GREENHOUSE CARBONATE RAMPS: IMPLICATIONS

9.1 Variable Down-Dip Sequence Record: the Rule rather than the Exception?

In the analyzed greenhouse Iberian Jurassic carbonate ramps, the stratigraphic record of orbitally driven cycles had different imprint in shallow (dominant eccentricity-related sequences) and deep ramp areas (dominant preccesion-related sequences). Comparison with cyclostratigraphic analysis performed in other greenhouse carbonate ramps may be useful to decipher whether the observed trend is the rule or the exception. Some of these works describe orbital-related sequences in specific areas of the ramp; others illustrate sequences encompassing inner-to outer ramp facies but do not describe clear lateral trends in the orbital-related sequences (Table 3).

Eccentricity-driven sea-level cycles have been interpreted as the main mechanism generating meter-scale sequences in shallow areas of greenhouse carbonate ramps of the Late Cambrian in USA (Osleger and Read, 1991), the Middle Ordovician ramp in Oman (Schalaich and Aigner, 2017) and Korea (Lee et al., 2001), and the Middle Triassic in Germany and Hungary (Götz and Török, 2008 and references therein). Milankovitch-related meter-scale sequences were also described in the Late Triassic in Spain (Calvet et al., 1990), but their timing is unknown due to poor age control. Cyclothems and megacycles encompassing inner-outer ramp facies have also been recorded in the Late Triassic of Italy and spectral analysis clearly indicate orbital imprint in their generation (Preto and Hinnov, 2003; Table 3). In the Late Cretaceous of the Iberian Basin studied by Gil et al. (2009), the hierarchical stacking of sequences and spectral analysis of facies indicate a dominance of eccentricity signal in the sequence record. Sequences related to short eccentricity and precession have been recorded in the Early Cretaceous of the Spanish Pyrenees (Bachmann and Willems, 1996).

Middle and outer carbonate ramp successions of the Late Jurassic in Germany (Pawellek and Aigner, 2003a,b; Ruf et al., 2005; Pross et al., 2006) also show a clear sequence record of eccentricity cycles. The mid- to outer ramp successions characterized in Germany (Pross et al., 2006) and in the latest Jurassic-earliest Cretaceous in Argentina (Kietzmann et al., 2015), are dominated by muddy sediments (limestone to marlstones).

Table 3 Recognition of Orbitally Driven Sequences and Orbital Cycles Obtained By Spectral Analysis in Greenhouse Carbonate Ramps Described in the Literature. The Gray Shading Indicates the More Prominent Sequences Recorded

Ramp area	Reference	Age, Location	Sequence (x) Spectral analysis (•)	lithological couplet (lc)	Orbitally driven sequences (x) and spectral cycles (•)			
					P	O	e	E
Inner	Schalaich and Aigner (2017)	Middle Jurassic, Oman	x				x 1–11.5 m	x
Inner to outer	Osleger and Read (1991)	Late Cambrian, USA	x				x 0.4–15 m	x
Inner to outer	Lee et al. (2001)	Middle Ordovician, Korea	x				x 1–14 m	
Inner to outer	Götz and Török (2008)	Middle Triassic, Germany, Hungary	x				x m-thick	x
Inner to outer	Preto and Hinnov (2003)	Late Triassic, Italy	x • facies		x ? cyclothems (dm–m) and megacycles			
					•	•	•	•
Inner to outer	Bachmann and Willens (1996)	Early Cretaceous, Spain	x • facies		x 5–9 m •		x •	
Inner to outer	Gil et al. (2009)	Late Cretaceous, Spain	x • facies		• 	x •	x 2.1–3.1 m •	x •
Middle to outer	Pawellek and Aigner (2003a,b)	Late Jurassic, Germany	x				x m-thick	x
Middle to outer	Kietzmann et al. (2015)	Latest Jurassic-earliest Cretaceous, Argentina	x • couplet thickness	limestone-marl	x (lc) 0.2–0.4 m		x •	x •
Outer	Ruf et al. (2005)	Late Jurassic, Germany	x				x ? 3–7 m	x 9–20
Outer	Pross et al. (2006)	Late Jurassic, Germany	x	marlstone-limestone-marly limestone		(not studied)	x 5–8 m	x
Outer	Wu et al. (2012)	Early Triassic, China	• MS & ARM	limestone-marl-mudstone millenial-scale	•	•	•	•
Outer	Cunha and Koutsoukos (2001)	Late Cretaceous, Brazil	• GR & sonic log	limestone-marlstone	•	•	•	•
Outer ramp-basin	Elrick and Hinnoz (2007)	Paleozoic, USA		limestone-marl/shale millenial-scale				
Outer ramp-basin	Payros et al. (2015), Martínez-Braceras et al. (2017). Dinarès-Turell et al. (2018)	Eocene, Spain	x • MS & color	limestone-marl	x (lc) dm–few m •	x (bundles) 2–6 m	• 	•
Outer ramp-basin	Betzler et al. (2000)	Miocene, Great Bahama Bank	x • GR	wackestone-packstone	x(lc) cm–m •	• 	x	

ARM, anhysteretic remanent magnetization; *GR*: gamma ray; *MS*: Magenetic susceptibility.

Pross et al. (2006) recognized eccentricity-related sequences reflecting sea-level changes driven by temperature changes (dominant marls during sea-level rises and cooler conditions *vs.* limestones during high sea level and warmer conditions). Kietzmann et al. (2015) describe eccentricity-related and precession-related sequences (limestone-marl couplets) driven by fluctuations in shallow-water carbonate production and basinward carbonate exportation.

Outer ramp limestones to mudstones successions described by Wu et al. (2012) in the Early Triassic in China, and by Cunha and Koutsoukos (2001) in the Late Cretaceous of Brazil do not show clear sequences related to the orbital cycles obtained using spectral analysis (Table 3). The most prominent cyclic features are limestone-marl/mudstone couplets thinner than the "spectral" precession cycles, which were controlled by changes in moonson activity (probably sub-Milankovitch cycles: Wu et al., 2012) or productivity cycles (Cunha and Koutsoukos, 2001). Elrick and Hinnov (2007) also indicate that the lithological couplets recorded in outer ramp-basin successions in several greenhouse Paleozoic ramps in USA reflect sub-Milankovitch related climate changes (aridity-humidity; variations of offshore transport). However, in the outer ramp-basin deposits of the Eocene of the Basque-Cantabrian Basin, preccesion-driven limestone-marl couplets are the most prominent sequence (related to pelagic productivity cycles or to dilution cycles, depending on the palaeogeographic location), and short-eccentricity bundles are also recognized (Payros et al., 2015; Martínez-Braceras et al., 2017). Spectral analysis of MS and color reflect a record of precession, short and long eccentricity cycles (Dinarès-Turell et al., 2018). In the Miocene of Great Bahama Bank, Betzler et al. (2000) also describe wackestone-packstone couplets linked to variations of exportation from shallow areas driven by precession cycles, and turbidite packages related to short eccentricity cycles.

In summary, the compiled data indicates that, although the spectral analyses can reflect a complete record of Milankovitch cycles in shallow to deep carbonate ramp areas, the sequence record of the orbital cycles varies with depth: (1) shallow (inner to middle) carbonate ramp areas are dominated by eccentricity-related, meter-thick facies sequences; (2) the middle-to outer ramp area is likely the domain potentially recording both eccentricity- and precession-related sequences; (3) the deep outer ramp-basin is characterized by precession-related sequences and/or sub-Milankovitch sedimentary cycles. Therefore, there is an apparent down-dip preservation trend from low-frequency (eccentricity) cycles to higher-frequency (precession to sub-Milankovitch) cycles, which is coherent with the down-dip trend proposed in the studied Jurassic ramps (Fig. 8B).

9.2 Precession-Related Sequences: Occurrence and Heterogeneity

The precession-related sequences are the smallest "building blocks" of the Milankovitch-band sequence record, so that their recognition is useful from both sedimentological and cyclostratigraphical points of views (Strasser et al., 2012). However, the compilation above reveals their unusual preservation in shallow ramp areas. The absence of the precession-related sequences may be linked to interrelated factors such as: (1) the low amplitude of precession-related sea-level cycles or seasonal changes and/or the setting not being sensitive enough to them (e.g., Osleger and Read, 1991; Gil et al., 2009), (2) phase relations of the interacting Milankovitch cycles (e.g., Osleger and Read, 1991), and (3) reduced accommodation space and internal processes of reworking by fair-weather waves and storms (Guo et al., 2018; present work). The link between lack of accommodation space and no preservation of small cycles was also proposed to explain the absence of sub-Milankovitch cycles in inner-middle areas of mixed siliciclastic-carbonate ramp deposits (La Rochelle platform, western France; Colombié et al., 2012), when compared to middle-outer ramp deposits (Carcel et al., 2010).

Regarding the character of precession-related sequences in greenhouse carbonate ramps, data summarized here indicate their great variability: they correspond to meter-thick facies sequences in shallow ramp (e.g., Bachmann and Willems, 1996); to dm-thick limestone-marl couplets (e.g., Kietzmann et al., 2015), ~1 m-thick bundles of beds and m-thick microbal-sponge facies sequences (e.g., Pawelleck and Aigner, 2003a,b; present work) in middle-outer ramp; and to cm-to m-thick wackestone-packstone couplets or limestone-marl couplets in outer ramp-basin (e.g., Betzler et al., 2000; Payros et al., 2015; Martínez-Braceras et al., 2017). The heterogeneity of the precession-related sequences along the carbonate ramp profile is due to the complex interaction of accommodation and processes controlling sediment production and accummulation, especially those recorded in the transitional area between the middle and outer ramp and the outer ramp and basin. Here, sediments can be accumulated by varied processes (in situ production, exportation from shallow areas, pelagic production), which may vary in down-dip direction (e.g., Aurell et al., 1998) but also in strike direction or depending on the paleogeographic location (e.g., Martínez-Braceras et al., 2017). Also, the possible out-of-phase response of these processes to the orbitally driven climate changes may hinder the recognition of the precession-related "building blocks." The heterogeneity of precession-related sequences linked to storm exportation, ranging from bundles of beds (this work) to bed couplets (e.g., Colombié et al., 2014; Kietzmann et al., 2015; Betzler et al., 2000), is an example of the complex interaction of climate and the sedimentary system: climate changes may translate in changes in storm intensity, but the variable effect of storm flows in the accumulation (carbonate productivity and export) but also erosion (cannibalism) of deposited sediments, may determine the great variability of the sequences recorded.

10. CONCLUSIONS

Orbitally driven insolation changes controlled the formation of high-frequency sequences of different order in the greenhouse Jurassic carbonate ramps of the Iberian Basin. The comparative review reported here shows that these sequences were recorded differently from shallow to deep ramp areas. A down-dip preferential record of low-frequency (eccentricity) sequences to high-frequency (precession) sequences is a consequence of the interplay between accomodation changes and internal processes controlling sediment production and accummulation above and below wave base levels (fair-weather and storm wave bases).

Comparison with cyclostratigraphic analysis performed in similar greenhouse carbonate ramps indicates that this variable down-dip sequence record is likely to be the rule in this kind of depositional systems. Of particular interest is the heretogeneous nature (e.g., bundles of beds, bed couplets) of the higher-frequency (precession) sequences recorded in middle to outer ramp areas, which can be attributed to the complex interaction of processes and factors controlling sediment production and accumulation in these domains. A well-know sedimentological context of the successions is required to fully understand the potential of these sedimentary settings to record these orbital signals.

More cyclostratigraphic studies focused on the analysis of high-frequency sequences in greenhouse carbonate ramps are required to confirm or refute the proposed down-dip trend on the preservation of orbitally controlled sequences. In any case, data and interpretations included here may be potentially useful for the identification, interpretation and comparison of orbitally controlled sequences in carbonate or mixed siliciclastic-carbonate ramps, developed in greenhouse or icehouse climates.

ACKNOWLEDGMENTS

This work has been supported by the project CGL2017−85038-P subsidized by the Spanish Ministry of Science and Innovation, the European Regional Development Fund and the project E18 of the Government of Aragón (*Aragosaurus: Recursos Geológicos y Paleoambientes*). We are grateful to Claude Colombie and Aitor Payros, whose useful revisions have notably improved the quality of this paper.

REFERENCES

Alnazghah, M.H., Bádenas, B., Pomar, L., Aurell, M., Morsilli, M., 2013. Facies heterogeneity at interwell-scale in a carbonate ramp, Upper Jurassic, NE Spain. Mar. Pet. Geol. 44, 140−163.

Aurell, M., Bádenas, B., 2004. Facies and depositional sequence evolution controlled by high-frequency sea-level changes in a shallow-water carbonate ramp (late Kimmeridgian, NE Spain). Geol. Mag. 141, 717−733.

Aurell, M., Bádenas, B., 2015a. Análisis comparado de secuencias de alta frecuencia en plataformas carbonatadas con subsidencia differential (Sinemuriense, Cordillera Ibérica). Rev. Soc. Geol. Esp. 28 (1), 77−90.

Aurell, M., Bádenas, B., 2015b. Facies architecture of a microbial-siliceous sponge dominated carbonate platform: the Bajocian of Moscardón (Middle Jurassic, Spain). In: Bosence, D.W.J., Gibbons, K.A., Le Heron, D.P., Morgan, W.A., Pritchard, T., Vining, B.A. (Eds.), Microbial Carbonates in Space and Time: Implications for Global Exploration and Production, Geol. Soc. Spec. Pub., 418, pp. 155−174.

Aurell, M., Bádenas, B., Bosence, D.W.J., Waltham, D.A., 1998. Carbonate production and offshore transport on a Late Jurassic carbonate ramp (Kimmeridgian, Iberian basin, NE Spain): evidence from outcrops and computer modeling. In: Wright, V.P., Burchette, T.P. (Eds.), Carbonate Ramps, Geol. Soc. Spec. Pub., 149, pp. 137−161.

Aurell, M., Robles, S., Bádenas, B., Quesada, S., Rosales, I., Meléndez, G., García-Ramos, J.C., 2003. Transgressive/regressive cycles and Jurassic palaeogeography of NE Iberia. Sed. Geol. 162, 239−271.

Aurell, M., Bádenas, B., Ipas, J., Ramajo, J., 2010. Sedimentary evolution of an upper Jurassic carbonate ramp (Iberian Basin, NE Spain). In: van Buchem, F., Gerdes, K., Esteban, M. (Eds.), Reference Models of Mesozoic and Cenozoic Carbonate Systems in Europe and the Middle East − Stratigraphy and Diagenesis, Geol. Soc. Spec. Pub., 329, pp. 87−109.

Aurell, M., Bádenas, B., Gasca, J.M., Canudo, J.I., Liesa, C., Soria, A.R., Moreno-Azanza, M., Najes, L., 2016. Stratigraphy and evolution of the Galve sub-basin (Spain) in the middle Tithonian-early Barremian: implications for the setting and age of some dinosaur fossil sites. Cretac. Res. 65, 138−162.

Bachmann, M., Willems, H., 1996. High frequency cycles in the upper Aptian carbonates of the Organyà basin. NE Spain. Geol. Rundsch. 85, 586−605.

Bádenas, B., Aurell, M., 2001. Proximal-distal facies relationship and sedimentary processes in a storm dominated carbonate ramp (Kimmeridgian, NW of the Iberian Ranges, Spain). Sed. Geol. 139, 319−342.

Bádenas, B., Aurell, M., 2010. Facies models of a shallow-water carbonate ramp based on distribution of non-skeletal grains (Kimmeridgian, Spain). Facies 56, 89−110.

Bádenas, B., Aurell, M., 2008. Kimmeridgian epeiric sea deposits of northeast Spain: sedimentary dynamics of a storm-dominated carbonate ramp. In: Pratt, B.R., Holmden, C. (Eds.), Dynamics of Epeiric Seas, Geol. Assoc. Can., Spec. Paper, 48, pp. 55−72.

Bádenas, B., Aurell, M., Rodríguez Tovar, F.J., Pardo-Izuzquiza, E., 2003. Sequence stratigraphy and bedding rhythms in an outer ramp limestone succession (Late Kimmeridgian, northeast Spain). Sed. Geol. 161, 153−174.

Bádenas, B., Salas, R., Aurell, M., 2004. Three orders of regional sea-level changes control facies and stacking patterns of shallow carbonates in the Maestrat Basin (Tithonian-Berriasian, NE Spain). Int. J. Earth Sci. 93, 144—162.

Bádenas, B., Aurell, M., Gröcke, D.R., 2005. Facies analysis and correlation of high-order sequences in middle-outer ramp successions: variations in exported carbonate in basin-wide $\delta^{13}C_{carb}$ (Kimmeridgian, NE Spain). Sedimentology 52, 1253—1276.

Bádenas, B., Aurell, M., Bosence, B., 2010. Continuity and facies heterogeneities of shallow carbonate ramp cycles (Sinemurian, Lower Jurassic, north-east Spain). Sedimentology 57, 1021—1048.

Berger, A., Loutre, M.F., Laskar, J., 1992. Stability of the astronomical frequencies over the Earth's history for paleoclimate studies. Science 255, 560—566.

Betzler, C., Pfeiffer, M., Saxena, S., 2000. Carbonate shedding and sedimentary cyclicities of a distally steepened carbonate ramp (Miocene, Great Bahama Bank). Int. J. Earth Sci. 89, 140—153.

Bosence, D.W.J., Procter, E., Aurell, M., Bel Kahla, A., Boudagher-Fadel, M., Casaglia, F., Cirilli, S., Mehdie, M., Nieto, L., Rey, J., Scherreiks, R., Soussi, M., Waltham, D., 2009. A tectonic signal in high-frequency, peritidal carbonate cycles? A regional analysis of Liassic platforms from western Tethys. J. Sedim. Res. 79, 389—415.

Boudagher-Fadel, M.K., Bosence, D.W.J., 2007. Early Jurassic benthic foraminiferal diversification and biozones in shallow-marine carbonates of western Tethys. Senckenberg. Lethaea 87, 1—39.

Boulila, S., Galbrun, B., Hinnov, L.A., Collin, P.Y., 2008. High-resolution cyclostratigraphic analysis from magnetic susceptibility in an upper Kimmeridgian (upper Jurassic) marl-limestone succession (La Méouge, Vocontian basin, France). Sed. Geol. 203, 54—63.

Boulila, S., de Rafélis, M., Hinnov, L.A., Gardin, S., Galbrun, B., Collin, P.Y., 2010. Orbitally forced climate and sea-level changes in the Paleoceanic Tethyan domain (marl—limestone alternations, Lower Kimmeridgian, SE France). Palaeogeogr. Palaeoclim. Palaeoecol. 292, 57—70.

Boylan, A., Waltham, D.A., Bosence, D., Bádenas, B., Aurell, M., 2002. Digital rocks: linking forward modelling to carbonate facies. Basin Res. 14, 401—415.

Brandano, M., Corda, L., Tomasseti, L., Testa, D., 2015. On the peritidal cycles and their diagenetic evolution in the lower Jurassic carbonates of the Calcare Massiccio formation (Central Apennines). Geol. Carpath. 66 (5), 393—407.

Brigaud, B., Durlet, C., Deconinck, J.F., Vincent, B., Pucéat, E., Thierry, J., Trouiller, A., 2009. Facies and climate/environmental changes recorded on a carbonate ramp: a sedimentological and geochemical approach on Middle Jurassic carbonates (Paris Basin, France). Sed. Geol. 222, 181—206.

Burgess, P.M., 2006. The signal and the noise: forward modelling of allocyclic and autocyclic processes influencing peritidal stacking patterns. J. Sedim. Res. 76, 962—977.

Calvet, F., Tucker, M.E., Henton, J.M., 1990. Middle Triassic carbonate ramp systems in the Catalan Basin, northeast Spain: facies, systems tracts, sequences and controls. In: Tucker, M.E., Wilson, J.L., Crevello, P.D., Sarg, J.R., Read, J.F. (Eds.), Carbonate Platforms: Facies Sequences and Evolution, Int. Assoc. Sedimentol. Spec. Publ., 9, pp. 79—108.

Carcel, D., Colombié, C., Giraud, F., Courtinat, B., 2010. Tectonic and eustatic control on a mixed siliciclastic-carbonate platform during the Late Oxfordian—Kimmeridgian (La Rochelle platform, western France). Sed. Geol. 223, 334—359.

Colombié, C., Schnyder, J., Carcel, D., 2012. Shallow-water marl limestone alternations in the Late Jurassic of western France: cycles, storm event deposits or both? Sed. Geol. 271—272, 28—43.

Colombié, C., Strasser, A., 2003. Depositional sequences in the Kimmeridgian of the Vocontian Basin (France) controlled by carbonate export from shallow-water platforms. Geobios 36, 675—683.

Colombié, C., Strasser, A., 2005. Facies, cycles and controls on the evolution of a keep-up carbonate platform (Kimmeridgian, Swiss Jura). Sedimentology 52, 1207—1228.

Colombié, C., Bádenas, B., Aurell, M., Götz, A., Bertholon, S., Boussaha, M., 2014. Feature and duration of metre-scale sequences in a storm-dominated carbonate ramp setting (Kimmeridgian, northeast Spain). Sed. Geol. 312, 94—108.

Comas-Rengifo, M.J., Gómez, J.J., Goy, A., Herrero, C., Perilli, N., Rodrigo, A., 1999. El Jurásico Inferior en la sección de Almonacid de la Cuba (sector central de la Cordillera Ibérica, Zaragoza, España). Cuad. Geol. Ibérica 25, 27—57.

Cortés, J.E., Gómez, J.J., Goy, A., 2009. Facies associations, sequence stratigraphy and timing of the earliest Jurassic peak transgression in central Spain (Iberian Range): correlation with other Lower Jurassic sections. J. Iber. Geol. 35 (1), 47—58.

Cunha, A.A.S., Koutsoukos, E.A.M., 2001. Orbital cyclicity in a Turonian sequence of the Cotinguiba formation, Sergipe basin, NE Brazil. Cretac. Res. 22, 529—548.

D'Argenio, B., Fischer, A.G., Premoli Silva, I., Weissert, H., Ferreri, V. (Eds.), 2004. Cyclostratigraphy. Approaches and Case Histories. SEPM Special Publ., 81 (Tulsa, Oklahoma, USA).

Dechamps, S., Da Silva, A.C., Boulvain, F., 2015. Magnetic susceptibility and facies relationship in Bajocian—Bathonian carbonates from the Azé caves, southeastern Paris Basin, France. In: Da Silva, A.C., Whalen, M.T., Hladil, J., Chadimova, L., Chen, D., Spassov, S., Boulvain, F., Devleeschouwer, X. (Eds.), Magnetic Susceptibility Application: A Window onto Ancient Environments and Climatic Variations, Geol. Soc. Spec. Pub., 414, pp. 1—13.

Dera, G., Brigaud, B., Monna, F., Laffont, R., Pucéat, E., Deconinck, J.-F., Pellenard, P., Joachimski, M.M., Durlet, C., 2011. Climatic ups and downs in a disturbed Jurassic world. Geology 39, 215—218.

Dercourt, J., Ricou, L.E., Vrielynck, B., 1993. Atlas: Tethys Palaeoenvironmental Maps. CCGM, Paris.

Dinarès-Turell, Martínez-Braceras, N., Payros, A., 2018. High-resolution integrated cyclostratigraphy from the oyambre section (Cantabria, N Iberian Peninsula): constraints for orbital tuning and correlation of middle Eocene Atlantic deep-sea records. Geochem. Geophys. Geosys. 19, 2017GC007367.

Dromart, G., Allemand, P., Garcia, J.-P., Robin, C., 1996. Variation cyclique de la production carbonatee au Jurassique le long d'un transect Bourgogne-Ardeche, Est-France. Bulletin de la Société Géologique de France 167 (3), 423—433.

Einsele, G., Ricken, W., 1991. Limestones—marls alternations: an overview. In: Einsele, G., Ricken, W., Seilacher, A. (Eds.), Cycles and Events in Stratigraphy. Springer-Verlag, Berlin, pp. 48—62.

Einsele, G., Ricken, W., Seilacher, A., 1991. Cycles and events in stratigraphy—basic concepts and terms. In: Einsele, G., Ricken, W., Seilacher, A. (Eds.), Cycles and Events in Stratigraphy. Springer-Verlag, Berlin, pp. 1—19.

Ellwood, B.B., Balsam, W.L., Roberts, H.H., 2006. Gulf of Mexico sediment sources and sediment transport trends from magnetic susceptibility measurements of surface samples. Mar. Geol. 230, 237—248.

Elrick, M., Hinnov, L.A., 2007. Millennial-scale paleoclimate cycles recorded in widespread Palaeozoic deeper water rhythmites of North America. Palaeogeogr. Palaeoclim. Palaeoecol. 243, 348—372.

Fezer, R., 1988. Die Oberjurassische Karbonatische Regressionsfazies im sudwestlichen Keltiberikum zwischen Griegos und Aras de Alpuente (Prov. Teruel, Cuenca, Valencia, Spanien). Arb. Inst. Geol. Palaont. Univ. Stuttg. 84, 1—119.

Fischer, A.G., D'Argenio, B., Silva, I.P., Weissert, H., Ferreri, V., 2004. Cyclostratigraphic approach to Earth's history: an introduction. In: D'Argenio, B., Fischer, A.G., Premoli Silva, I., Weissert, H., Ferreri, V. (Eds.), Cyclostratigraphy Approaches and Case Histories, SEPM, Spec. Publ., 81, pp. 5—16.

Gil, J., García-Hidalgo, J.F., Mateos, R., Segura, M., 2009. Orbital cycles in a late Cretaceous shallow platform (Iberian ranges, Spain). Palaeogeogr. Palaeoclim. Palaeoecol. 274, 40—53.

Ginsburg, R.N., 1971. Landward movement of carbonate mud: new model for regressive cycles in carbonates (abstract). AAPG 55 (2), 340. Annual Meeting, meeting abstracts.

Goldhammer, R.K., Dunn, P.A., Hardie, L.A., 1990. Depositional cycles, composite sea-level changes, cycle stacking patterns, and the hierarchy of stratigraphic forcing: examples from Alpine Triassic platform carbonates. Geol. Soc. Am. Bull. 102, 535—562.

Gómez, J.J., Fernández-López, S.R., 2006. The Iberian Middle Jurassic carbonate-platform system: synthesis of the palaeogeographic elements of its eastern margin (Spain). Palaeogeogr. Palaeoclim. Palaeoecol. 236, 190—205.

Gómez, J.J., Goy, A., 2005. Late Triassic and early Jurassic palaeogeographic evolution and depositional cycles of the Western Tethys Iberian platform system (Eastern Spain). Palaeogeogr. Palaeoclim. Palaeoecol. 222, 77—94.

Götz, A., Török, A., 2008. Correlation of Tethyan and Peri-Tethyan long-term and high-frequency eustatic signals (Anisian, middle Triassic). Geol. Carpath. 59 (4), 307—317.

Gradstein, F.M., Ogg, J.G., Smith, A.G., 2004. A Geological Time Scale 2004. Cambridge University Press.

Gradstein, F.M., Ogg, J.G., Schmitz, M.D., Ogg, G.M., 2012. The Geologic Time Scale 2012. Elsevier, Oxford.

Guo, C., Daizhao, C., Song, Y., Zhou, X., Ding, Y., Zhang, G., 2018. Depositional environments and cyclicity of the early Ordovician carbonate ramp in the western Tarim Basin (NW China). J. Asian Earth Sci. 158, 29—48.

Hardenbol, J., Thierry, J., Farley, M.B., Jacquin, T., De Gracianski, P.C., Vail, P.R., 1998. Mesozoic and Cenozoic sequence chronostratigraphic framework of European basins. In: De Gracianski, P.C., Hardenbol, J., Jacquin, T., Vail, P.R. (Eds.), Mesozoic and Cenozoic Sequence Stratigraphy of European Basins, SEPM Spec. Public., 60 charts 1—8.

Hebbeln, D., Samankassou, E., 2015. Where did ancient carbonate mounds grow—In bathyal depths or in shallow shelf waters? Earth-Sci. Rev. 145, 56—65.

Herbert, T.D., 1994. Reading orbital cycles distorted by sedimentation: models and examples. IAS Spec. Publ. 19, 483—507.

Hilgen, F.J., Schwarzacher, W., Strasser, A., 2004. Concepts and definitions in cyclostratigraphy (second report of the cyclostratigraphy working group). SEPM. Spec. Publ. 81, 303—305.

Holz, M., 2015. Mesozoic paleogeography and paleoclimates — a discussion of the diverse greenhouse and hothouse conditions of an alien world. J. South Am. Earth Sci. 61, 91—107.

Husinec, A., Read, J.F., 2007. The late Jurassic Tithonian, a greenhouse phase in the middle Jurassic—early Cretaceous "cool" mode: evidence from the cyclic Adriatic platform, Croatia. Sedimentology 54, 317—337.

Jones, B., Desrochers, A., 1992. Shallow platform carbonates. In: Walker, R.G., James, N.P. (Eds.), Facies Models: Response to Sea Level Change, Geol. Assoc. Can. St. John's, pp. 277—301.

Kemp, D.B., Sadler, P.M., 2014. Climatic and eustatic signals in a global compilation of shallow marine carbonate accumulation rates. Sedimentology 61, 1286—1297.

Kietzmann, D.A., Palma, R.M., Iglesia Llanos, M.P., 2015. Cyclostratigraphy of an orbitally-driven Tithonian—Valanginian carbonate ramp succession, southern Mendoza, Argentina: implications for the Jurassic—Cretaceous boundary in the Neuquén basin. Sed. Geol. 315, 29—46.

Lambeck, K., Esat, T.M., Potter, E.K., 2002. Links between climate and sea level for the past three million years. Nature 419 (6903), 199–206.

Laskar, J., Robutel, P., Joutel, F., Gastineau, M., Correia, A.C.M., Levrard, B., 2004. A long term numerical solution for insolation quantities of the Earth. Astron. Astrophys. 428 (1), 261–285.

Laya, J.C., Tucker, M.E., Pérez-Huerta, A., 2013. Metre-scale cyclicity in Permian ramp carbonates of equatorial Pangea (Venezuelan Andes): implications for sedimentation under tropical Pangea conditions. Sed. Geol. 292, 15–35.

Lee, Y., Hyeong, K., Yoo, C.H., 2001. Cyclic sedimentation across a middle ordovician carbonate ramp (Duwibong formation), Korea. Facies 44, 61–74.

Leinfelder, R.R., Schmid, D.U., 2000. Mesozoic reefal thrombolites and other microbialites. In: Riding, R.E., Awramik, S.M. (Eds.), Microbial Sediments. Springer-Verlag, Berlin, pp. 289–294.

Leinfelder, R.R., Schmid, D.U., Nose, M., Werner, W., 2002. Jurassic reef patterns-The expression of a changing globe. In: Kiessling, W., Flügel, E., Golonka, J. (Eds.), Phanerozoic Reef Patterns, SEPM Spec. Public, 72, pp. 465–520.

Long, D.G.F., 2007. Tempestite frequency curves: a key to late Ordovician and early Silurian subsidence, sea-level change, and orbital forcing in the Anticosti foreland basin, Quebec, Canada. Can. J. Earth Sci. 44, 413–431.

Martínez-Braceras, N., Payros, A., Miniati, F., Arostegi, J., Franceschetti, G., 2017. Contrasting environmental effects of astronomically driven climate change on three Eocene hemipelagic successions from the Basque-Cantabrian Basin. Sedimentology 64 (4), 960–986.

Mitchum, R.M., Van Wagoner, J.C., 1991. High-frequency sequences and their stacking patterns: sequence-stratigraphic evidence of high-frequency eustatic cycles. Sed. Geol. 70 (2), 131–160.

Munnecke, A., Westphal, H., 2004. Shallow-water aragonite recorded in bundles of limestone-marl alternations — the Upper Jurassic of SW Germany. Sed. Geol. 64, 191–202.

Nose, M., 1995. Vergleichende Faziesanalyse und Palökologie koral- lenreicher VerXachungsabfolgen des Iberischen Oberjura. Pro Wl 8, 1–237.

Ogg, J.G., Ogg, G., Gradstein, F.M., 2016. A Concise Geologic Time Scale. Elsevier B.V.

Olivier, N., Pittet, B., Werner, W., Hantzpergue, P., Gaillard, G., 2008. Facies distribution and coral-microbialite reef development on a low-energy carbonate ramp (Chay Peninsula, Kimmeridgian, western France). Sed. Geol. 205, 14–33.

Osete, M.L., Gómez, J.J., Pavón-Carrasco, F.J., Villalaín, J.J., Palencia, A., Ruíz Martínez, V.C., Heller, F., 2011. The evolution of the Iberia during the Jurassic from paleomagnetic data. Tectonophysics 502, 105–120.

Osleger, D., 1991. Subtidal carbonate cycles: implications for allocyclic vs. autocyclic controls. Geology 19, 917–920.

Osleger, D., Read, D.V., 1991. Relation of eustasy to stacking patterns of meter scale carbonate cycles, Late Cambrian U.S.A. J. Sediment. Pet. 61 (7), 1225–1252.

Pawellek, T., Aigner, T., 2003a. Stratigraphic architecture and gamma ray logs of deeper ramp carbonates (Upper Jurassic, SW Germany). Sed. Geol. 159, 203–240.

Pawellek, T., Aigner, T., 2003b. Apparently homogenous "reef"-limestones built by high-frequency cycles: Upper Jurassic, SW-Germany. Sed. Geol. 160, 259–284.

Payros, A., Dinarès-Turell, J., Monechi, S., Orue-Etxebarria, X., Ortiz, S., Apellaniz, E., Martínez-Braceras, N., 2015. The Lutetian/Bartonian transition (middle Eocene) at the Oyambre section (northern Spain): implications for standard chronostratigraphy. Palaeogeogr. Palaeoclimatol. Palaeoecol. 440, 234–248.

Pomar, L., Hallock, P., 2008. Carbonate factories: a conundrum in sedimentary geology. Earth Sci. Rev. 87, 134–169.

Pomar, L., Haq, B.U., 2016. Decoding depositional sequences in carbonate systems: concepts vs experience. Glob. Planet. Change 146, 190–225.

Pomar, L., Morsilli, M., Hallock, P., Bádenas, B., 2012. Internal waves, an under-explored source of turbulence events in the sedimentary record. Earth Sci. Rev. 111, 56—81.

Pratt, B.R., James, N.P., Cowan, C.A., 1992. Peritidal carbonates. In: Walker, R.G., James, N.P. (Eds.), Facies Models: Response to Sea Level Change, Geol. Assoc. Can. St. John's, pp. 303—322.

Preto, N., Hinnov, L.A., 2003. Unravelling the origin of shallow-water cyclothems in the Upper Triassic Dürrenstein Fm. (Dolomites, Italy). J. Sediment. Res. 73, 774—789.

Price, G.D., 1999. The evidence and implications of polar ice during the Mesozoic. Earth Sci. Rev. 48, 183—210.

Pross, J., Link, E., Ruf, M., Aigner, T., 2006. Delineating sequence stratigraphic patterns in deeper ramp carbonates: quantitative palynofacies data from the Upper Jurassic (Kimmeridgian) of Southwest Germany. J. Sediment. Res. 76, 524—538.

Read, J.F., 1982. Carbonate platforms of passive (extensional) continental margins: types, characteristics and evolution. Tectonophysics 81, 195—212.

Read, J.F., Osleger, D., Elrick, M., 1991. Two-dimensional modeling of carbonate ramp sequences and component cycles. In: Franseen, E.K., Watney, W.L., Kendall, C.G.S.C., Ross, W. (Eds.), Sedimentary Modeling: Computer Simulations and Methods for Improved Parameter Definition, Kansas Geol. Surv. Bull., 233, pp. 473—488.

Ruf, M., Link, E., Pross, J., Aigner, T., 2005. Integrated sequence stratigraphy: Facies, stable isotope and palynofacies analysis in a deeper epicontinental carbonate ramp (Late Jurassic, SW Germany). Sed. Geol. 175, 391—411.

Ruhl, M., Hesselbo, S., Hinnov, L., Jenkyns, H., Xu, W., Riding, J., Storm, M., Minisini, D., Ullmann, C., Leng, M., 2016. Astronomical constraints on the duration of the Early Jurassic Pliensbachian stage and global climatic fluctuations. Earth Planet. Sci. Lett. 455, 149—165.

Salas, R., Guimerà, J., Mas, R., Martín-Closas, C., Meléndez, A., Alonso, A., 2001. Evolution of the Mesozoic central Iberian Rift System and its Cainozoic inversion (Iberian chain). In: Ziegler, P.A., Cavazza, W., Robertson, A.H.F., Crasquin-Soleau, S. (Eds.), Peri-tethys Memoir 6: Peri-tethyan Rift/Wrench Basins and Passive Margins, Paris, Memoir. Mus. Natl. Hist., 186, pp. 145—185.

Sames, B., Wagreich, M., Wendler, J.E., Haq, B.U., Conrad, C.P., Melinte-Dobrinescu, M.C., Hu, X., Wendler, I., Wolfgring, E., Yilmaz, I.O., Zorina, S.O., 2016. Short-term sea-level changes in a greenhouse world — A view from the Cretaceous. Palaeogeogr. Palaeoclimatol. Palaeoecol. 441, 393—411.

San Miguel, G., Aurell, M., Bádenas, B., 2017. Occurrence of high-diversity metazoan- to microbial-dominated bioconstructions in a shallow Kimmeridgian carbonate ramp (Jabaloyas, Spain). Facies 63, 1—21.

Schalaich, M., Aigner, T., 2017. Facies and integrated sequence stratigraphy of an Epeiric carbonate ramp succession: Dhruma formation, Sultanate of Oman. Depositional. Rec. 3 (1), 92—132.

Schlager, W., 2010. Ordered hierarchy versus scale invariance in sequence stratigraphy. Int. J. Earth Sci. 99 (Suppl 1), S139—S151.

Schwarzacher, W., Fischer, A.G., 1982. Limestone—shale bedding and perturbations of the Earth's orbit. In: Einsele, G., Seilacher, A. (Eds.), Cyclic and Event Stratification. Springer-Verlag, Berlin, pp. 72—95.

Sequero, C., Bádenas, B., Muñoz, A., 2017. Sedimentología y cicloestratigrafía de las calizas de plataforma abierta de la Fm. Río Palomar (Pliensbachiense inferior; Cuenca Ibérica). Rev. Soc. Geol. Esp. 30 (1), 71—84.

Spencer, G.H., Tucker, M.A., 2007. A proposed integrated multisignature model for peritidal cycles in carbonates. J. Sedimen. Res. 77, 797—808.

Stanton, R.J., 2006. Nutrient models for the development and location of ancient reefs. Geol. Alp. 3, 191—206.

Strasser, A., Samankassou, E., 2003. Carbonate Sedimentation Rates Today and in the Past: Holocene of Florida Bay, Bahamas, and Bermuda vs. Upper Jurassic and Lower Cretaceous of the Jura Mountains (Switzerland and France). Geol. Croat. 56 (1), 1—18.

Strasser, A., Védrine, S., 2009. Controls on facies mosaics of carbonate platforms: a case study from the Oxfordian of the Swiss Jura. IAS Spec. Publ. 41, 199—213.

Strasser, A., Pittet, B., Hillgärtner, H., Pasquier, J.B., 1999. Depositional sequences in shallow carbonate-dominated sedimentary systems: concepts for a high-resolution analysis. Sed. Geol. 128, 201—221.

Strasser, A., Hilgen, F.J., Heckel, P.H., 2006. Cyclostratigraphy—concepts, definitions, and applications. Newsl. Stratigr. 42 (2), 75—114.

Strasser, A., Vedrine, S., Stienne, N., 2012. Rate and synchronicity of environmental changes on a shallow carbonate platform (Late Oxfordian, Swiss Jura Mountains). Sedimentology 59, 185—211.

Tresch, J., Strasser, A., 2011. Allogenic and autogenic processes combined in the formation of shallow-water carbonate sequences (Middle Berriasian, Swiss and French Jura Mountains). Swiss J. Geosci. 104 (2), 299—322.

Tucker, M.E., Garland, J., 2010. High-frequency cycles and their sequence stratigraphic context: orbital forcing and tectonic controls on Devonian cyclicity. Belg. Geol. Belg. 13 (3), 213—240.

Vail, P.R., Audemard, F., Bowman, S.A., Eisner, P.N., Pérez-Cruz, C., 1991. The stratigraphic signatures of tectonics, eustasy and sedimentology—an overview. In: Einsele, G., Ricken, W., Seilacher, A. (Eds.), Cycles and Events in Stratigraphy. Springer-Verlag, Berlin, pp. 617—659.

Val, V., Bádenas, B., Aurell, M., Rosales, I., 2017. Cyclostratigraphy and chemostratigraphy of a bioclastic storm-dominated carbonate ramp (late Pliensbachian, Iberian Basin). Sed. Geol. 355, 93—113.

van de Schootbrugge, B., McArthur, J.M., Bailey, T.R., Rosenthal, Y., Wright, J.D., Miller, K.G., 2005. Toarcian oceanic anoxic event: An assessment of global causes using belemnite C isotope records. Paleoceanography 20, PA3008.

Waltham, D., 2015. Milankovitch period uncertainties and their impact on Cyclostratigraphy. J. Sediment. Res. 85, 990—998.

Weedon, G.P., 2003. Time-series Analysis and Cyclostratigraphy. Cambridge Univ. Press, Cambridge.

Westphal, H., Halfar, J., Freiwald, A., 2010. Heterozoan carbonates in subtropical to tropical settings in the present and past. Int. J. Earth Sci. 99, 153—169.

Wu, H., Zhang, S., Feng, Q., Jiang, G., Li, H., Yang, T., 2012. Milankovitch and sub-Milankovitch cycles of the early Triassic Daye Formation, South China and their geochronological and paleoclimatic implications. Gondwana Res. 22, 748—759.

Yang, W., Lehrmann, D.J., Hu, X.F., 2014. Peritidal carbonate cycles induced by carbonate productivity variations: A conceptual model for an isolated Early Triassic greenhouse platform in South China. Journal of Palaeogeography 3 (2), 115—126.

Zachos, J., Pagani, M., Sloan, L., Thomas, E., Billups, K., 2001. Trends, rhythms, and aberrations in global climate 65 Ma to present. Science 292, 686—693.

CHAPTER EIGHT

Astronomical Calibration of the Tithonian — Berriasian in the Neuquén Basin, Argentina: A Contribution From the Southern Hemisphere to the Geologic Time Scale

Diego A. Kietzmann[*,§,1], **Maria Paula Iglesia Llanos**[*,§] and **Melisa Kohan Martínez**[*,§]

[*]Universidad de Buenos Aires, Facultad de Ciencias Exactas y Naturales, Departamento de Ciencias Geológicas, Ciudad Universitaria, Pabellón 2, Ciudad Autónoma de Buenos Aires, Argentina
[§]CONICET-Universidad de Buenos Aires, Instituto de Geociencias Básicas, Ambientales y Aplicadas de Buenos Aires (IGeBA), Ciudad Universitaria, Pabellón 2, Ciudad Autónoma de Buenos Aires, Argentina
[1]Corresponding author: E-mail: diegokietzmann@gl.fcen.uba.ar

Contents

Abstract

Detailed cyclostratigraphical analyses have been made from five Tithonian–Berriasian sections of the Vaca Muerta Formation, exposed in the Neuquén Basin, Argentina. The Vaca Muerta Formation is characterized by decimetre-scale rhythmic alternations of

Stratigraphy & Timescales, Volume 3
ISSN 2468-5178
https://doi.org/10.1016/bs.sats.2018.07.003

marlstones and limestones, showing a well-ordered hierarchy of cycles, where elementary cycles, bundles of cycles and superbundles have been recognized. According to biostratigraphic data, elementary cycles have a periodicity of ~21 ky, which correlates with the precession cycle of Earth's axis. Spectral analysis based on time series of elementary cycle thicknesses allows us to identify frequencies of ~400 ky and ~90–120 ky, which we interpret as the modulation of the precessional cycle by the Earth's orbital eccentricity. Correlation between studied sections allowed us to estimate a minimum duration for each Andean ammonite zone. Moreover, cyclostratigraphic data allowed us to build the first continuous floating astronomical time scale for the Tithonian — Berriasian, which is anchored to the geological time scale through magnetostratigraphy. We estimated a minimum duration of 5.67 myr for the Tithonian and 5.27 myr for the Berriasian. The resulted durations of some polarity chrones are also different with respect to the GTS2016, however such differences could be due to condensation or discontinuities not detected in the studied sections.

1. INTRODUCTION

In contrast to most geological systems, the Jurassic/Cretaceous transition is characterized by the absence of a significant faunal turnover, as well as by the remarkable increase of faunal provincialism. The uncertainties in the inter-regional biostratigraphic correlation constitute a major classic problem for this time interval across the world, and absolute data are still scarce and conflicting with each other (e.g., Remane, 1991; Wimbledon, 2008; Wimbledon et al., 2011, 2013; Pálfy et al., 2000; Pálfy, 2008; Vennari et al., 2014; Wimbledon, 2017). Cyclostratigraphy has become an important tool in measuring Jurassic - Cretaceous geologic time and establishing floating astronomical time scales (Ogg and Hinnov, 2012a; b). Almost the entire Jurassic and Cretaceous have been calibrated using astronomical cycles, especially the long-term and short-term excentricity cycle. However, the Tithonian (uppermost Upper Jurassic) remains still somewhat uncertain, and should be even carefully reviewed and compared with magnetostratigraphy and absolute dating (e.g., Pálfy et al., 2000; Pálfy, 2008; Huang et al., 2010; Ogg and Hinnov, 2012a).

Many important works involving the Tithonian — Berriasian were dedicated to demonstrate the presence of Milankovitch cycles in the sedimentary record; among them, it is important to mention the contributions of Molinie and Ogg (1992) for the Callovian–Valanginian of North Pacific, Melnyk et al. (1994) for the Kimmeridgian-Portlandian of the Wessex Basin, Park and Fürsich (2001) for the Tithonian of Solnhofen (Germany), Husinec and Read (2007) for the Tithonian of the Adriatic platform, and Kietzmann et al. (2011) for the Tithonian of the Neuquén Basin. Nevertheless, only a few works have carried out detailed cyclostratigraphic studies, as in the case of Strasser (1994), Strasser et al. (2004), Strasser and Hillgärtner (1998) for the Berriasian of the Jura Mountains, Weedon et al. (1999, 2004) and Huang et al. (2010) for the Kimmeridge Clay Formation, and Kietzmann et al. (2015) for the Tithonian — Berriasian in the Neuquén Basin.

The Tithonian — Berriasian interval in the Neuquén Basin (Fig. 1), is represented by the Vaca Muerta Formation, a thick rhythmic alternation of dark bituminous shales, marlstones and limestones deposited as result of a rapid and widespread Paleo-pacific marine transgression (Legarreta and Uliana, 1991, 1996). It is famous for its important oil and gas resources, as well as its rich fossil content and temporal continuity along several hundreds of meters thick. This unit contains nine Andean ammonite zones with an excellent biostratigraphic resolution, but their autochthony prevents a straightforward correlation with the Tethys, and prompts the occurrence of different correlation schemes between Andean and Tethyan zones (e.g., Leanza, 1981, 1996; Aguirre Urreta, 2001; Riccardi, 2008, 2015; Vennari et al., 2014). Different research groups based on diverse stratigraphic disciplines have recently attempted to fix the Jurassic - Cretaceous boundary in the Neuquén Basin (Vennari et al., 2014; Kietzmann et al., 2015; Iglesia Llanos et al., 2017; Ivanov and Kietzmann, 2017; Kietzmann, 2017; López Martinez et al., 2018; Kietzmann and Iglesia Llanos, 2018), but there are still many inconsistencies between them and the problem can only be solved with more detailed multidisciplinary studies (Kietzmann et al., 2018). At the moment the most conclusive data for this time interval in the Neuquén Basin are provided by magnetostratigraphy (Iglesia Llanos et al., 2017; Kohan Martínez et al., 2018). The aim of this work is to contribute to the calibration of the Tithonian - Berriasian using high resolution cyclostratigraphic data.

2. GEOLOGICAL SETTING

2.1 Neuquén Basin

The Neuquén Basin was a retro-arc basin developed in Mesozoic times in the Pacific margin of South America (Fig. 1A), where different tectonic regimes exerted a first-order control in basin development and sedimentary evolution (Legarreta and Uliana, 1991, 1996). An extensional

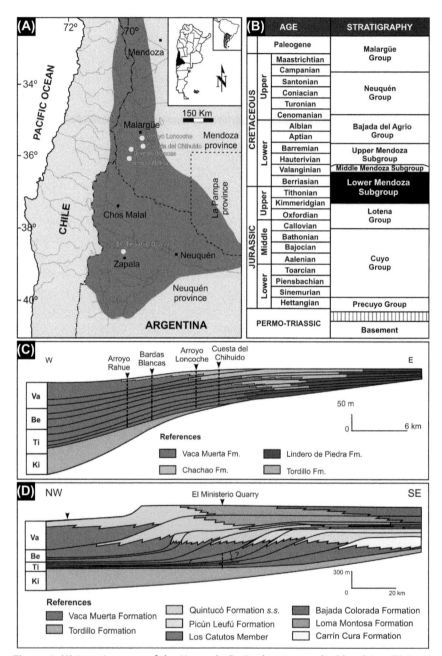

Figure 1 (A) Location map of the Neuquén Basin showing studied localities, (B) Stratigraphic chart for the Neuquén Basin; (C–D) Litostratigraphic subdivision of the Lower Mendoza Group in the Southern Mendoza area (C), and Central Neuquén area (D), showing the location of the studied sections.

regime was established during Late Triassic—Early Jurassic, which generated a series of narrow, isolated depocentres controlled by large transcurrent fault systems filled mainly with continental deposits of the Precuyano Cycle (Fig. 1B, Manceda and Figueroa, 1993; Vergani et al., 1995). Thermal subsidence with localized tectonic events characterized the Early Jurassic to Late Cretaceous interval (Vergani et al., 1995). Depocentres were filled by continental and marine siliciclastic, carbonate and evaporitic sediments (Cuyo, Lotena, and Mendoza Groups). Marine sequences developed throughout the basin during Late Jurassic—Early Cretaceous, are included in the Mendoza Group (Legarreta and Gulisano, 1989). Finally, a compressive deformation regime was established during the Late Cretaceous, and continued throughout the Cenozoic, although alternating with extensional events (Ramos and Folguera, 2005; Ramos, 2010).

2.2 Lower Tithonian — Lower Valanginian Interval

The Mendoza Group was divided into three main shallowing-upward sedimentary cycles: Lower Mendoza Subgroup (Lower Tithonian—Lower Valanginian), Middle Mendoza Subgroup (Lower Valanginian), and Upper Mendoza Subgroup (Lower Valanginian — Lower Barremian) (Legarreta and Gulisano, 1989; Leanza, 2009). The Lower Mendoza Subgroup corresponds to a Lower Tithonian — Lower Valanginian shallowing-upward carbonate ramp system (Fig. 1C and D), in which most distal facies are included into the Vaca Muerta Formation.

Biostratigraphy is well defined based on ammonites (Riccardi, 2008; Riccardi et al., 2011; Aguirre-Urreta et al., 2011). Microfossils, such as calcareous nannofossils, radiolarians, organic and calcareous dinoflagellates, calpionellids and saccocomids (Quattrocchio et al., 2003; Bown and Concheyro, 2004; Ballent et al., 2004, 2011; Kietzmann and Palma, 2009; Ivanova and Kietzmann, 2017; Kietzmann, 2017; López Martinez et al., 2018), still need thorough reviews to accurately establish the stratigraphic position of the most important bioevents (Fig. 2). Detailed micropaleontological studies in the southern Mendoza area of the Neuquén Basin reveal a relatively rich micropaleontological assemblage of 24 known species of calcareous dinoflagellate cysts (Ivanova and Kietzmann, 2017), that allowed to distinguish 6 calcareous dinoflagellate cyst zones, including the Early Tithonian *Carpistomiosphaera tithonica* to the Early Valanginian *Carpistomiosphaera valanginiana* Zones. Likewise, 18 poorly preserved known species of calpionellids were identified by Kietzmann (2017), allowing the determination of the *Chitinoidella* and *Crassicollaria* calpionellid Standard

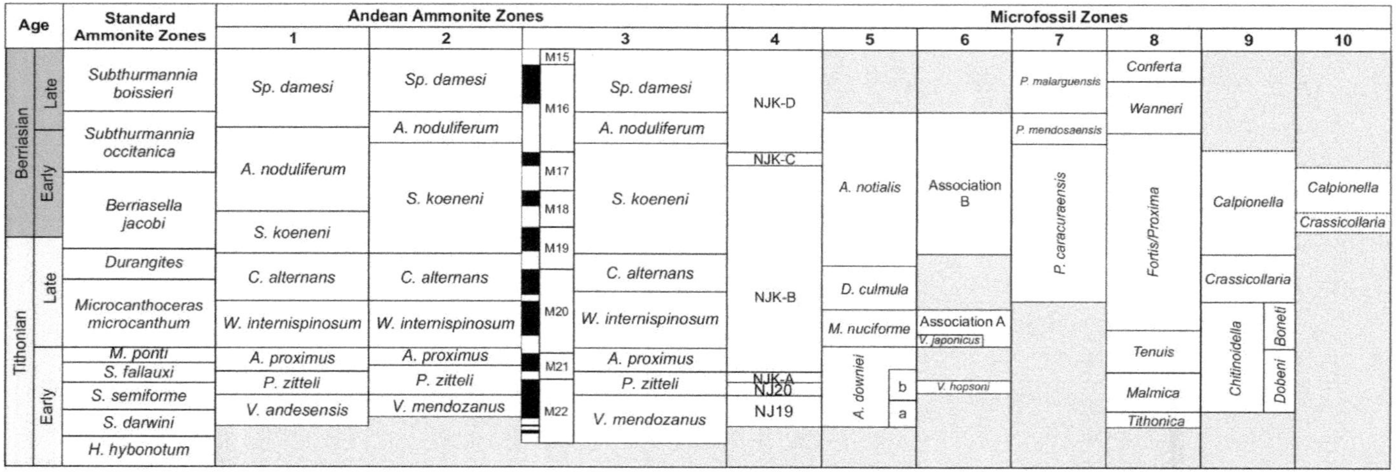

Figure 2 Biostratigraphic subdivision of the Tithonian – Berriasian in the Neuquén Basin. (1–3) Correlation of Andean ammonite zones with ammonite standard zones: (1) biostratigraphic correlation according to Vennari et al. (2014) and Vennari (2016), (2) biostratigraphic correlation according to Riccardi (2015) (3) magnetostratigraphic calibration after Iglesia Llanos et al. (2017). (4–9), Distribution of microfossil zones according to their coretaliton to Andean ammonite zones (in this case we have used the magnetostratigrahic calibration, which is the one that presents lower uncertainties and correlates quite well with Riccardi's (2015) biostratigraphic proposal): (4) Nannofossil zones after Ballent et al. (2011); (5) Organic-walled dinoflagellates zones after Volkheimer et al. (2011); (6) radiaolaria associations after Ballent et al. (2011); (7) Microcoprolite associations after Kietzmann and Palma (2014); (8) Calcareous dinoflagellate zones after Ivanova and Kietzmann (2017) and Kietzmann et al. (2018); (9) calpionellid zones after Kietzmann (2017) and Kietzmann et al. (2018); (10) calpionellid zones López Martínez et al. (2017) correlated with Vennari' et al. (2014) ammonite zones.

Zones (Fig. 2). Andean Ammonite zones are mostly assemblage zones; therefore the boundary between zones is established as the first occurrence of species of the association (Fig. 3).

Magnetostratigraphy published by Iglesia Llanos et al. (2017) recognized 11 reverse and 10 normal polarity zones, which were calibrated based on the correlation between Andean and Tethyan ammonite zones, and the Geomagnetic Polarity Time Scale (GPTS) compiled by Ogg and Hinnov (2012a, b), indicating that the deposition of the Vaca Muerta Formation took place during the M22r.2r to M15r Subchrons. From base to top, the *Virgatosphinctes andesensis* (ex *mendozanus*) Zone comprises a set of reverse, normal, reverse and normal polarities, which we interpret to span the M22r.2r to M22n Subchrons in the GPTS. The following *Pseudolissoceras zitteli* Zone bears normal, reverse and normal polarities that would correspond to M22n to the base of M21n Subchrons. The *Aulacosphinctes proximus* Zone comprises a set of normal and reverse polarities that are correlated with M21n to M20r Subchrons. The *Windhauseniceras internispinosum* Zone bears a reverse and normal polarity that is correlated with the M20r and M20n.2n Subchron. Above, the *Corongoceras alternans* Zone includes normal, reverse, normal, reverse and normal polarities which are interpreted to correspond to the upper M20n.2n to M19n Subchrons. The *Substeueroceras koeneni* Zone comprises normal, reverse, normal, reverse, normal, reverse, normal and reverse polarities that are correlated with M19n to M16r Subchrons. Further above, the *Argentiniceras noduliferum* Zone includes a dominant reverse with a minor opposite polarity, which is correlated with M16r Subchron. Finally, the *Spiticeras damesi* Zone comprises reverse, normal and reverse polarities that are interpreted to correspond to M16r to M15r Subchrons (Fig. 2).

Sequence stratigraphic framework was established by Kietzmann et al. (2014a, b), who recognized two hierarchies of depositional sequences: (1) Composite depositional sequences (CS) for large-scale sequences, and (2) high-frequency depositional sequences (HFS) for those of small scale. Because of their average duration, composite sequences are considered to be equivalent to third order sequences, while high-frequency sequences are considered as fourth order sequences. In this paper we will consider only composite sequences, because it is important to eliminate these trends for spectral analysis. High-frequency sequences are too close to the super-bundle hierarchy, and therefore their removal could generate fictitious frequencies. The studied interval includes 4 trasgressive-regressive composite depositional sequences. The first composite depositional sequence (CS-1) starts with a regional flooding surface at the boundary between the Vaca

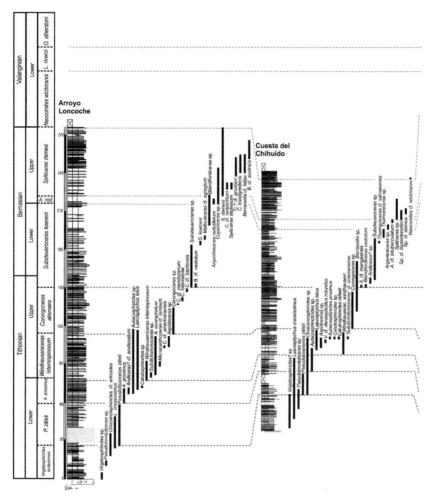

Figure 3 Ammonite distribution and zone correlation between the five studied sections.

Muerta Formation and the continental facies of the Tordillo Formation, in the Lower Tithonian *Virgatosphinctes andesenis* Zone. The upper boundary is marked by other regional flooding surface (SB-1) located within the lowermost Upper Tithonian *W. internispinosum* Zone, which is also the base of the second composite depositional sequence (CS-2). Sequence boundary SB-2 lies within the uppermost Upper Tithonian *S. koeneni* Zone. Sequence boundary SB-3 that separates composite depositional sequences CS-3 and CS-4 coincides with the lower part of the *S. damesi* Zone (Upper Berriasian), while SB-4 lies in the upper part of this zone.

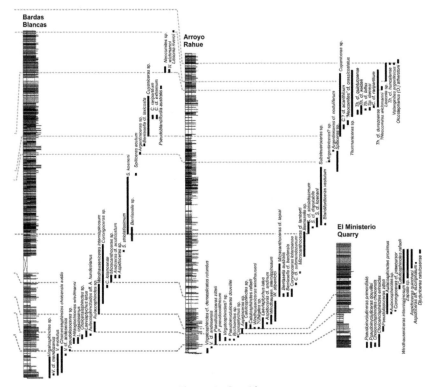

Figure 3 Cont'd

3. METHODOLOGY

Five detailed sedimentological sections from the Vaca Muerta Formation were selected from more than 30 measured sections described bed-by-bed in the Neuquén Basin. They include the Arroyo Loncoche (~290 m), Cuesta del Chihuido (~185 m), Bardas Blancas (~230 m), Arroyo Rahue (~340 m), and El Ministerio Quarry (~100 m) sections (Figs. 1 and 7). Detailed facies and sequence stratigraphic analysis of these sections were published by Kietzmann et al. (2008, 2011, and 2014a, b). This study differs from that of Kietzmann et al. (2015) in three respects. Firstly, during paleomagnetic sampling (Iglesia Llanos et al., 2017) we corrected some stratigraphic intervals due to structural problems that were not considered in the previous work. Secondly, new findings of ammonite allowed correcting the boundaries between some ammonite zones. Finally, the addition of a new section with an expanded Tithonian interval (El Ministerio Quarry; Kohan Martinez et al., 2018) has allowed reinterpreting the cyclostratigraphy of Kietzmann et al. (2015).

Cyclostratigraphic analysis is based on the differentiation of dm-scale carbonate/siliciclastic lithofacies couplets or elementary cycles, bundles and superbundles differentiated in the field. Time series were constructed using elementary cycle thickness. Lower frequencies are searched using spectral analysis with the software POWGRAF2 (Pardo-Igúzquiza and Rodríguez-Tovar, 2004) using Blackman-Tukey method. Spectral analyses are based in the following primary premises: (1) the sections consist of a succession of elementary cycles showing similar thickness within the range of several decimetres, and they are considered of similar time duration; (2) the number of cycles is large enough to obtain statistically significant results. According to Weedon (2003) the sampling density must be at least 12 times the lowest frequency sought. In this case is at least 25 times longer than the low-frequency eccentricity cycle; and (3) vertical changes in facies and bed thickness throughout the sections could be related with periodic climate factors or with other non-periodic factors as basin evolution.

Data set were corrected prior to spectral analysis, substracting the mean value and trends generated by changes in sea level (composite depositional sequences in Kietzmann et al., 2014a,b, 2016a,b), allowing data centering and variance stabilization. Poorly-cyclic intervals are not incorporated into the spectral analysis. However, these "non-cyclic" intervals need to be also considered for interpreting the long term variations as well as for estimations of the total number of cycles in the stratigraphic sections. In fact, Kohan Martínez et al. (2018), using time series of magnetic susceptibility demonstrate that "non-cyclic" intervals contain a similar cyclic pattern.

4. CYCLOSTRATIGRAPHIC ANALYSIS

4.1 Hierarchy of Cycles

A well-ordered hierarchy of cycles, including elementary cycles, bundles and superbundles has been recognized within the Vaca Muerta Formation. Elementary cycles have a relatively regular thickness in the order of 20 to 40 cm, so that they can be regarded as temporarily equivalent units. Three types of elementary cycles are recognized: limestone/marlstone, marlstone/marlstone and marlstone/mudstone (Figs. 4 and 5).

Each elementary cycle consists of two hemicycles of similar thickness. Limestone/marlstone cycles begin with a limestone hemicycle, wich pass transitionally to a marlstone hemicycle. On unweathered surface, limestones have a net basal boundary and show enrichment in clays upward, resulting in a transitional passage to marlstones. The sedimentological feature of each hemicycle varies according to the facies in which occurs. Marlstone/marlstone and

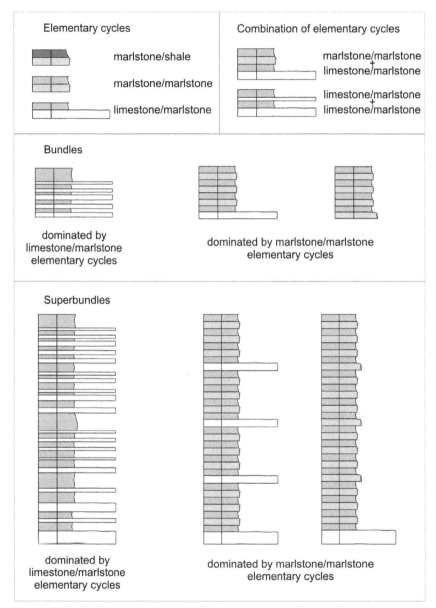

Figure 4 Hierachy and types of cycles in the Vaca Muerta Formation.

marlstone/mudstone elementary cycles show variations in the content of carbonate, allowing detecting them in weathering profile. Elementary cycles are grouped into sets of 4—5 elementary cycles (bundles), and these are grouped into sets of 4—5 bundles (superbundles). Such stacking pattern is a diagnostic criterion to identify the influence of orbital forcing

Figure 5 (A) Outcrops example of a cyclic interval dominated by limestone/marlstone elementary cycles in Cuesta del Chihuido section (Upper Tithonian). (B) Outcrops example of a cyclic interval dominated by marlstone/marlstone elementary cycles in Arroyo Rahue section (Upper Berriasian).

(e.g., Goldhammer et al., 1990; Schwarzacher, 1993; Lehmann et al., 1998; Raspini, 2001; Anderson, 2004; Strasser et al., 2004, among others). The 5:1 ratio (5 elementary cycles per bundles) is commonly attributed to high-frequency eccentricity (95 ky and 125 ky), while the 4:1 ratio (4 bundles per superbundles) as low-frequency eccentricity (405 ky).

Bundles and superbundles start with a thick elemental cycle and show higher proportion of marls towards the top. This characteristic stacking pattern is clearly shown on the field in stratigraphic sections with low sedimentation rate, such as the Cuesta del Chihuido section (Fig. 5A). Bundles and superbundles can be dominated by marlstone/marlstone elementary cycles; so that the proportion of marlstone exceeds the proportion of limestones, or be dominated by limestone/marlstone elementary cycles, i.e., the proportion of limestones is greater or similar to the proportion of marlstones (Figs. 4 and 5). This descriptive division correlates with facies and system tracts, so that variations in the proportion of marlstones or limestones have no real genetic connotation respect to the mechanisms which transfers the orbital signal to the sedimentary record, but with the proximal–distal trend within the carbonate ramp.

Basinal to outer ramp deposits are dominated by marlstone/marlstone elementary cycles. Bundles or superbundles can occasionally start with a limestone/marlstone elementary cycle, by contrast, outer ramp to middle ramp deposits, are dominated by limestone/marlstone elementary cycles (Figs. 4 and 5).

4.2 Elementary Cycle Periodicity and Spectral Analysis

According to ammonite biostratigraphic data, the Vaca Muerta Formation spans the Early Tithonian to Early Valanginian (Leanza and Hugo, 1977). However, it should be noted that the Lower Tithonian is not complete in the Vaca Muerta Formation, because the biostratigraphic correlation with the *H. hybonotum* Standard Zone has not be found (Leanza, 1996; Riccardi, 2015; Vennari, 2016). On the other hand, detailed magnetostratigraphic correlations by Iglesia Llanos et al. (2017) indicate that the Vaca Muerta Formation spans the M22r.2r to M15r Subchrons (uppermost *H. hybonotum* to *S. boissieri* Standard Zones). These data point to a time interval of *ca.* 10 myr (Gradstein et al., 2004) and 11 myr (Gradstein et al., 2012; Ogg et al., 2016). Absolute ages from the Vaca Muerta Formation are still scarce and and inconsistent with the currently accepted geological scale; detrital zircon data from the Tordillo Formation, which underlies the Vaca Muerta Formation, indicate a maximum U—Pb LAM-MC-ICP-MS age of 144 Ma (Naipauer et al., 2015), and Vennari et al. (2014) obtained recently a ID-TIMS age of 139.55 ± 0.18 Ma for the middle part of the *A. noduliferum* ammonite Zone. In fact, Ogg et al. (2016) indicate that one or two new dates would suggest eventually that the Late Jurassic age model for M-sequence should be shifted about 3 myr younger than the GTS2012. Therefore, until new data confirm or refute these ages, absolute age data will not be considered yet.

The Vaca Muerta Formation at Arroyo Loncoche and Cuesta del Chihuido sections comprises the Lower Tithonian to Upper Berriasian *V. andesensis* to lower *Neocomites wichmanni* Zones (Fig. 7), reaching about

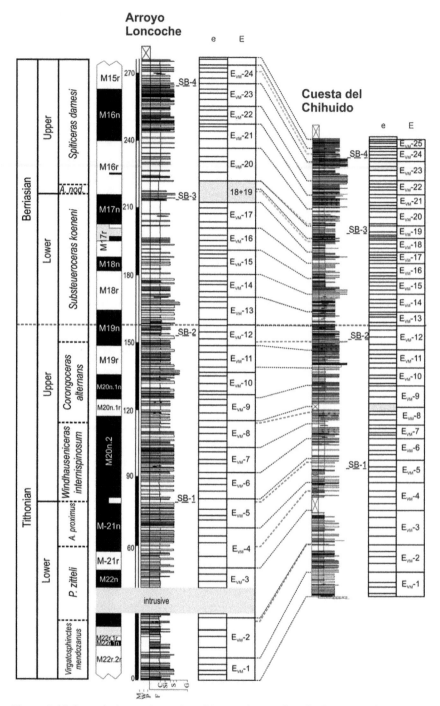

Figure 6 High resolution cyclostratigraphic correlation of studied stratigraphic sections of the Vaca Muerta Formation, using low-frequency eccentricity cycle (E) and high-frequency eccentricity cycle (e). Jurassic − Cretaceous boundary according to the last proposal of the Berriasian Working Group (Wimbledon, 2017), ammonite zones after Kietzmann et al. (2014a), magnetostratigraphy from Arroyo Loncoche section after Iglesia Llanos et al. (2017), magnetostratigraphy from El Ministerio Quarry section after Kohan Martínez et al. (2018).

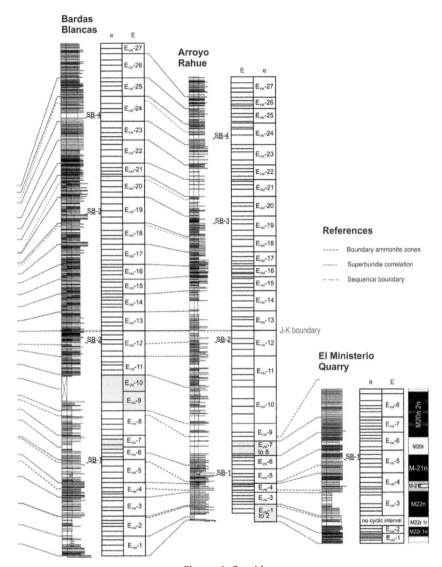

Figure 6 Cont'd

10 myr. The Arroyo Loncoche section contains 487 elementary cycles, which are grouped in 98 bundles and 24 superbundles, while the Cuesta del Chiuhido section include 495 elementary cycles, 101 bundles and 25 superbundles. The Bardas Blancas and Arroyo Rahue sections span the Lower Tithonian to Lower Valanginian *V. andesensis* Zone to the *Olcosthephanus (O) atherstoni* Zones (Fig. 6). The Bardas Blancas section contains 501 elementary cycles, which are grouped in 106 bundles and 27 superbundles, while the Arroyo Rahue section includes 510 elementary cycles,

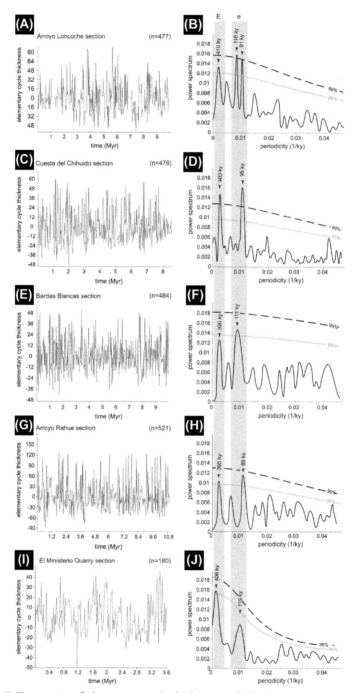

Figure 7 Time series of elementary cycle thickness and Blackman-Tuckey spectrum of studied localities, showing the presence of the low-frequency (E) and high-frequency (e) excentricity cycle. (A–B) Arroyo Loncoche section, (C–D) Cuesta del Chihuido section, (E–F) Bardas Blancas section, (G–H) Arroyo Rahue section, (I–J) El Ministerio Quarry section.

Table 1 Estimation of Elementary Cycle Periodicity

Stratigraphic section	Estimated time (Ma)	Elementary cycles	Periodicity (ky)
Arroyo Loncoche	10	487	20.53
Cuesta del Chihuido	10	495	20.20
Bardas Blancas	11	501	21.9
Arroyo Rahue	11	510	21,56
El Ministerio Quarry	3.5	168	20.83

108 bundles and 27 superbundles. In El Ministerio Quarry section, comprises the Lower to Upper Tithonian *V. andesensis* to *W. internispinosum* Zones, including 168 elementary cycles, 38 bundles and 8 superbundles. Dividing the time-length of these sections by the number of cycles, elementary cycles in this area have duration of ∼21 ky (Table 1), which can be attributed to the precessional cycle of the Earth. Data from other Jurassic examples are consistent with this average value (Berger, 1978; Berger et al., 1992; Strasser et al., 2006; Hinnov and Hilgen, 2012).

Spectra obtained from time series of elementary cycle thickness are very consistent with each other (Fig. 7). It is important to note that time series are aliased for the precesional periodicity, because the elementary cycle is the unit of time series and is higher than the Nyquist frequency, so the Fourier analysis does not detect it. For the lowest frequencies, at Arroyo Loncoche section Blackman-Tuckey spectra show a peak above the 95% confidence level, and two peaks above the 99% confidence level (Fig. 7A and B). The first one has a periodicity of 410 ky which is consistent with the low-frequency eccentricity cycle. The other two peaks have periodicities of 118 cycles ky and 91 ky. These periodicities can be also attributed to the high-frequency eccentricity cycle. Similar results are observed in the other four sections. At Cuesta del Chihuido section a first 403 ky peak and a second single peak of 95 ky were obtained (Fig. 8C and D), which appear in Blackman-Tuckey spectra from Bardan blancas section at 390 ky and 111 ky (Fig. 7e and f), 395 ky and 89 ky at Arroyo Rahue section (Fig. 7G and H), and 406 ky and 119 ky at El Ministerio Quarry section (Fig. 7I and J).

Some poorly defined peaks below confidence levels could be assigned to the obliquity signal, which would indicate that the obliquity cycle was not an important forcing for the latitudes of the Neuquén Basin. In fact, the obtained periodicities fit well with the so called precession-eccentricity syndrome (PES) defined by Fischer et al. (2004), which is characteristic of mid and low latitudes (Berger and Loutre, 1994).

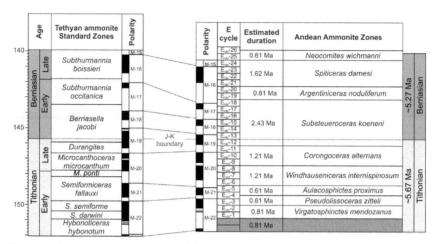

Figure 8 Astronomical calibration of the Tithonian − Berriasian in the Neuquén Basin, using 405 ky low-frequency eccentricity cycles, and magnetostratigraphic correlation with the GTS2016 (Ogg et al., 2016). Gray-shaded interval is not represented in the Vaca Muerta Formation and the illustrated cycles are taken from Huang et al. (2010). According to our cyclostratigraphic data we estimated a minimum duration of 5.67 myr for the Tithonian, and 5.27 myr for the Berriasian. These results are similar to those proposed in the GTS 2004.

4.3 Orbital Calibration of Tithonian − Berriasian in the Neuquén Basin

Using high and low-frequency eccentricity cycles identified in the four studied stratigraphic sections, we built a floating astronomical scale for the Tithonian−Berriasian of the Neuquén Basin, which allowed us estimate the minimum time durations of Andean ammonite zones and related magnetozones. There are some differences in the number of high-frequency eccentricity cycles between sections, so longer duration of biozones of the four sections is used for building the astronomical timescale (Fig. 6).

The most important difference in the number of cycles between ammonite biozones is shown in the *A. noduliferum* Zone at Arroyo Loncoche and Cuesta del Chihuido sections, where it has the scale of two high-frequency eccentricity cycles, while in Bardas Blancas and Arroyo Rahue sections up to eighth cycles can be identified. This interval corresponds to oyster-biostrome middle carbonate ramp facies, showing evidence of recurrent erosion and low sedimentation rates (Kietzmann et al., 2008, 2011; 2014a). Another interval where the cycles are condensed is the lower part of the Arroyo Rahue section, consisting of basinal to distal outer ramp facies where sedimentation rates were low (Kietzmann et al., 2014a).

According to our cyclostratigraphic data the minimum durations of ammonite zones within the Vaca Muerta Formation are summarized in Fig. 8. The *V. andesensis* Zone contains two low-frequency eccentricity cycles, so that considering the 405 ky periodicity, its minimum duration is 0.81 myr. This ammonite zone comprises the M22r.2r to M22n Subchrons (Iglesia Llanos et al., 2017; Kohan Martinez et al., 2017) which correlates to the uppermost *H. hybonotum* and lowermost *S. darwini* Standard Zones.

The *P. zitteli* Zone shows 1.5 low-frequency eccentricity cycles with a minimum duration of 0.61 myr, and according to Iglesia Llanos et al. (2017) spans the M22n to M21n Subchrons (upper *S. darwini* to lower *S. fallauxi* Standard Zones). This zone also contains the FO of *Polycostella beckmannii* (see Kietzmann et al., 2011) which occurs at the upper M22n Subchron (Casellato, 2010).

The *A. proximus* Zone shows 1.5 low-frequency eccentricity cycles with a minimum duration of 0.61 myr, and comprises the M21n and M20r Subchrons (upper *S. fallauxi* to *M. ponti* Standard Zones; Iglesia Llanos et al., 2017; Kohan Martinez et al., 2017). This zone also contains the *L. dobeni* Subzone of the *Chitinoidella* Zone (Kietzmann, 2017).

The *W. internispinosum* Zone has three low-frequency eccentricity cycles with a minimum duration of 1.21 myr, and bears the M20 Chron, which is consistent with the correlation to the *M. microcanthum* Standard Zone (Iglesia Llanos et al., 2017; Kohan Martinez et al., 2017). This zone contains the *Ch. boneti* Subzone of the *Chitinoidella* Zone (Kietzmann, 2017), supporting its Late Tithonian age.

The *C. alternans* Zone has also three low-frequency eccentricity cycles with a minimum duration of 1.21 myr, and comprises the M20n to M19n Subchrons (upper *M. microcanthum* Standard Zone and lower part of the "*Durangites*" Standard Zone; Iglesia Llanos et al., 2017). This interval also contains the *Crassicollaria* calpionellid Zone (Kietzmann, 2017).

The *S. koeneni* Zone presents seven low-frequency eccentricity cycles, so it would have a minimum duration of ~2.43 myr, comprises the M19n to the lowermost part of M16r Subchrons (upper part of "*Durangites*" to *S. occitanica* Standard Zones; Iglesia Llanos et al., 2017). This zone contains the *Stomiosphaerina proxima* calcareous dinoflagellate Zone, as well as elements of the *Calpionella* Zone (Kietzmann, 2013 in Gonzalez Tomassini et al., 2015; Ivanova and Kietzmann, 2017). Taking into account that recently the Berriasian Working Group has voted the Jurassic/Cretaceous boundary at the base of the *Calpionella* Standard Calpionellid Zone in the middle part of magnetosubzone M19n.2n (e.g., Ogg et al., 2016;

Wimbledon, 2017), the Jurassic/Cretaceous boundary should be placed within the lower third of the *S. koeneni* Zone.

Further above, the *A. noduliferum* Zone shows two low-frequency eccentricity cycles with a minimum duration of ∼0.81 myr, and includes a dominant reverse, which is correlated with M16r Subchron (upper *S. occitanica* to lower *S. boissieri* Standard Zones; Iglesia Llanos et al., 2017). Finally, the *S. damesi* Zone contains four low-frequency eccentricity cycles, with a minimum duration of ∼1.62 myr, and spans the M16r to M15r Subcrons (*S. boissieri* Standard Zone; Iglesia Llanos et al., 2017).

Based on the number of low-frequency eccentricity cycles within the Vaca Muerta Formation we estimated a minimum duration of 4.86 myr for the Tithonian, and 5.27 myr for the Berriasian in the Neuquén Basin (Fig. 8). The basal part of the Tithonian is not represented in the Vaca Muerta Formation, however, taking into account ammonite zones durations reported by Huang et al. (2010) we interpret that about two low-frequency eccentricity cycles are missing, and therefore, the estimated minimum duration for the Tithonian is 5.67 myr.

5. A COMPLETE FLOATING ASTRONOMICAL TIME SCALE FOR THE TITHONIAN — BERRIASIAN

The latest Late Jurassic represents an epoch with scarce number of data in terms of absolute ages and calibration points, and latest Jurassic tie-points are ⁴⁰Ar/³⁹Ar ages (Pálfy, 2008; Gradstein et al., 2004, 2012). This led to numerous time duration proposals over the years for the Tithonian, ranging from 5 to 8.7 myr, and 3 to 7 myr for the Berriasian. In the latest two geologic time scales (GTS2004 ans GTS2012) the Tithonian spans 5.3 and 7.1 myr, while the Berrasian 5.3 and 5.6 myr, respectively. Recently, Ogg et al. (2016) updated the GTS2012 incorporating the last proposal of the Jurassic — Cretaceous boundary at the base of the *Calpionella* Zone, so the duration of the Tithonian changed to 6.4 myr, and that of the Berriasian to 6.3 myr.

Our cyclostratigraphic data suggest a time span of 5.67 myr for the Tithonian and 5.27 myr for the Berriasian. However, if we take into consideration that the Tithonian most likely spans 7.1 myr (Ogg and Hinnov, 2012a), it would imply that *c.* four low-frequency eccentricity cycles (405 ky) are missing either by omission or condensation in the Neuquén Basin. Iglesia Llanos et al. (2017) indicate the transgressive surfaces as possible positions where time might be missing. However, there is no sedimentolog-ical evidence to correctly evaluate this possibility.

An accurate astromomical time scale (ATS) for the Tithonian—Berriasian is possible from the comparison and combination of available studies (Fig. 9). The lowermost Lower Tithonian was studied by Weedon et al. (1999, 2004) using magnetic-susceptibility measures made on exposures, core material and down boreholes from the Kimmeridge Clay Formation (Kimmeridgian—Tithonian). From the recognition of the obliquity (38 ky) and precession (20 ky) cycles, these authors proposed a duration of 1.35 myr for the *A. autissiodorensis* Boreal Zone and 0.53 myr for the *P. elegans* Boreal Zone, indicating that the interval equivalent to the *H. hybonotum* Standar Zone is longer than 0.9 myr. Weedon et al. (2004) studied the type section of the Kimmeridge Clay Formation from the measurement of three inde-pendent variables (magnetic susceptibility, photoelectric factor and total gamma ray), obtained a cyclostratigraphic calibration based on the 38 ky obliquity cycle. These authors achieved *ca.* 1.1 myr for the *A. autissiodoren-sis* Boreal Zone and *ca.* 3.9 myr for the *P. elegans* to *Virgatosphinctes fittoni* Boreal Zones (~ *H. hybonotum* to *M. ponti* Standard Ammonite Zones).

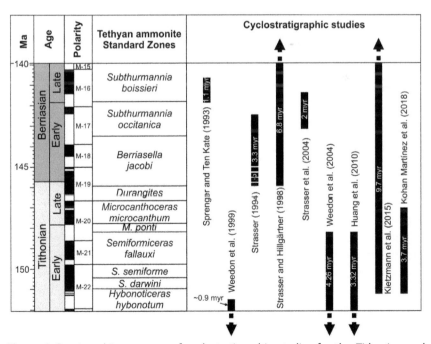

Figure 9 Stratigraphic coverage of cyclostratigraphic studies for the Tithonian and Berriasian. All studies were calibrated to the low-frequency eccentricity cycle of 405 ky.

This means that the equivalent interval to the *H. hybonotum* to *M. ponti* Standard Ammonite Zones would span *ca.* 4.3 myr. Later, Huang et al. (2010) refined the astromomical time model of Weedon et al. (2004) for the Kimmeridge Clay Formation, using the stable 405 ky low-frequency eccentricity cycle obtained from %TOC and FMS microconductivity measures. They obtained a *ca.* 3.3 myr duration for the *P. elegans* to *V. fittoni* Boreal Ammonite Zones (∼ *H. hybonotum* to *M. ponti* Standard Ammonite Zones), very similar to the duration obtained in this work for the same stratigraphic interval.

The Upper Tithonian was only studied by Kietzmann et al. (2015) and Kohan Martínez et al. (2018) from elementary cycle thickness and magnetic susceptibility measurements from the Vaca Muerta Formation, and recalibrated in this study using the 405 ky low-eccentricity cycle. The *W. internispinosum, C. alternans* and lower third of the *S. koeneni* Andean Zones span 2.825 myr.

The Berriasian includes the studies from Strasser (1994), Strasser and Hillgärtner (1998), and Strasser et al. (2004), who studied the shallow-lagoonal and peritidal carbonates of the French Jura Mountains, interpreting these deposits are driven by 20 ky, 100 ky and 400 ky cycles and represents ∼ 6.8 myr from the combination of 405 ky cycles and the correlation with eustatic curves of Haq et al. (1987) and Hardenbol et al. (1998). On the other hand, Sprenger and Ten Kate (1993) studied the Upper Berriasian of Spain, estimating 1.2 myr for the Calpionellopsis Standard Calpionellid Zone, and 3.1 myr for the Berriasian. The ∼ 6.8 myr duration of the Berriasian that arises from the combination of cyclostratigraphic data from Strasser (1994), Strasser and Hillgärtner (1998) and Strasser et al. (2004) differs in ∼ 1.5 myr with the 5.27 myr obtained in the Neuquén Basin. In fact, Kietzmann et al. (2015) detected a discontinuity within the *S. koeneni* Zone that could be the cause of a shorter M17r subchron with respect to the GTS2016 (Fig. 8). Also, the correlation of a 300 km regional seismic line (Gonzalez et al., 2016) and outcropping sections in the northern Neuquén Basin (Kietzmann et al., 2016a,b) indicates that there is an important forced regression in the Lower Berriasian that does not seem to have expression in the outcrops (B1 seismic horizon).

New absolute age data they are being processed, which together with forthcoming magnetostratigraphic studies of all the studied stratigraphic sections, and systematic studies of microfossils, will allow a better calibration of the Tithonian — Berriasian, and the anchoring of the floating astronomical time scale provided in this work.

6. CONCLUSIONS

The Vaca Muerta Formation is characterized by decimetre-scale rhythmic alternations of marlstones and limestones, showing a well-ordered hierarchy of cycles, including elementary cycles, bundles and superbundles. Based on spectral analysis, elementary cycles have a periodicity of ∼21 ky, which correlates with the precession cycle of Earth's axis; bundles and superbundles have periodicities of ∼90–120 ky and ∼400 ky, which we interpret as the modulation of the precessional cycle by the Earth's orbital eccentricity.

Cyclostratigraphic data allowed us to build a floating astronomical time scale for the Tithonian — Berriasian in the Neuquén Basin, which is anchored to the geological time scale through magnetostratigraphy. We obtained the following correlation and minimum time duration of ammonite zones: The *V. andesensis* Zone (M22r.2r to M22n Subchrons - uppermost *H. hybonotum* and lowermost *S. darwini* Standard Zones) spans 0.81 myr. The *P. zitteli* Zone (M22n to M21n Subchrons - upper *S. darwini* to lower *S. fallauxi* Standard Zones) has a minimum duration of 0.61 myr. The *A. proximus* Zone (M21n and M20r Subchrons - upper *S. fallauxi* to *M. ponti* Standard Zones) show a minimum duration of 0.61 myr. The *W. internispinosum* Zone (M20 Chron - *M. microcanthum* Standard Zone) spans 1.21 myr. The *C. alternans* Zone (M20n to M19n Subchrons - upper *M. microcanthum* to lower part of the "*Durangites*" Standard Zones) has a minimum duration of 1.21 myr. The *S. koeneni* Zone (M19n to the lowermost part of M16r Subchrons - upper part of "*Durangites*" to *S. occitanica* Standard Zones) would have a minimum duration of 2.43 myr. The *A. noduliferum* Zone (M16r Subchron - upper *S. occitanica* to lower *S. boissieri* Standard Zones) spans 0.81 myr. Finally, the *S. damesi* Zone (M16r to M15r Subcrons - *S. boissieri* Standard Zone) show a minimum duration of 1.62 myr.

The resulted durations of some polarity chrones are different with respect to the GTS2016, however such differences could be due to condensation or discontinuities not detected in the studied sections.

The basal part of the Tithonian is not represented in the Vaca Muerta Formation, nevertheless, the combination of the 405 ky calibrated ATS from Huang et al. (2010) and that presented in this work, we estimated a minimum duration of 5.67 myr for the Tithonian and 5.27 myr for the Berriasian.

ACKNOWLEDGMENTS

We are especially grateful to Dr. A.C. Riccardi (Universidad Nacional de La Plata y Museo, Argentina) and Dr. H.A. Leanza (Museo de Ciencias Naturales Bernandino Rivadavía, Argentina) for the helpful discussions regarding the biostratigraphy of the Vaca Muerta Formation. We thank the Pacheco family (Arroyo Loncoche) and Patricio Zapata (Lonco of the Comunidad Millaqueo, Sierra de la Vaca Muerta) for allowing us to work in their terrains. This research has been done under the framework of the PICT-2015-0206 and PICT-2016-3762 projects supported by Agencia Nacional de Promoción Científica y Tecnológica. We are grateful to Dr. J. Ogg and Dr. H. Leanza for their comments and suggestions that improved the original manuscript.

REFERENCES

Aguirre Urreta, M.B., 2001. Estratigrafía y bioestratigrafía del Jurásico Superior-Cretácico Inferior marino de la Cuenca Neuquina-Aconcagüina, Argentina y Chile. J. Iber. Geol. 27, 71–90.

Aguirre-Urreta, B., Lazo, D.G., Griffin, M., Vennari, V.V., Parras, A.M., Cataldo, C., Garberoglio, R., Luci, L., 2011. Megainvertebrados del Cretácico y su importancia bioestratigráfica. In: Leanza, H.A., Arregui, C., Carbone, O., Danieli, J.C., Vallés, J.M. (Eds.), Geología y Recursos Naturales de la Provincia del Neuquén. Asociación Geológica Argentina, Buenos Aires, pp. 465–488.

Anderson, E.J., 2004. The cyclic hierarchy of the "Purbeckian" Sierra del Pozo Section, Lower Cretaceous (Berriasian), southern Spain. Sedimentology 51, 455–477.

Ballent, S.C., Ronchi, D.I., Angelozzi, G.N., 2004. Microfósiles calcáreos tithonianos (Jurásico superior) en el sector oriental de la cuenca Neuquina, Argentina. Ameghiniana 41, 13–24.

Ballent, S., Concheyro, A., Náñez, C., Pujana, I., Lescano, M., Carignano, A.P., Caramés, A., Angelozzi, G., Ronchi, D., 2011. Microfósiles mesozoicos y cenozoicos. In: Leanza, H.A., Arregui, C., Carbone, O., Danieli, J.C., Vallés, J.M. (Eds.), Geología y Recursos Naturales de la Provincia del Neuquén. Asociación Geológica Argentina, Buenos Aires, pp. 489–528.

Berger, A., 1978. Long-term variations of daily insolation and Quaternary climatic changes. J. Atmos. Sci. 35, 2362–2367.

Berger, A., Loutre, M.F., 1994. Astronomical forcing through geological time. In: de Boer, P.L., Smith, D.G. (Eds.), Orbital Forcing and Cyclic Sequences, vol. 19. International Association of Sedimentologists, pp. 15–24. Special Publication.

Berger, A., Loutre, M.F., Laskar, J., 1992. Stability of the astronomical frequencies over the Earth's history for paleoclimate studies. Science 255, 560–566.

Bown, P., Concheyro, A., 2004. Lower cretaceous calcareous nannoplankton from the Neuquén basin, Argentina. Mar. Micropaleontol. 52, 51–84.

Casellato, C.E., 2010. Calcareous nannofossil biostratigraphy of Upper Callovian – Lower Berriasian successions from the Southern Alps, North Italy. Rivista Italiana di Paleontologia e Stratigrafia 116, 357–404.

Fischer, A.G., D'Argenio, B., Premoli Silva, I., Weissert, H., Ferreri, V., 2004. Cyclostratigraphic approach to Earth's history: an introduction. In: D'Argenio, B., Fischer, A.G., Premoli Silva, I., Weissert, H., Ferreri, V. (Eds.), Cyclostratigraphy: Approaches and Case Histories, vol. 81. Society of Economic Paleontologists and Mineralogists, pp. 5–16. Special Publication.

Goldhammer, R.K., Dunn, P.A., Hardie, L.A., 1990. Depositional cycles, composite sea-level changes, cycle stacking patterns, and the hierarchy of stratigraphic forcing. Geol. Soc. Am. Bull. 102, 535–562.

González, G., Vallejo, D., Kietzmann, D.A., Marchal, D., Desjardins, P., González Tomassini, F., Gómez Rivarola, L., Domínguez, F., 2016. Transecta Regional de la Formación Vaca Muerta Integración de sísmica, registros de pozos, coronas y afloramientos. IAPG-AGA, Buenos Aires, p. 252.

González Tomassini, F., Kietzmann, D.A., Fantín, M.A., Crousse, L.C., Reinjenstein, H.M., 2015. Estratigrafía y análisis de facies de la Formación Vaca Muerta en el área de El Trapial, Cuenca Neuquina, Argentina. Petrotecnia 2015/2, 78—89.

Gradstein, F.M., Ogg, J.G., Smith, A.G., 2004. A Geologic Time Scale. Cambridge University Press, Cambridge, p. 589.

Gradstein, F.M., Ogg, J.G., Schmitz, M.D., Ogg, G.M., 2012. The Geologic Time Scale. Elseier, Oxford, p. 1144.

Haq, B.U., Hardenbol, J., Vail, P.R., 1987. Chronology of fluctuating sea level since the Triassic. Science 235, 1156—1167.

Hardenbol, J., Thierry, J., Farley, M.B., Jacquin, T., De Graciansky, P.C., Vail, P.R., 1998. Cretaceous chronostratigraphy. In: De Graciansky, P.C., Hardenbol, J., Jacquin, T., Vail, P.R., Farley, M.B. (Eds.), Sequence Stratigraphy of European Basins, vol. 60, pp. 3—13. Spec. Pubi. Soc. Sed.Geol.

Hinnov, L., Hilgen, F., 2012. Cyclostratigraphy and astrochronology. In: Gradstein, F.M., Ogg, J.G., Schmitz, M.D., Ogg, G.M. (Eds.), The Geologic Time Scale. Elsevier, Oxford, pp. 63—84.

Huang, C., Hesselbo, S.P., Hinnov, L., 2010. Astrochronology of the late Jurassic Kimmeridge clay (Dorset, England) and implications for Earth system processes. Earth Plnetary Sci. Lett. 289, 242—255.

Husinec, A., Read, J.F., 2007. The late Jurassic tithonian, a greenhouse phase in the middle Jurassic—early cretaceous "cool" mode: evidence from the cyclic adriatic platform, Croatia. Sedimentology 54, 317—337.

Iglesia Llanos, M.P., Kietzmann, D.A., Kohan Martínez, M., Palma, R., 2017. Magnetostratigraphy of the upper Jurassic lower cretaceous from Argentina: implications for the J-K boundary in the Neuquén basin. Cretac. Res. 70, 189—208.

Ivanova, D.K., Kietzmann, D.A., 2017. Calcareous dinoflagellate cysts from the tithonian - valanginian Vaca Muerta Formation in the southern Mendoza area of the Neuquén basin, Argentina. J. S. Am. Earth Sci. 77, 150—169.

Kietzmann, D.A., 2017. Chitinoidellids from the early tithoniane - early valanginian Vaca Muerta Formation in the northern Neuquén basin, Argentina. J. S. Am. Earth Sci. 76, 152—164.

Kietzmann, D.A., Iglesia Llanos, M.P., 2018. Comment on "tethyan calpionellids in the Neuquén basin (argentine andes), their significance in defining the Jurassic/cretaceous boundary and pathways for tethyan-eastern pacific connections" by R. López-Martínez, B. Aguirre-urreta, M. Lescano, a. Concheyro, V. Vennari and V. Ramos. J. S. Am. Earth Sci. 84, 444—447.

Kietzmann, D.A., Palma, R.M., 2009. Microcrinoideos saccocómidos en el Tithoniano de la Cuenca Neuquina. ¿Una presencia inesperada fuera de la región del Tethys? Ameghiniana 46, 695—700.

Kietzmann, D.A., Palma, R.M., 2014. Early Cretaceous crustacean microcoprolites from Sierra de la Cara Cura, Neuquén Basin, Argentina: taphonomy, environmental distribution, and stratigraphic correlation. Cretac. Res. 49, 214—228.

Kietzmann, D.A., Palma, R.M., Bressan, G.S., 2008. Facies y microfacies de la rampa tithoniana-berriasiana de la Cuenca Neuquina (Formación Vaca Muerta) en la sección del arroyo Loncoche — Malargüe, provincia de Mendoza. Rev. la Asoc. Geol. Argent. 63, 696—713.

Kietzmann, D.A., Martín-Chivelet, J., Palma, R.M., López-Gómez, J., Lescano, M., Concheyro, A., 2011. Evidence of precessional and eccentricity orbital cycles in a Tithonian source rock: the mid-outer carbonate ramp of the Vaca Muerta Formation, Northern Neuquén Basin, Argentina. AAPG (Am. Assoc. Pet. Geol.) Bull. 95, 1459—1474.

Kietzmann, D.A., 2013. Estudio bioestratigráfico integrado basado en calciesferas y calpionéllidos, Formación Vaca Muerta, Pozos ETxp-2001, ETxp-2002, ETxp.-2003 y ETxp-2006. Unpublished report.

Kietzmann, D.A., Palma, R.M., Martín-Chivelet, J., López-Gómez, J., 2014a. Sedimentology and sequence stratigraphy of a Tithonian-Valanginian carbonate ramp (Vaca Muerta Formation): a misunderstood exceptional source rock in the Southern Mendoza area of the Neuquén Basin, Argentina. Sediment. Geol. 302, 64—86.

Kietzmann, D.A., Ambrosio, A., Suriano, J., Alonso, S., Vennari, V.V., Aguirre-Urreta, M.B., Depine, G., Repol, D., 2014b. Variaciones de facies en las secuencias basales de la Formación Vaca Muerta en su localidad tipo (Sierra de la Vaca Muerta), Cuenca Neuquina. In: IX Congreso de Exploración y Desarrollo de Hidrocarburos TT2, pp. 299—317.

Kietzmann, D.A., Palma, R.M., Iglesia Llanos, M.P., 2015. Cyclostratigraphy of an orbitally-driven tithonian-valanginian carbonate ramp succession, southern Mendoza, Argentina: implications for the Jurassic-Cretaceous boundary in the Neuquén basin. Sed. Geo. 315, 29—46.

Kietzmann, D.A., Ambrosio, A., Suriano, J., Alonso, S., González Tomassini, F., Depine, G., Repol, D., 2016a. The vaca muerta-quintuco system (tithonian — valanginian) in the Neuquén basin, Argentina: a view from the outcrops in the chos malal fold and thrust belt. AAPG (Am. Assoc. Pet. Geol.) Bull. 100, 743—771.

Kietzmann, D.A., Ambrosio, A., Alonso, S., Suriano, J., 2016b. Puerta Curaco. In: González, G., Vallejo, D., Kietzmann, D.A., Marchal, D., Desjardins, P., González Tomassini, F., Gómez Rivarola, L., Domínguez, F. (Eds.), Transecta regional de la Formación Vaca Muerta, Integración y correlación de sismica, perfilaje de pozos, coronas y afloramineto. IAPG-AGA, Buenos Aires, pp. 219—232.

Kietzmann, D.A., Iglesia Llanos, M.P., Ivanova, D.K., Kohan Martinez, M., Sturlesi, M.A., 2018. Toward a multidisciplinary chronostratigraphic calibration of the Jurassic-Cretaceous transition in the Neuquén Basin. Rev. la Asoc. Geol. Argent. 75, 175—187.

Kohan Martínez, M., Kietzmann, D.A., Iglesia Llanos, M.P., Leanza, H.A., Luppo, T., 2018. Magnetostratigraphy and cyclostratigraphy of the tithonian interval from the vaca muerta formation, southern Neuquén basin, Argentina. J. S. Am. Earth Sci. 85, 209—228.

Leanza, H.A., 1981. Faunas de ammonites del Jurásico y Cretácico inferior de América del Sur, con especial consideración de la Argentina. In: Volkheimer, W., Musacchio, E. (Eds.), Cuencas Sedimentarias de América del Sur. Asociación Geológica Argentina, Buenos Aires, pp. 559—597.

Leanza, H.A., 1996. Advances in the ammonite zonation around the Jurassic/Cretaceous boundary in the andean realm and correlation with tethys. In: Jost Wiedmann Symposium, Abstracts, Tübingen, pp. 215—219.

Leanza, H.A., 2009. Las principales discordancias del Mesozoico de la Cuenca Neuquina según observaciones de superficie. Rev. del Museo Argent. de Ciencias Nat. 11, 145—184.

Leanza, H.A., Hugo, C.A., 1977. Sucesión de amonites y edad de la Formación Vaca Muerta y sincrónicas entre los Paralelos 35o y 40o l.s. Cuenca Neuquina-Mendocina. Rev. Asoc. Geol. Argent. 32, 248—264.

Legarreta, L., Gulisano, C.A., 1989. Análisis estratigráfico de la Cuenca Neuquina (Triásico Superior-Terciario Inferior). In: Chebli, G.A., Spalletti, L.A. (Eds.), Cuencas Sedimentarias Argentinas, Serie Correlación Geológica, vol. 6, pp. 221—243. Tucumán.

Legarreta, L., Uliana, M.A., 1991. Jurassic–cretaceous marine oscillations and geometry of back arc basin, central Argentina Andes. In: McDonald, D.I.M. (Ed.), Sea Level Changes at Active Plate Margins: Process and Product, vol. 12. International Association of Sedimentologists, pp. 429–450. Special Publication.

Legarreta, L., Uliana, M.A., 1996. The Jurassic succession in west central Argentina: stratal patterns, sequences, and paleogeographic evolution. Palaeogeogr. Palaeoclimatol. Palaeoecol. 120, 303–330.

Lehmann, C., Osleger, D.A., Montañez, I.P., 1998. Controls on cyclostratigraphy of lower cretaceous carbonates and evaporites, cupido and coahuila platforms, Northeastern Mexico. J. Sediment. Res. 68, 1109–1130.

López-Martínez, Aguirre-Urreta, B., Lescano, M., Concheyro, A., Vennari, V., Ramos, V., 2018. Tethyan calpionellids in the Neuquén basin (Argentine Andes), their significance in defining the Jurassic/Cretaceous boundary and pathways for Tethyan-Eastern Pacific connections. J. S. Am. Earth Sci. (in press).

Manceda, R., Figueroa, D., 1993. La inversión del rift mesozoico de la faja fallada y plegada de Malargüe. Provincia de Mendoza. In: 12 Congreso Geologico Argentino y 2 Congreso de Exploración de Hidrocarburos, Actas, vol. 3, pp. 219–232. Mendoza.

Melnyk, D.H., Smith, D.G., Amiri-Garroussi, K., 1994. Filtering and frequency mapping as tools in subsurface cyclostratigraphy, with examples from the Wessex Basin, UK. In: de Boer, P.L., Smith, D.G. (Eds.), Orbital Forcing and Cyclic Sequences, vol. 19. International Association of Sedimentologists, pp. 35–46. Special Publication.

Molinie, A.J., Ogg, J.G., 1992. Milankovitch cycles in upper Jurassic and lower Cretaceous radiolarites of the Equatorial pacific: spectral analysis and sedimentation rate curves. Proc. Ocean. Drill. Program, Sci. Results 129, 529–547.

Naipauer, M., Tunik, M., Marques, J.C., Rojas Vera, E., Vujovich, G.I., Pimentel, M., Ramos, V.A., 2015. U-Pb detrital zircon ages of Upper Jurassic continental successions: implications for the provenance and absolute age of the Jurassic–Cretaceous boundary in the Neuquén Basin. In: Sepúlveda, S.A., Giambiagi, L.B., Moreiras, S.M., Pinto, L., Tunik, M., Hoke, G.D., Farías, M. (Eds.), Geodynamic Processes in the Andes of Central Chile and Argentina, vol. 399. Geological Society, London, pp. 131–154. Special Publication.

Ogg, J.G., Hinnov, L., 2012a. The Jurassic period. In: Gradstein, F.M., Ogg, J.G., Schmitz, M., Ogg, G. (Eds.), The Geologic Time Scale 2012. Elsevier, Amsterdam, pp. 731–792.

Ogg, J.G., Hinnov, L., 2012b. The Cretaceous period. In: Gradstein, F.M., Ogg, J.G., Schmitz, M., Ogg, G. (Eds.), The Geologic Time Scale 2012. Elsevier, Amsterdam, pp. 793–854.

Ogg, J.G., Ogg, G.M., Gradstein, F.M., 2016. A Concise Geologic Time Scale. Elsevier, Amsterdam, 243 p.

Pálfy, J., 2008. The quest for refined calibration of the Jurassic time-scale. PGA (Proc. Geol. Assoc.) 119, 85–95.

Pálfy, J., Smith, P.L., Mortensen, J.K., 2000. A U-Pb and 40Ar/39Ar time scale for the Jurassic. Can. J. Earth Sci. 37, 923–944.

Pardo-Igúzquiza, E., Rodríguez-Tovar, F., 2004. POWGRAF 2: a program for graphical spectral analysis in cyclostratigraphy. Comput. Geosci. 30, 533–542.

Park, M.H., Fürsich, F., 2001. Cyclic nature of lamination in the Tithonian Solnhofen Plattenkalk of southern Germany and its palaeoclimatic implications. Int. J. Earth Sci. 90, 847–854.

Quattrocchio, M.E., Martínez, M.A., García, V.M., Zavala, C.A., 2003. Palinoestrtigrafia del tithoniano-Hauteriviano del centro-oeste de la Cuenca Neuquina, Argentina. Rev. Española Micropaleontol. 354, 51–74.

Ramos, V.A., 2010. The tectonic regime along the Andes: present-day and Mesozoic regimes. Geol. J. 45, 2—25.

Ramos, V.A., Folguera, A., 2005. Tectonic evolution of the Andes of Neuquén: constraints derived from the magmatic arc and Foreland deformation. In: Veiga, G.D., Spalletti, L.A., Howell, J.A., Schwarz, E. (Eds.), The Neuquén Basin, Argentina: A Case Study in Sequence Stratigraphy and Basin Dynamics, vol. 252. Geological Society of London, pp. 15—35. Special Publications.

Raspini, A., 2001. Stacking pattern of cyclic carbonate platform strata: lower Cretaceous of southern Appennines, Italy. J. Geol. Soc. 158, 353—366.

Remane, J., 1991. The Jurassic-Cretaceous boundary: problems of definition and procedure. Cretac. Res. 12, 447—453.

Riccardi, A.C., 2008. The marine Jurassic of Argentina: a biostratigraphic framework. Episodes 31, 326—335.

Riccardi, A.C., 2015. Remarks on the Tithonian—Berriasian ammonite biostratigraphy of west central Argentina. Volumina Jurassica 13, 23—52.

Riccardi, A.C., Damborenea, S.E., Manceñido, M.O., Leanza, H.A., 2011. Megainvertebrados jurásicos y su importancia geobiológica. In: Leanza, H.A., Arregui, C., Carbone, O., Danieli, J.C., Vallés, J.M. (Eds.), Geología y Recursos Naturales de la Provincia del Neuquén. Asociación Geológica Argentina, Buenos Aires, pp. 441—464.

Schwarzacher, W., 1993. Cyclostratigraphy and the Milankovitch Theory. Elsevier, Amsterdam, pp. 1—224.

Sprenger, A., Ten Kate, W.G., 1993. Orbital Forcing of Calcilutite-marl Cycles in Southeast Spain and an Estimate for the Duration of the Berriasian Stage. Geological Society of America, pp. 807—818. Bulletin 105.

Strasser, A., 1994. Milankovitch cyclicity and high-resolution sequence stratigraphy in lagoonal-peritidal carbonates (upper Tithonian—lower Berriasian, French Jura Mountains). In: de Boer, P.L., Smith, D.G. (Eds.), Orbital Forcing and Cyclic Sequences, vol. 19. International Association of Sedimentologists, pp. 285—301. Special Publication.

Strasser, A., Hillgärtner, H., 1998. High-frequency sea-level fluctuations recorded on a shallow carbonate platform (Berriasian and lower valanginian of mount Salève, French Jura). Eclogae Geol. Helv. 91, 375—390.

Strasser, A., Hillgärtner, H., Pasquier, J.B., 2004. Cyclostratigraphic timing of sedimentary processes: an example from the Berriasian of the Swiss and French Jura Mountains. In: D'Argenio, B., Fischer, A.G., Premoli Silva, I., Weissert, H., Ferreri, V. (Eds.), Cyclostratigraphy: Approaches and Case Histories, vol. 81. Society of Economic Paleontologists and Mineralogists, pp. 135—151. Special Publication.

Strasser, A., Hilgen, F., Heckel, P., 2006. Cyclostratigraphy - Concepts, definitions, and applications. Newslett. Stratigr. 42, 75—114.

Vennari, V.V., 2016. Tithonian ammonoids (Cephalopoda, Ammonoidea) from the Vaca Muerta Formation, Neuquén basin, west-central Argentina. Palaeontographica A 306, 85—165.

Vennari, V.V., Lescano, M., Naipauer, M., Aguirre-Urreta, B., Concheyro, A., Schaltegger, U., Armstrong, R., Pimentel, M., Ramos, V.A., 2014. New constraints on the Jurassic-Cretaceous boundary in the High Andes using high-precision U-Pb data. Gondwana Res. 26, 374—385.

Vergani, G.D., Tankard, A.J., Belotti, H.J., Welkink, H.J., 1995. Tectonic evolution and paleogeography of the Neuquén basin, Argentina. In: Tankard, A.J., Suarez Soruco, R., Welsink, H.J. (Eds.), Petroleum Basins of South America, vol. 62. American Association of Petroleum Geologists, Memoir, Tulsa, pp. 383—402.

Volkheimer, W., Quattrocchio, Martínez, M., Prámparo, M., Scafati, L. and Melendi, D. Palinobiotas fósiles. In: Leanza, H.A., Arregui, C., Carbone, O., Danieli, J.C., Vallés, J.M. (Eds.), Geología y Recursos Naturales de la Provincia del Neuquén. Asociación Geológica Argentina, Buenos Aires, pp. 579—590.

Weedon, G.P., Jenkyns, H.C., Coe, A.L., Hesselbo, S.P., 1999. Astronomical calibration of the Jurassic time scale from cyclostratigraphy in British mudrock formations. Philos. Trans.: Math., Phys. Eng. Sci. 357, 1787—1813.

Weedon, G., 2003. Time—Series Analysis and Cyclostratigraphy. Examining stratigraphic record of environmental cycles. Cambridge University Press, New York, pp. 1—259.

Weedon, G.P., Coe, A.L., Gallois, R.W., 2004. Cyclostratigraphy, orbital tuning and inferred productivity for the type Kimmeridge clay (late jurassic), southern England. J. Geol. Soc., London 161, 655—666.

Wimbledon, W.A.P., 2008. The Jurassic-Cretaceous boundary: an age-old correlative enigma. Episodes 31, 423—428.

Wimbledon, W.A.P., 2017. Developments with fixing a Tithonian/Berriasian (J/K) boundary. Volumina Jurassica XV, 181—186.

Wimbledon, W.A.P., Casellato, C.E., Reháková, D., Bulot, L.G., Erba, E., Gardin, S., Verreussel, R.M.C.H., Munsterman, D.K., Hunt, C., 2011. Fixing a basal Berriasian and Jurassic/Cretaceous (J/K) boundary - is there perhaps some light at the end of the tunnel? Rivista Italiana di Paleontologia e Stratigrafia 117, 295—307.

Wimbledon, W.A.P., Reháková, D., Pszczółkowski, A., Casellato, C.E., Halásová, E., Frau, C., Bulot, L.G., Grabowski, J., Sobień, K., Pruner, P., Schnabl, P., Čížková, K., 2013. An account of the bio- and magnetostratigraphy of the upper tithonian—lower berriasian interval at le chouet, drôme (SE France). Geol. Carpathica 64, 437—460.

CHAPTER NINE

Paleocene-Eocene Calcareous Nannofossil Biostratigraphy and Cyclostratigraphy From the Neo-Tethys, Pabdeh Formation of the Zagros Basin (Iran)

Seyed Hamidreza Azami*, Erik Wolfgring*, Michael Wagreich*, [1] and Mohamad Hosein Mahmudy Gharaie[§]
*University of Vienna, Department of Geodynamics and Sedimentology, Vienna, Austria
[§]Ferdowsi University of Mashhad, Department of Geology, Mashhad, Iran
[1]Corresponding author: E-mail: Michael Wagreich michael.wagreich@univie.ac.at

Contents

Stratigraphy & Timescales, Volume 3
ISSN 2468-5178
https://doi.org/10.1016/bs.sats.2018.08.006

357

Abstract

The Pabdeh Formation in the Zagros Basin, Iran, records cyclic pelagic sedimentation from the middle Paleocene to the middle Eocene in a Tethyan setting. The cyclic successions consist of deeper-water pelagic to hemipelagic shale, marl (stone) and limestone, with a predominantly shaly lower part, and a marl-limestone upper part. Nannofossil biostratigraphy indicates standard zones CNP8 - NP6 to CNE8 - NP14. The Paleocene-Eocene Thermal Maximum (PETM) interval is indicated by a distinct nannofossil assemblages and a significant negative carbon isotope excursion. Sediment accumulation rates are in general around 6–68 mm/ka. Cyclic signals investigated include fluctuations in carbonate content. Power spectra using LOWSPEC (Robust Locally-Weighted Regression Spectral Background Estimation) and EHA (Evolutive Harmonic Analysis) indicate the presence of twentyone 405 ka cycles from the base of the PETM up to the top of the studied section. Orbital tuning to the established Laskar target curve is in accordance with the general Paleocene-Eocene cyclostratigraphy and yields insights into Paleogene chronostratigraphy.

1. INTRODUCTION

The Paleogene System records the transition from the major greenhouse climate phase of the Cretaceous to cooler climates and the onset of large ice sheets (Hay et al., 2004). This general climate evolution was punctuated by short-lived climate events termed hyperthermals (Abels et al., 2012; Zachos et al., 2008) such as the Paleocene-Eocene Thermal Maximum (PETM). The PETM event constitutes one of the most abrupt and short-lived global warming events of the past 100 million years (Bains et al., 1999; Zachos et al., 2008). This event was associated with a severe shoaling of the ocean calcite compensation depth and with a pronounced negative carbon isotope excursion recorded in carbonate and organic materials, reflecting a massive release of ^{13}C–depleted carbon and widespread dissolution of seafloor carbonates (Zachos et al., 2003, 2005). The PETM with abrupt warming of around 10,000 years and other hyperthermals during the Paleogene serve as analogs of the recent anthropogenically-induced climate warming and associated global change (e.g. Bowen and Zachos, 2010). High-resolution stratigraphic correlation and a stable Paleogene time scale are thus critical for evaluation of rates of past global change to relate to recent global warming in the Anthropocene (Waters et al., 2016).

The PETM event, more exactly the base of the CIE as the initiation of basal Eocene, has been used to define the base of the Ypresian Stage (early Eocene), and thus the base of the Eocene Series, with the Global Boundary Stratotype Section and Point (GSSP) "golden spike" at Dababiya, Egypt (Aubry et al., 2007). Various event scenarios were put forward for the PETM and other Paleogene hyperthermals (e.g. Pagani et al., 2006; Zachos et al., 2005) including light carbon from dissociation of methane hydrates (Dickens et al., 1997) and/or thermal combustion or oxidation of sedimentary organic matter (Svenson et al., 2004). The following early to mid-Eocene time interval indicates a longer-term climate warming trend (e.g. Kelly et al., 2005) to the Early and Middle Eocene Climatic Optimum (EECO/MECO) punctuated by short-term hyperthermal events, with astrochronological dating and timing of events in fast progress (Gradstein et al., 2012; Westerhold et al., 2012, 2017).

This chapter uses a case study of the Paleocene-Eocene time interval from a low-latitude Tethyan pelagic section at Paryab, Zagros Mountains, Iran. Based on nannofossil biostratigraphy and zonation, an evalution of cyclostratigraphy and astrochronology of the late Paleocene and the early Eocene are discussed, using primarily the stable long eccentricity term of Milankovitch cyclicity (e.g. Kent et al., 2018) the stable "metronome" of the Phanerozoic (Gradstein et al., 2012).

2. GEOLOGICAL SETTING

The Zagros Mountains are tectonically a part of the Alpine–Himalayan mountain belt, and formed after convergence between Eurasia and Gondwana, followed by collision of the Eurasian and African-Arabian plates and, finally, the closure of the Neo-Tethys Ocean in southwest Iran (e.g. Sengör, 1990).

The Zagros Basin is the second largest basin in the Middle East with an area of about 553,000 km^2, which extends from Turkey, northeastern Syria and northeaster Iraq throughout northwestern to southeastern Iran, forming nowadays the Zagros fold—thrust belt. The basin resulted from the continental collision between the Arabian margin and the Eurasian plate (e.g. Barbarian and King, 1981; Falcon, 1961) after Neo-Tethys subduction to the north, with the study area situated at the active margin during Late Cretaceous to Paleogene times (e.g. Saura et al., 2011).

During the Paleocene to Eocene, the Zagros Basin was situated on the eastern continental margin of the Tethys and provides a Cenozoic carbonate platform margin and a deeper-water pelagic depositional area characterized by alternating carbonates and fine-grained siliciclastics (Sengör, 1990).

Paleocene-Eocene sections in the Ilam province at Paryab were sampled in detail (Fig. 1A and B). The cyclic limestone-marl successions of the Paleocene—lower Eocene Pabdeh Formation were deposited in a deeper-water bathyal marine environment in the (closing) oceanic area of the Neo-Tethys, and consist of deep-water pelagic to hemipelagic shale, marl (stone) and limestone (Fig. 2). Stratigraphic data for the Paleogene Pabdeh Formation were reported by Senemari (2014), Khorassani et al. (2015), and Rabbani et al. (2015).

Above the gray shales of the Gurpi Formation follows a purple shale unit at the base of the Pabdeh Formation (Fig. 2). This unit was identified as an interval of pelagic oceanic red bed (ORB, Wang et al., 2011) by Khorassani et al. (2015). Above this shaly lower part a cyclic limestone-marl succession follows, overlain by limestones of the Asmari Formation (Figs. 2 and 3).

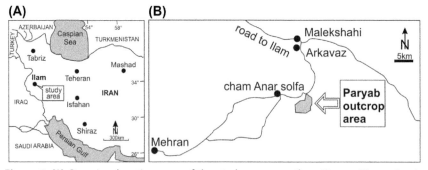

Figure 1 (A) Overview location map of the study area near Ilam, Zagros Mountains, in northwest Iran. (B) Inset detailed road map of the studied outcrops at Paryab.

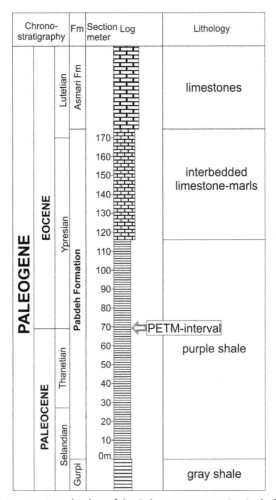

Figure 2 Overview stratigraphic log of the Paleogene succession including the studied part of the Pabdeh Formation in the Zagros Mountains, near Ilam, northwest Iran.

3. MATERIAL AND METHODS

A total of 394 samples for biostratigraphy and geochemistry analyses was taken from the 172.4 m thick section (Fig. 3) including the shaly lower part and marly (limestone–rich) upper part of the Pabdeh Formation in the section at the village of Paryab, NW Iran (N 33°15′14″, E 46°37′3.2″). However, due to outcrop conditions, sampling intervals varied within the section, above 2 m distance in the lower shaly section part up to 60 m,

Figure 3 Field photograph of the main outcrop section measured in the Pabdeh Formation (172 m thickness). To the left, gray shales of the Gurpi Formation underlie purple shales of the lower part of the Pabdeh Formation, including the PETM interval. The upper part of the Pabdeh Formation to the right comprises distinctive limestone-marl cycles.

and a more denser sampling interval of below 50 cm in the middle part of the section including the shaly PETM. The upper part of the section, displaying marl–limestone cycles from 116 m onwards, was sampled bed-by-bed for every marl and limestone bed, resulting in a mean sampling distance of 40 cm.

3.1 Nannofossil Biostratigraphy

100 samples were processed using the standard smear-slide technique to prepare nannofossil samples for light microscopy (Bown, 1998). Small amount of decanted in 50 mL distilled water, then placed for 30 s in an ultrasonic bath. Next, the materials were to be settled for 1 min. After pouring out the upper solution and settling for 1 h, the supernatant was poured off and the residue was diluted with distilled water and used for slide. Small amount of material scraped on to cover slip, drop of water added, mixed and smeared using a toothpick, dried on a hotplate, and the cover slip was attached with Canada balsam to the slide and also one drop was dried on the filter for SEM examination. Biostratigraphic age determination is based on qualitative examination using 100x oil immersion light microscope. Standard nannofossil zonations of Martini (1971) and Agnini et al. (2014) are applied.

3.2 Carbonate Contents and Total Organic Carbon

The carbonate content was measured by Müller-Gastner-Bomb devices (Müller and Gastner, 1971) using diluted hydrochlorid acid on 273 samples. An error range for individual measurements of 1% $CaCO_3$ is reported based on frequently run calibration samples. The mean sampling distance of around 63 cm corresponds to a time resolution of one sample each c. 37 ka.

3.3 Whole Rock Carbonate Stable Carbon Isotopes

This study includes δ^{13}C data for the PETM interval of the Pabdeh Formation of the study area (a carbon and oxygen isotope curve of the whole section is in preparation). 44 powdered samples were analyzed for stable carbon and oxygen isotopes using a Thermo Fisher DeltaPlusXL mass spectrometer equipped with a GasBench II following the procedure of Spötl and Vennemann (2003) at the University of Innsbruck. The δ^{13}C values are corrected according to the NBS19 standard and reported in per mil (‰) relative to the Vienna-PeeDee Belemnite (V-PDB) standard; analytical precision was at 0.05% for δ^{13}C.

3.4 Orbital Cyclicity and Astronomical Calibration

The program package R (R Core Team, 2016) with the software package "astrochron" (Meyers, 2014) was applied for statistical analyses. The carbonate content signature (%CaCO$_3$) was used to search for harmonic frequencies between 69.1 and 172.5 m. Due to irregular sampling intervals data were interpolated (using piecewise linear interpolation).

Peaks in power spectra were calculated with the functions LOWSPEC (Robust Locally-Weighted Regression Spectral Background Estimation) and EHA (Evolutive Harmonic Analysis), to visualize the evolution of spectral density through the section. Window sizes of 20 m (raw data) and 2 Ma (tuned record) were applied. The record was tuned using frequency-domain minimal tuning (see Meyers, 2014). Tuning was performed upon tracing the frequency drift of what is believed to be the 405 ka signal and adjusting the hypothetical sedimentation rate accordingly in the %CaCO$_3$ signature. Significant frequencies were extracted using a bandpass filter with a cosine-tapered window. Extracted frequencies were subsequently correlated to the Laskar orbital solution LA2011b (Laskar et al., 2011).

4. RESULTS

4.1 Nannofossil Biostratigraphy

Nannofossil biostratigraphy indicates standard zones NP6 to NP14 of Martini (1971) and CNP8 to CNE8 of Agnini et al. (2014), respectively, for the studied section of the Pabdeh Formation at Paryab. The section includes the following identified nannofossil zones (for light microscope photographs of selected markers see Fig. 4).

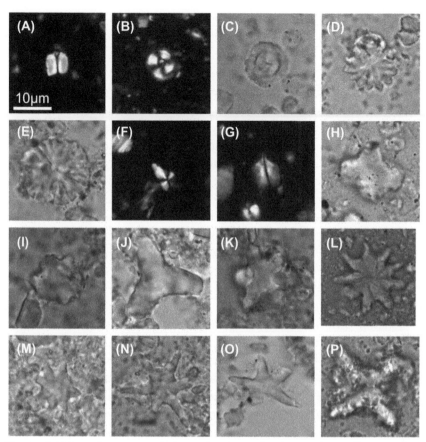

Figure 4 Light microscope photographs (magnification 1000× oil immersion) of selected nannofossil markers from the Paryab section (all same scale as Fig. 4A). (A) *Fasciculithus tympaniformis* Hay & Mohler in Hay et al., 1967 (68.2 m, crossed nicols); (B) *Heliolithus* cf. *cantabriae* Perch-Nielsen, 1971 (2.5 m, crossed nicols); (C) *Heliolithus kleinpellii* Sullivan, 1964 (2.5 m, normal light); (D) *Discoaster mohleri* Bramlette & Percival, 1971 (47.5 m, normal light); (E) *Discoaster multiradiatus* Bramlette & Riedel, 1954 (68.2 m, normal light); (F) *Sphenolithus anarrhopus* Bukry & Bramlette, 1969 (77.5 m, crossed nicols); (G) *Fasciculithus richardii* Perch-Nielsen, 1971 (68.2 m, crossed nicols); (H) *Rhomboaster cuspis* (69.1 m, normal light); (I) *Tribrachiatus* cf. *contortus* (Stradner, 1958) Bukry, 1972 (81.2 m, normal light); (J) *Tribrachiatus orthostylus* Shamrai, 1963 (97.5 m, normal light); (K) *Rhomboaster bramlettei* (Brönnimann & Stradner, 1960) Bybell & Self-Trail, 1995 (68.5 m, normal light); (L) *Discoaster nobilis* Martini, 1961 (62.5 m, normal light); (M) *Discoaster araneus* Bukry, 1971 (70.9 m, normal light); N *Discoaster lodoensis* Bramlette & Riedel, 1954 (142.5 m, normal light); (O) *Discoaster* cf. *sublodoensis* Bramlette & Sullivan, 1961 (158.8 m, normal light); (P) *Nannotetrina cristata* (Martini, 1958) Perch-Nielsen, 1971 (172.4 m, normal light).

4.1.1 Nannofossil Zone CNP8 - NP6

The interval from the base of the section (0 m) to 47.5 m records nannofossil standard zone CNP8 (Agnini et al., 2014: *Heliolithus cantabriae* Zone) and NP6 (Martini, 1971; Perch–Nielsen, 1985; *Heliolithus kleinpellii* Zone). *H. kleinpellii* (Fig. 4C) is present throughout that interval in the Paryab section, together with various *Fasciculithus* (*F. tympaniformis, F.* cf. *bitectus, F. ulii, F. janii*) and rare *Sphenolithus anarrhopus*. The presence of *H. kleinpellii* indicates NP6 of Martini (1971). Although different morpho-types of *H. kleinpellii* are present, no typical *H. cantabriae* (Fig. 4B) could be found, therefore, the middle to upper part of CNP8 is indicated according to Agnini et al. (2014).

The assemblages are predominated by common *Coccolithus pelagicus, Sphenolithus* spp. (mostly *S. primus*), *Cruciplacolithus* spp. (*C. frequens, C.* cf. *asymetricus, C. intermedius*) and few *Toweius eminens, T. pertusus, Ericsonia subpertusa* and *Operculodinella* sp. Rare *Neochiastozygus* (*N. distentus, N. saepes*), *Chiasmolithus* (*C. consuetus*), *Ellipsolithus macellus, Prinsius* sp., *Zygodiscus* spp. are present.

Reworked specimens from Upper Cretaceous (e.g. *Lucianorhabdus cayeuxii, Arkhangelskiella cymbiformis, Cribrosphaerella ehrenbergii*) and Lower Paleocene (e.g. *Placozygus fibuliformis*) make up a few percent of the assemblages.

4.1.2 Nannofossil Zone CNP9/10 - NP7/8

The interval from 47.5 m to 62.5 m is assigned to nannofossil standard zones NP7/8 (Martini, 1971; Perch–Nielsen, 1985: *Discoaster mohleri* Zone and *Discoaster nobilis* Zone) and CNP9/10 (Agnini et al., 2014: *D. mohleri* Zone and *Discoaster backmanii* Zone). *D. mohleri* (Fig. 4D) is consistently present in this interval. and *D. nobilis* (Fig. 4L) occurs rarely. *C. pelagicus* is again abundant; other common genera include *Fasciculithus* spp. (Fig. 4A), *Sphenolithus* spp., *Toweius* spp. Further evolved *Fasciculithus* like *F. clinatus* can be found.

Within the rather low resolution study of this section interval no distinct base for *D. nobilis* (very rare from 62.0 m onwards) or *D. backmanii* (not identified unambiguously) could be found, therefore we join the two zones into one combined nannofossil zone. *H. kleinpellii* has its top within this zone at 50.0 m.

4.1.3 Nannofossil Zone CNP11 - NP9a

The interval from 62.5 m to 67.9 m is assigned to nannofossil standard zones NP9 (Martini, 1971; Perch–Nielsen, 1985: *Discoaster multiradiatus* Zone) and

CNP11 (Agnini et al., 2014: *D. multiradiatus/Fasciculithus richardii* Concurrent Range Zone). *D. multiradiatus* (Fig. 4E) appears at 62.5 m. Nannofossil standard zone NP9 was further subdivided into 2 subzones (e.g. Aubry et al., 2007), with NP9a defined by the base of *D. multiradiatus,* and the base of *Rhomboaster* spp. defining the base of NP9b; thus, in the zonal scheme of Aubry et al. (2007) this interval can be assigned to subzone NP9a.

4.1.4 Nannofossil Zone CNE1 - NP9b

Nannofossil zone CNE1 (Agnini et al., 2014: *F. tympaniformis* Zone) or subzone NP9b ranges from 67.9 to 69.1 m in the Paryab section. The *F. richardii* group (Fig. 4G) disappears at 67.9 m. The first morphotypes of *Tribrachiatus/Rhomboaster* appear at 69.1 m, *Rhomboaster* spp., mainly *Rhomboaster cuspis* (Fig. 4H). *Discoaster araneus* (Fig. 4M) also has its base in this interval as well as *Spenolithus* cf. *moriformis.* The disappearance of common *F. tympaniformis* (Fig. 4A) at 69.1 m is coeval with the base of *Rhomboaster* spp.

4.1.5 Nannofossil Zone CNE2 - NP10

The interval from 69.1 m to 76.2 m is assigned to nannofossil standard zone CNE2 (Agnini et al., 2014: *T. eminens* Partial Range Zone). This correlates to the lower part of NP10 (Martini, 1971; Perch-Nielsen, 1985: *Tribrachiatus contortus* Zone), corresponding most probably to NP10a of Aubry et al. (2007). Subzones NP10a, NP10b, NP10c, NP10d which are defined by the *Tribrachiatus* lineage of *T. bramlettei* - *T. digitalis* - *T. contortus* - *T. orthostylus* (Aubry et al., 2007), could not all be identified unambiguously. *Discoaster diastypus* appears at 73.2 m in this zone. The first *Tribrachiatus* in our study, *T. contortus* (Fig. 4I), appears at 76.2 m and may thus define the base of NP10d of Aubry et al. (2007).

4.1.6 Nannofossil Zone CNE3 — NP10d/NP11

The interval from 76.2 m to 87.6 m ranges into nannofossil standard zone CNE3 (Agnini et al., 2014: *T. orthostylus* Zone). This correlates to the uppermost part of NP10 (NP10d) and the whole NP11 zone (*Discoaster binodosus* Zone) of Martini (1971). The first *T. orthostylus* (Fig. 4J) occurs rarely at 76.2 m, still together with *R. cuspis. Rhomboaster bramletteii* (Fig. 4K) has its first sporadic occurrence at 78.6 m, *Discoaster barbadiensis* at 79.2 m. *T. digitalis* and *Coccolithus bownii* start at 81.6 m, *Sphenolithus radians* and *D. binodosus* start consistently at 82.6 m. The last occurrence of

T. contortus at 83.6 m defines the top of NP10 which thus covers an interval up to 83.6 m; however, the range of *T. contortus* overlaps completely with (rare) *T. orthostylus.*

4.1.7 Nannofossil Zone CNE4 - NP12

The interval from 87.6 m to 119.6 m is assigned to nannofossil standard zone CNE4 (Agnini et al., 2014: *Discoaster lodoensis/T. orthostylus* Concurrent Range Zone) and NP12 (*T. orthostylus* Zone) of Martini (1971). The first *Discoaster* sp. aff *D. lodoensis* (Fig. 4N) occurs at 87.6 m defining the base of CNE4 (base common *D. lodoensis* of Agnini et al., 2014) and NP12; however, this species is rather rare throughout the section.

Common *C. pelagicus* and *Sphenolithus* spp. (mainly *S. radians*) characterize the assemblages, besides *Discoasters* (e.g. *D. barbadiensis, D. salisburgensis, D. mahmoudi*), *Toweius gammation*, and rare *Chiasmolithus* (*C.* cf. *californicus*).

4.1.8 Nannofossil Zone CNE5 - NP13

Nannofossil zone CNE5 (Agnini et al., 2014: *Reticulofenestra dictyoda* Partial Range Zone) ranges from 119.6 m to 138.7 m, and correlates to NP13 (*D. lodoensis* Zone). The last occurrence of *T. orthostylus* is at 119.6 m, *D. lodoensis* is still rarely present.

Besides *Discoaster* spp. (*D. salisburgensis, D. kuepperi, D. barbadiensis, D. lodoensis*), *S. radians* dominates the assemblages, whereas *C. pelagicus* decreases significantly.

4.1.9 Nannofossil Zone CNE6 - NP14

The interval from 138.7 m to 158.8 m is assigned to nannofossil standard zones CNE6 (Agnini et al., 2014: *Discoaster sublodoensis/D. lodoensis* Concurrent Range Zone) and the lower part of NP14 (*D. sublodoensis* Zone) of Martini (1971). The first five-rayed *Discoaster* cf. *sublodoensis* (Fig. 4O) occurs sporadically at 138.7 m together with *D. lodoensis*. The top of *D. lodoensis* is around 158.8 m.

4.1.10 Nannofossil Zone CNE7 - NP14

The interval from 158.8 to 172.0 m is assigned to nannofossil standard zones CNE7 (Agnini et al., 2014: *D. barbadiensis* Partial Range Zone) and the middle part of NP14 (*D. sublodoensis* Zone) of Martini (1971). The top of *D. lodoensis* is recorded at 158.8 m, defining the base of CNE7.

4.1.11 Nannofossil Zone CNE8 - NP14

The top of the Paryab section from 172.0 m to 172.4 m is assigned to nan-nofossil standard zone CNE8 (Agnini et al., 2014: *Nannotetrina cristata* Zone) and the uppermost part of NP14 (*Nannotetrina fulgens* Zone) of Martini (1971). *Nannotetrina cristata* (Fig. 4P), the first *Nannotetrina* species recorded, occurs from 172 m onwards.

Blackites spp. (*Blackites spinosus, Blackites stylus*) and *Reticulofenestra* (mainly *R. dictyoda* and *R. bisecta*) are present. In addition, *Sphenolithus* spp. show an evolutionary step to forms that may lead to *Sphenolithus furcalithoides* higher up.

4.2 Onset of the PETM CIE

A prominent negative carbon isotope excursion starts in the Paryab section at 69.1 m, indicated by a shift of up to $-2‰$ from values around $2.0‰$ to $-0.2‰$. Carbon isotope values then fluctuate between 0% and $1‰$ without showing distinct peaks or plateaus (Fig. 5 inset).

4.3 Carbonate Content Signature

The carbonate content signature fluctuates mostly between 50% and 70% from 0 to 125 m. We document a slight increase to values that fluctuate around 70% between 125 and 132m, that is followed by distinct drop in %CaCO$_3$ at 142 m. The top of the section once again shows slightly higher %CaCO3 levels that oscillate around 70%. Albeit irregular sampling intervals, a cyclic pattern is clearly visible in the data (Fig. 6).

4.4 Cyclostratigraphic Evaluation

The LOWSPEC analysis finds several significant spectral peaks above the 80% confidence interval (CI) according to the criteria outlined in Meyers (2012) (see Figs. 7 and 8). Significant harmonics that clearly meet the requirements were identified at 36.12 m, 5.49 m, and 3.49 m and in the frequency band between 2.47 and 1.89 m cycle thickness. The evolution of significant harmonics visualized in the EHA shows two distinct harmonic frequencies that are present throughout almost the whole section. Seemingly continuous signals can be followed from 70 m at 0.1 cycles per/m to 140 m at 0.07 cycles/m and from 70 m with 0.18 cycles/m to 135 m with 0.35 cycles/m. Other signals visible in this analysis seem to be discontinuous and show a patchy distribution. The spectral resolution between 70 and 120 m suffers from an undersampled interval in the shales present in the older segments of the section. The

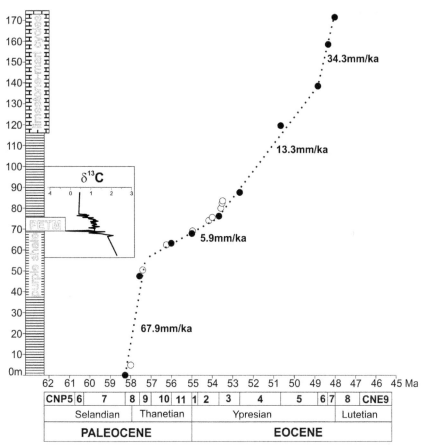

Figure 5 Age-depth plot of the Paryab section using the nannofossil zonal age model of Agnini et al. (2014). Inset shows $\delta^{13}C$ carbon isotope curve around the PETM interval. Black circles indicate nannofossil zonal markers, white circles secondary nannofossil biomarkers recorded in the section. Numbers refer to sediment accumulation rates in mm/ka.

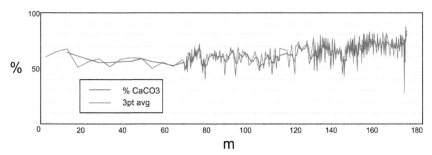

Figure 6 Carbonate content (%CaCO$_3$) of the Paryab section between 0 and 172.43 m. A 3-point average is calculated and given in red.

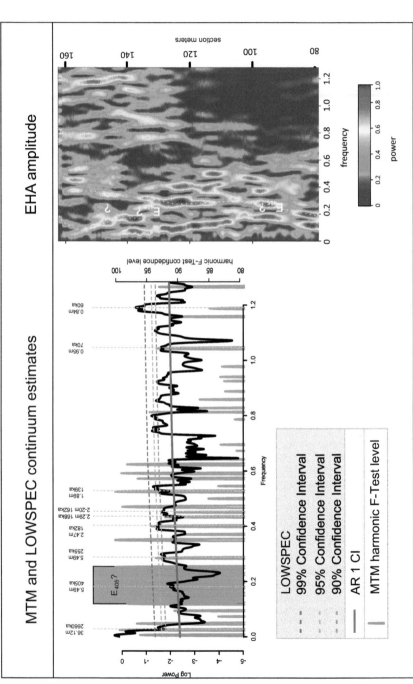

Figure 7 LOWSPEC and EHA analyses of the Paryab section. Significant peaks at 36.12 m, 5.49 m, 3.45 m and a frequency band from 2.47 to 1.89 m as well as two significant signals at 0.95 and 0.84 m are visualized in the LOWSPEC analysis. A green bar marks the hypothetical position of the frequency band corresponding to the stable 405 ka cycle. The EHA shows two distinct frequency bands starting at 0.1 cycles/m and 0.18 cycles/m that indicates a

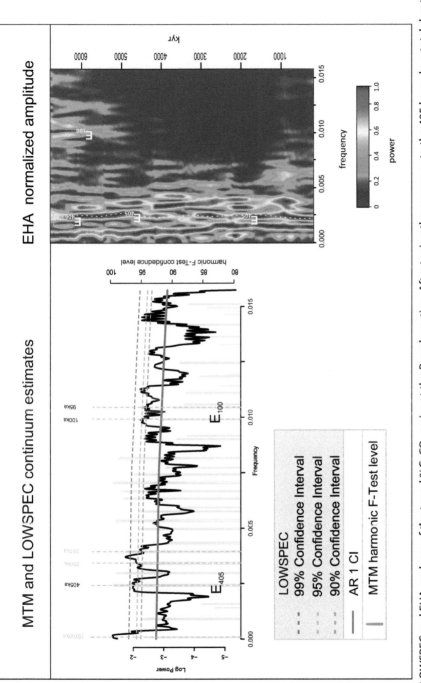

Figure 8 LOWSPEC and EHA analyses of the tuned %CaCO$_3$ series at the Paryab section. After tuning the sequence to the 405 ka cycle, a total duration of 8.566 Ma can be inferred. The temporal location of peaks in this analysis is given in black for signals that can be linked to an orbital target and in gray for signals that do not match orbital targets in the LOWSPEC analysis. The EHA shows a continuous 405 ka eccentricity signal as well as faint signals for the 100 ka eccentricity that is present in the uppermost parts of the section. We also record a significant signal that represents durations from 800 to 1600 ka (between 0.0006 and 0.0012 frequency).

resolution of the topmost intervals in the EHA shows several faint signals between 0.5 and 1.5 cycles/m. Fig. 7 shows the results of spectral analyses of %CaCO$_3$ raw data.

5. DISCUSSION

5.1 The Base of the PETM

From 68.8 m to 69.1 m in the Paryab section, carbon isotope values fall from 1.68‰ to −0.23‰ (see inset δ^{13}C curve in Fig. 5) depicting the beginning of the distinct negative carbon isotope excursion (CIE). This co-incides in the same sample with the base of nannofossil zone CNE2 (Agnini et al., 2014: *F. tympaniformis* Zone). The disappearance of the *F. richardii* group appears 1.2 m below this level, at 67.9 m. The first *R. cuspis* and *D. araneus* appear also at this level, as well as the disappearance of common *F. tympaniformis,* attesting to the base of the PETM interval (Aubry et al., 2007).

Lithologically, there is no visible sign of changes in sedimentation as known from other sites, e.g. carbonate dissolution or black shale deposition. Carbonate contents fluctuate around the PETM interval from 47% to 72%, without a clear-cut regularity. Although the onset of the PETM-CIE can be defined precisely, the following interval does not show a typical PETM δ^{13}C excursion pattern, but fluctuations, followed by a longer term plateau phase upwards (Azami et al., in prep.).

5.2 Nannoplankton Biostratigraphy

Nannofossil zones recorded at Paryab include the interval from CNP8 (NP6) to CNE8 (NP14). This indicates the base of the section (CNP8) in the late Paleocene, at around the Selandian/Thanetian boundary (Agnini et al., 2014). The upper part of CNP8 already ranges into the Thanetian, as well as CNP9/10 (NP7/8) and CNP11 (NP9).

The base of the Eocene is defined by the onset of the PETM CIE, which correlates to the last appearance of the *F. richardii* group and the first appearance of *Rhomboaster* spp (Aubry et al., 2007; Agnini et al., 2014). Nannofossil zone CNE1 starts at 69.1 m in the Paryab section with the first occurrence of *R. cuspis.* Due to reworking the last occurrence of the *F. richardii* group is not unambiguous to define but seems to correspond to roughly the same level except single occurrences further upsection that are interpreted as reworked. *D. araneus* also appears at this level as well as *Spenolithus* cf. *moriformis.* This marks both the base of the PETM as indicated by the negative CIE, and the base of the Eocene and the base of the Ypresian (early Eocene). Thus, this horizon could be precisely defined in the Paryab section.

Higher up-section, *Sphenolithus moriformis* and *D. diastypus* appear, and $\delta^{13}C$ values increase again. *T. contortus* and *T. orthostylus* have their first occurrence at 75.6 m, followed by *R. bramletteii*, *D. barbadiensis*, *Sphenolithus* cf. *radians*, and *D. binodosus*.

Assemblages at Paryab are typically Tethyan/low-latitude and of warm-water character, with lots of *Discoaster* and other warm-water species (e.g. Bown, 1998; Agnini et al., 2014). The nannoplankton assemblages allow the identification of the PETM interval due to the large turnover in assemblages at the base of the PETM interval (Aubry et al., 2007). This short-term marine warming was followed by a long-term cooling after the early-middle Eocene. Though thermal variation from warmer ocean waters during the PETM to the cooler ocean water temperatures afterward is likely to have decreased the effects of Milankovitch cyclicity on the sedimentation pattern, however, significant facies change from shales to shale—marl cycles can be identified in the section after the PETM interval, and may be regarded as an expression to higher productivity of calcareous plankton organisms in the aftermath of the event (Kelly et al., 2005).

5.3 Cyclostratigraphy and Astronomical Tuning

The interval subject to cyclostratigraphic investigation was chosen because of the relatively dense sample positions, the presence of a probably undisturbed succession between 69.1 and 172 m and the astronomically calibrated onset of the PETM at 55.93 Ma (e.g., Westerhold et al., 2012, 2017), that is evident in the lowermost segment of this interval. Despite a difference in lithological properties, data from shales and overlying limestone/marl rhythmites were not analyzed separately as sedimentation rates inferred from biostratigraphic data suggest to be almost uniform from 50 m onwards (Fig. 5). Still, the analysis suffers from a rather poor sampling resolution in the shaly interval between ~85 and 110 m: An undersampled interval is best reflected in the EHA spectrum of the % $CaCO_3$ raw data (see Fig. 7).

The untuned %$CaCO_3$ signature (Fig. 7) is hard to interpret as the stable 405 ka is not well expressed and shorter frequencies tend to be either masked by the dominant long-term trend or too close to the Rayleigh frequency to be considered a robust signal. Upon lowering the threshold for significant peaks in the LOWSPEC analyses to 80%, a signal that is likely to represent the 405 ka cycle was identified.

Also evident in the raw data are signals that might correspond to the long 2.4 Ma- and the short ~ 100 ka eccentricity cycle. Other shorter signals evident in the LOWSPEC analysis of %CaCO₃ raw data are either discontinuous and cannot be attributed an astronomical target or suffer from the poor sample resolution (particularly in the older segments of the section). Nevertheless, signals that correspond to wavelengths representing 150–260 ka could be related to precession and obliquity amplitude modulations (see Hinnov, 2000).

The %CaCO₃ signature was tuned to the possible 405 ka signal (Fig. 8 and Table 2). Significant signals visible in spectral analyses of the tuned record include two short eccentricity terms of 100 and 90 ka durations and a 291 ka harmonic frequency that cannot be attributed an orbital target. The EHA of the tuned timeseries also shows evidence for prominent harmonic signals below the 405 ka frequency band that represent durations from 800 to 1900 ka. As these signals are not present in the robust LOWSPEC

Table 1 Age Estimates of Nannofossil Bioevents and Standard Zones Used in this Study, Based on Agnini et al. (2014), Including Position in the Paryab Section

Bioevent	Base zone	Age (Ma)	Position (m)
Base *Nannotetrina cristata*	CNE8	47.99	172.
Top *Discoaster lodoensis*	CNE7	48.37	158.8
Base *Discoaster sublodoensis* (5-rayed)	CNE6	48.96	138.7
Top *Tribrachiatus orthostylus*	CNE5	50.66	119.6
Base *Discoaster lodoensis*	CNE4	52.64	87.6
Top *Tribrachiatus contortus*		53.49	83.6
Base *Sphenolithus radians*		53.53	82.6
Top *Discoaster multiradiatus*		53.58	80.2
Base *Tribrachiatus orthostylus*	CNE3	53.67	76.2
Base *Tribrachiatus contortus*		54.00	76.2
Base *Discoaster diastypus*		54.13	73.2
Top *Fasciculithus tympaniformis*	CNE2	54.71	69.1
Base *Rhomboaster* spp.		54.99	69.1
Top *Fasciculithus richardii* group	CNE1	55.00	1 67.9
Base *Discoaster multiradiatus*	CNP11	56.01	62.5
Base *Discoaster nobilis*		56.25	62.0
Top *Heliolithus kleinpellii*		57.42	50.0
Base *Discoaster mohleri*	CNP9	57.57	47.5
Base *Heliolithus kleinpellii*		58.03	5.0
Base *Heliolithus* cf. *cantabriae*	CNP8	58.27	0.0?

Table 2 Comparison of Harmonic Signals Detected in LOSWSPEC Analyses of Tuned %CaCO$_3$ and Raw Data

Raw data		Tuned data		
cycles/m	ka	cycles/ka	ka	Orbital target in ka
		0.000009	c. 10700	
36.12	2660			E2400
5.49	405	0.0024	405	E405
3.45	255	0.0034	291	
2.47	182	0.0039	253	
2.29	168			E120?
2.20	162	0.0099	100	E100
1.89	139	0.0104	95	E90
0.95	70			
0.84	60			

analyses, we tend to interpret them as harmonic artifacts that originate from orbital tuning and represent multiples of harmonic frequencies.

The tuned %CaCO$_3$ sequence from the onset of the PETM at 69.1 m to the top of the Paryab section spans a total of 8.566 Ma and records twenty one 405 ka eccentricity cycles. In Fig. 8 the tuned %CaCO$_3$ signature was correlated to the LA2011B (Laskar et al., 2011) solution (as used in Westerhold et al., 2012).

The duration of the orbitally tuned %CaCO$_3$ data was compared to estimates for the Eocene nannofossil zonations (see Agnini et al., 2014): The base of the Lutetian and nannofossil zone CNE8 (Agnini et al., 2014) at 170 m can be dated with c. 47.5 Ma (this cyclostratigraphic solution finds an age value of 47.455 Ma for 170 m). Agnini et al. (2014) give an estimated age of 47.99 Ma for the base of this zone. Thus, our age estimate for the base of CNE8 at the Paryab section is approximately one 405 ka eccentricity cycle off the calibrated age for this zone. The reasons for that could be flaws in our cyclostratigraphic model, gaps in the record at Paryab section or diachroneity of nannofossil markers. In addition, differences in the exact timing of the onset of the PETM are plausible, e.g. Agnini et al. (2014) use a different age of 55.0 Ma for their base of the PETM CIE (and base of CNE1) than the tuned data by Westerhold et al. (2007, 2012, 2017).

5.4 Chronostratigraphy

The nannofossil zonal scheme and the CIE allow to interpret age models for the Paryab section and to give an age–depth plot (Fig. 5) for the succession. We apply the age model of Agnini et al. (2014) using their dating of

CNP and CNE zones based on magnetostratigraphic correlations mainly. According to the Agnini et al. (2014) age model (Table 1), the base of the section is at 58.27 Ma (base CNP8), base CNE1 at 55.00 and CNE2 at 54.99 correlating to the PETM base CIE, and base CNE8 at the top of the section at 47.99 Ma, According to this age model, the section spans about 10.28 Ma (base uncertainty at maximum of 0.7 Ma). The Eocene part of the section records a time span of c. 7 Ma, from 55.00 Ma to 47.99 Ma.

Modifying ages with the astronomically calibrated onset of the PETM at 55.53 Ma (option 1 of Westerhold et al., 2012) and more recently of 55.93 Ma (Westerhold et al., 2017; error range according to Westerhold et al., 2007: $\pm$ 0.02 Ma) results in a possible duration of the Eocene part of c. 7.5 Ma or 7.9 Ma, respectively. According to an alternative age model of Gradstein et al. (2012) the base of the Paryab section would be at 59.54 Ma (base NP6), up to the middle part of NP14 with the base of *Nannotetrina* spp. at 47.73 Ma. Thus, the investigated interval records a time duration of c. 11.81 Ma (with an uncertainty of c. 0.5 Ma, as an unknown lower part of NP6 may be missing).

Using the age-depth plot based on the Agnini et al. (2014) age model indicates continuous deposition without major gaps, with varying sediment accumulation rates from 68 mm/ka in the lowermost part (CNP8-CNP9) of the section, a middle part around the PETM interval (CNP10-CNE2) with 5.9 mm/ka, an interval (CNE3-CNE6) with to 13.3 mm/ka, and an uppermost part (CNE7-CNE8) with 34.3 mm/ka.

Furthermore, the continuous and tuned Paryab section allows investigating definitions and correlations of stages of the Paleogene System (Fig. 9). The studied section starts in the Selandian and ends in the Lutetian, and thus can contribute to the refinement of the bases of the Thametian, Ypresian and Lutetian. All of these stages are defined by GSSPs (Global Boundary Stratotype Section and Point) and conform the "golden spike" principle in chronostratigraphic definitions (e.g. Remane et al., 1996).

5.4.1 The Base of the Thanetian
The Thanetian Stage comprises the youngest stage of the Paleocene Series with a type locality in Southern England (Gradstein et al., 2012). The base of the Thanetian is defined by a GSSP in the Zumaia section, Itzurun Beach, Basque Country, northern Spain (Schmitz et al., 2011). The primary marker for the base of the Thanetian is a magnetic reversal, i.e. the base of

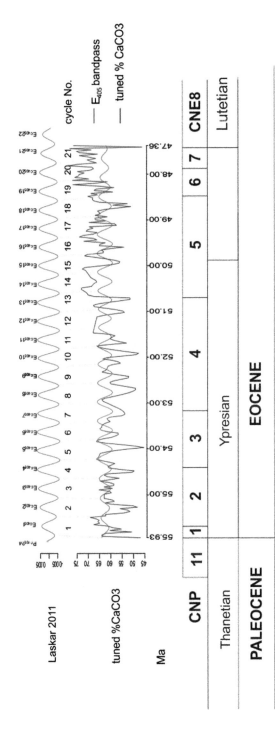

Figure 9 The tuned %CaCO₃ signature is correlated to the Laskar orbital solution 2011. The nomenclature of 405 ka eccentricity cycles corresponds to that of Westerhold et al. (2008). The %CaCO₃ signature was bandpassed to extract the frequencies corresponding to the 405 ka eccentricity signal. This solution records twentyone 405 ka eccentricity cycles from 69.1–172.43 m that correspond to a total duration of 8.56 Ma (from the onset of the PETM at 55.93 Ma to the top of the section at 47.36 Ma). Paleocene-Eocene chronostratigraphy and nannofossil zones of Agnini et al. (2017) are modified according to astrochronology.

magnetochron C26n (the C26r/C26n reversal) in the Zumaia section (Dinarès-Turell et al., 2007). According to Schmitz et al. (2011) this golden spike corresponds to a distance of eight precession cycles above the base of the clay-rich interval associated with the Mid-Paleocene Biotic Event (MPBE), characterized by calcareous nannofossil and foraminifer assemblage changes (Dinarès-Turell et al., 2007). Generally, nannofossil assemblages of the GSSP interval in nannofossil zones CNP8 and NP6 are characterized by a diversification of the genera *Heliolithus* and *Discoaster*. Especially the first appearance and radiation of *Discoaster* is remarkable, providing one of the most stratigraphically significant calcareous nannofossil groups throughout the Paleogene. The base of *D. mohleri* (= base of CNP9 of Agnini et al., 2014) provides the nearest nannofossil bioevent, c. 5 m above the GSSP level in the Zumaia section. Cyclostratigraphy (Dinarès-Turell et al., 2007) indicates for magnetochron C26r (upper part of the Selandian) a duration of 2.793 Ma (133 precession cycles), for chron C26n (basal part of the Thanetian) a short duration of 0.231 Ma (11 precession cycles) and for chron C25r a duration of 1.449 Ma (69 precession cycles). Gradstein et al. (2012) indicated a numerical age of 59.11 Ma for the base of C26n from their radio-isotopic age model, and 59.24 Ma by means of their astronomic age model for the base of the Thanetian.

At Paryab, the base of the Thanetian cannot be precisely defined by its primary marker due to lack of magnetostratigraphic analysis. The Selandian/Thanetian boundary correlates into CN8/NP6, between the base of *H. kleinpellii* at 5.0 m and the base of *D. mohleri* at 47.5 m. Thickness evidence from Zumaia may infer a position somewhere in the middle part of NP6, probably around 20 m in the Paryab section (Fig. 5). However, no sign of any lithological or facies change can be recognized within this shaly interval as would be suggested by the MPBE, a hyperthermal event present in several Tethyan sections and the Zumaia GSSP, but not identified in the studied Paryab section.

5.4.2 The Base of the Ypresian

The Ypresian Stage comprises the lowermost stage of the Eocene, with the Paleocene/Eocene boundary at the base; therefore, the base of the Ypresian is equivalent to the base of the Eocene, and, consequently, relates to the lower Eocene. The classical stratotype is situated in western Belgium (Gradstein et al., 2012). The modern GSSP was defined in the Dababiya section near Luxor, Egypt (Aubry et al., 2007). The marker event chosen was the onset of the PETM CIE, i.e. the onset of the 2.5 to 4‰ negative carbon isotope excursion. Correlations to secondary marker include: (1) the lower to middle part of magnetochron C24r — according to Westerhold et al. (2008) at C24r.36, c. 1.4 Ma above the base of C24r, (2) a position within nannofossil zone NP9 marked by the occurrence of the *Rhomboaster-D. araneus* assemblage (Aubry et al., 2007) or the calcareous nannofossil excursion taxa (CNET, Agnini et al., 2014; Westerhold et al., 2015). The base of the Ypresian thus correlates to the top of the *F. richardii* group, the base of *Rhomboaster* spp. and the subsequent top of *F. tympaniformis*/spp. (Aubry et al., 2007; Agnini et al., 2014). The age of the base of the Ypresian, and thus the onset of the PETM interval, is still in discussion. Gradstein et al. (2012) indicate a numerical age of 56.0 Ma or 55.96 Ma, with the base of *Rhomboaster* spp. at 55.96 Ma, whereas Agnini et al. (2014) give an age of 55.0 Ma for the top of the *F. richardii* group, and 54.99 for the base of *Rhomboaster* spp. Westerhold et al. (2012, 2015, 2017) combining astrochronology and available geochronological data report most recently an age of the PETM base at 55.93 Ma (option 2 in Westerhold et al., 2012). All of these ages derived from cyclostratigraphy have error ranges of 0.01 to 0.04 Ma (Westerhold et al., 2007, 2012).

The studied section at Paryab provides a high resolution record around the onset of the PETM and thus the Thanetian/Ypresian and Paleocene/Eocene boundary. The base of the CIE is, however, not defined unambiguously — a first slight decrease of values start at 68.2 m (from 1.40 to 0.91‰) followed by a slight increase (δ^{13}C of 1.23 and 1.68‰) and a final abrupt decrease to $-0.23‰$ from 68.8 to 69.1 m. We take this level as the actual onset of the CIE, keeping in mind the small bias introduced by using the onset of a chemostratigraphic signal/peak – a small uncertainty in correlations due to sample spacing and analytical errors already noted by Aubry et al. (2007). At 69.1 m, as *Rhomboaster* spp. and other CNET taxa (Aubry et al., 2007; Agnini et al., 2014) occur. Applying the astrochronological solution of Laskar et al. (2011) and our tuned %CaCO$_3$ record we infer the Thanetian/Ypresian boundary (Fig. 9) at a position within the Ec$_{405}$1 cycle of Westerhold et al. (2008). However, using the Agnini et al. (2014) age of 55 Ma for the base of the Ypresian, this datum may already fall into Ec$_{405}$3, giving still a possible correlation uncertainty of 2 long eccentricity cycles.

5.4.3 The Base of the Lutetian

The Lutetian Stage comprises the second stage of the Eocene, above the Ypresian, defined first within the Paris Basin (Gradstein et al., 2012). The

GSSP of the base of the Lutetian is defined in the Gorrondatxe section, Basque Country, Spain (Molina et al., 2011). The primary marker chosen is the base of *Blackites inflatus* within NP14 and at the base of CNE8 (Agnini et al., 2014), between the top of *D. sublodoensis* and the base of *Nannotetrina cristata*. This corresponds to a maximum flooding event and 39 precession cycles (c. 820 ka) above the base of magnetochron C21r. The lowest occurrence of *Nannotetrina cristata* is approximately 115 ka younger than the GSSP level, and the base of *D. sublodoensis* more than 900 ka older (Molina et al., 2011).

Gradstein et al. (2012) report an age of 47.8 Ma for the base of the Lutetian. Agnini et al. (2014) have the base of *B. inflatus* and *Nannotetrina cristata* at the same level, at 47.99 Ma.

At Paryab, the base of CNE8 was inferred to the base of *Nannotetrina cristata* at the top of the section, at 172.0 m. Thus, we interpret this level as the base of the Lutetian (with a possible error of c. 115 ka correlated to the GSSP section of Molina et al., 2011). A high resolution correlation based on nannofossil biostratigraphy to the GSSP is not possible due to the rareness of the genus *Blackites*, and the lack of *B. inflatus* within the Paryab section. Using the astrochronological solution of Laskar et al. (2011) and our tuned % $CaCO_3$ record (Fig. 9) we infer the Ypresian/Lutetian boundary to be positioned in the $Ec_{405}21$ cycle of Westerhold et al. (2008), with a calculated duration for the Ypresian of 8.566 Ma.

6. CONCLUSIONS

The Paryab section, NW Iran, records a continuous pelagic-hemipelagic sedimentation during the Paleogene within the Neo-Tethys. The section of the Pabdeh Formation sheds light on late Paleocene to early Eocene stratigraphy. Nannofossil biostratigraphy constrains the studied section to the standard nannofossil zones CNP8 to CNE8 (NP6 to NP14), indicating Selandian, Thentian, Ypresian to basal Lutetian ages. The Paleocene-Eocene Thermal Maximum (PETM) interval was identified by a significant negative carbon isotope excursion of c. 2‰. An age–depth plot using ages for nannofossil zones indicates continuous deposition, with varying sediment accumulation rates from 68 mm/ka in the lowermost part of the section to 6 mm/ka around the PETM interval. Power spectra analysis indicates the presence of twentyone 405 ka cycles from the base of the PETM up to the top of the studied section. Orbital tuning to the established Laskar target curve is in accordance with the

Paleocene-Eocene cyclostratigraphic models. The base of the Thanetian, Ypresian and Lutetian are present within the studied sections and can be further refined, giving a duration 8.57 Ma for the Ypresian.

ACKNOWLEDGMENTS

We thank UNESCO IGCP 609 *Climate-environmental deteriorations during greenhouse phases: Causes and consequences of short-term Cretaceous sea-level changes*, and the Austrian Academy of Sciences, International Research Programs, for financial support. We also thank Sabine Hruby-Nichtenberger and Maria Meszar (all University of Vienna) for sample preparation and lab analyses.

REFERENCES

Abels, H.A., Clyde, W.C., Gingerich, P.D., et al., 2012. Terrestrial carbon isotope excursions and biotic change during Palaeogene hyperthermals. Nat. Geosci. 5, 326—329.

Agnini, C., Fornaciari, E., Raffi, I., Catanzariti, R., Pälike, H., Backman, J., Rio, D., 2014. Biozonation and biochronology of Paleogene calcareous nannofossils from low and middle latitudes. Newsl. Stratigr. 47, 131—181.

Aubry, M.-P., Ouda, K., Dupuis, C., et al., 2007. Global standard stratotype-section and point (GSSP) for the base of the Eocene series in the Dababiya section (Egypt). Episodes 30, 271—286.

Bains, S., Corfield, R.M., Norris, R.D., 1999. Mechanisms of climate warming at the end of the Paleocene. Science 285, 724—727.

Barbarian, M., King, G.C.P., 1981. Towards a paleogeography and tectonic of Iran. Can. J. Earth Sci. 18, 210—265.

Bowen, G.J., Zachos, J.C., 2010. Rapid carbon sequestration at the termination of the Palaeocene—Eocene thermal maximum. Nat. Geosci. 3, 866—869.

Bown, P.R. (Ed.), 1998. Calcareous Nannofossil Biostratigraphy. British Micropalaeontological Soc. Publ. Ser., Chapman and Hall, London.

Dickens, G.R., Castillo, M.M., Walker, J.C.G., 1997. A blast of gas in the latest Paleocene: simulating first-order effects of massive dissociation of oceanic methane hydrate. Geology 25, 259—262.

Dinarès-Turell, J., Baceta, J.I., Bernaola, G., Orue-Etxebarria, X., Pujalte, V., 2007. Closing the mid-Palaeocene gap: toward a complete astronomically tuned Palaeocene Epoch and Selandian and Thanetian GSSPs at Zumaia (Basque Basin, W. Pyrenees). Earth Planet Sci. Lett. 262, 450—467.

Falcon, N.L., 1961. Major earth-flexuring in the Zagros mountain of southwest Iran. J. Geol. Soc. London 117, 367—376.

Gradstein, F.M., Ogg, J.G., Schmitz, M.D., Ogg, G.M. (Eds.), 2012. The Geological Time Scale 2012. Elsevier, Amsterdam.

Hay, W.W., Flögel, S., Söding, E., 2004. Is the initiation of glaciation on Antarctica related to a change in the structure of the ocean? Glob. Planet. Change 45, 23—33.

Hinnov, L.A., 2000. New perspectives on orbitally forced stratigraphy. Annu. Rev. Earth Planet Sci. 28, 419—475.

Kelly, D.C., Zachos, J.C., Bralower, T.J., Schellenberg, S.A., 2005. Enhanced terrestrial weathering/runoff and surface ocean carbonate production during the recovery stages of the Paleocene- Eocene thermal maximum. Paleoceanography 20, 1—11.

Kent, D.V., Olsen, P.E., Rasmussen, C., et al., 2018. Empirical evidence for stability of the 405-kiloyear Jupiter—Venus eccentricity cycle over hundreds of millions of years. Proc. Natl. Acad. Sci. Unit. States Am. https://doi.org/10.1073/pnas.1800891115.

Khorassani, M.P.K., Ghasemi-Nejad, E., Wagreich, M., Hadavi, F., Richoz, S., Harami, R.M., 2015. Biostratigraphy and geochemistry of upper Paleocene - Lower Eocene Oceanic Red Beds from the Zagros Mountains, SW Iran. J. Earth Sci. Climatic Change 6, 302. https://doi.org/10.4172/2157-7617.1000302.

Laskar, J., Gastineau, M., Delisle, J.B., Farres, A., Fienga, A., 2011. Strong chaos induced by close encounters with Ceres and Vesta. Astron. Astrophys. 532, L4.

Martini, E., 1971. Standard Tertiary and Quaternary calcareous nannoplankton zonation. In: Farinacci, A. (Ed.), Proceedings 2nd International Conference Planktonic Microfossils Roma, Rome 2, pp. 739—785.

Meyers, S.R., 2012. Seeing red in cyclic stratigraphy: spectral noise estimation for astrochronology. Paleoceanography 27, PA3228. https://doi.org/10.1029/2012PA002307.

Meyers, S.R., 2014. Astrochron: An R Package for Astrochronology. http://cran.r-project.org/package=astrochron.

Molina, E., Alegret, L., Apellaniz, E., et al., 2011. The global stratotype section and point (GSSP) for the base of the Lutetian stage at the Gorrondatxe section, Spain. Episodes 34, 86—108.

Müller, G., Gastner, M., 1971. The "Karbonat-Bombe", a simple device for the determination of the carbonate content in sediments, soils, and other materials. N. Jahrb. Mineral., Mh. 10, 466—469.

Pagani, M., Caldeira, K., Archer, D., Zachos, J.C., 2006. An ancient carbon mystery. Science 314, 1556—1557.

Perch-Nielsen, K., 1985. Cenozoic calcareous nannofossils. In: Bolli, H.M., Sanders, J.B., Perch-Nielsen, K. (Eds.), Plankton Stratigraphy. Cambridge University Press, Cambridge, pp. 427—554.

Rabbani, J., Ghasemi-Nejad, E., Ashori, A., Vahidinia, M., 2015. Quantitative palynostratigraphy and palaeoecology of Tethyan Paleocene—Eocene red beds in north of Zagros sedimentary basin, Iran. Arab. J. Geosci. 8, 827—838.

Remane, J., Bassett, M.G., Cowie, J.W., Gohrbandt, K.H., Lane, H.R., Michelsen, O., Wang, N., 1996. Revised guidelines for the establishment of global chronostratigraphic standards by the International Commission on Stratigraphy (ICS). Episodes 19, 77—81.

R Development Core Team, 2016. R: A Language and Environment for Statistical Computing. R Foundation for Statistical Computing, Vienna.

Saura, E., Vergés, J., Homke, S., et al., 2011. Basin architecture and growth folding of the NW Zagros early foreland basin during the Late Cretaceous and early Tertiary. J. Geol. Soc. London 168, 235—250.

Schmitz, B., Pujalte, V., Molina, E., Monechi, S., Orue-Etxebarria, X., Speijer, R.P., Alegret, L., Apellaniz, E., Arenillas, I., Aubry, M.-P., Baceta, J.I., Berggren, W.A., Bernaola, G., Caballero, F., Clemmensen, A., Dinarès-Turell, J., Dupuis, C., Heilmann-Clausen, C., Hilario Orús, A., Knox, R., Martín-Rubio, M., Ortiz, S., Payros, A., Petrizzo, M.R., von Salis, K., Sprong, J., Steurbaut, E., Thomsen, E., 2011. The global stratotype sections and points for the bases of the selandian (middle Paleocene) and Thanetian (upper Paleocene) stages at Zumaia Spain. Episodes 34, 220—243.

Senemari, S., 2014. Diversity changes among calcareous nannofossil assemblages across the Paleocene/Eocene Boundary in the Zagros (Southwest Iran). J. Tethys 2, 45—54.

Şengör, A.M.C., 1990. A new model for late Paleozoic- Mesozoic tectonic evolution of Iran and implications for Oman. In: Robertson, A.H.F., Searle, M.P., Ries, A.C. (Eds.), The Geology and Tectonics of the Oman Region, Geol. Soc. Spec. Publ., vol. 49, pp. 797—831.

Spötl, C., Vennemann, T., 2003. Continuous -flow isotope ratio mass spectrometric analysis of carbonate minerals. Rapid Commun. Mass Spectrom. 17, 1004—1006.

Svensen, H., Planke, S., Malthe-Sorenssen, A., et al., 2004. Release of methane from a volcanic basin as a mechanism for initial Eocene global warming. Nature 429, 542–545.

Wang, C., Hu, X., Huang, Y., Wagreich, M., Scott, R.W., et al., 2011. Cretaceous oceanic red beds as possible consequence of oceanic anoxic events. Sediment. Geol. 235, 27–37.

Waters, C.N., Zalasiewicz, J., Summerhayes, C., et al., 2016. The Anthropocene is functionally and stratigraphically distinct from the Holocene. Science 351, 6269 aad2622.

Westerhold, T., Röhl, U., Laskar, J., Bowles, J., Raffi, I., Lourens, L.J., Zachos, J.C., 2007. On the duration of magnetochrons C24r and C25n and the timing of early Eocene global warming events: implications from the Ocean Drilling Program Leg 208 Walvis Ridge depth transect. Paleoceanography 22, PA2201. https://doi.org/10.1029/2006PA001322, 2007.

Westerhold, T., Röhl, U., Raffi, I., Fornaciari, E., Monechi, S., Reale, V., Bowles, J., Evans, H., 2008. Astronomical calibration of the Paleocene time. Palaeogeogr. Palaeoclimatol. Palaeoecol. 257, 377–403.

Westerhold, T., Röhl, U., Laskar, J., 2012. Time scale controversy: accurate orbital calibration of the early Paleogene. Geochem., Geophys., Geosyst. 13, Q06015. https://doi.org/10.1029/2012GC004096.

Westerhold, T., Röhl, U., Frederichs, T., Bohaty, S.M., Zachos, J.C., 2015. Astronomical calibration of the geological timescale: closing the middle Eocene gap. Clim. Past 11, 1181–1195.

Westerhold, T., Röhl, U., Friedrichs, T., Agnini, C., Raffi, I., Zachos, J.C., Wilkens, R.H., 2017. Astronomical calibration of the Ypresian timescale: implications for seafloor spreading rates and the chaotic behavior of the solar system. Clim. Past 13, 1129–1152.

Zachos, J.C., Wara, M.W., Bohaty, S., Delaney, M.L., Petrizzo, M.R., Brill, A., Bralower, T.J., Premoli-Silva, I., 2003. A transient rise in tropical sea surface temperature during the Paleocene–Eocene thermal maximum. Science 302, 1551–1554.

Zachos, J.C., Röhl, U., Schellenberg, et al., 2005. Rapid acidification of the ocean during the Paleocene–Eocene thermal maximum. Science 308, 1611–1615.

Zachos, J.C., Dickens, G.R., Zeebe, R.E., 2008. An early Cenozoic perspective on greenhouse warming and carbon-cycle dynamics. Nature 451, 279–283.

CPI Antony Rowe
Chippenham, UK
2018-10-16 17:42